BRAIN-COMPUTER INTERFACES

Brain-Computer Interfaces

An International Assessment of Research and Development Trends

by

THEODORE W. BERGER
University of Southern California, Los Angeles, CA, USA

JOHN K. CHAPIN
State University of New York, Downstate Medical Center, Brooklyn, NY, USA

GREG A. GERHARDT
University of Kentucky, Lexington, KY, USA

DENNIS J. MCFARLAND
Wadsworth Centre, Albany, NY, USA

JOSÉ C. PRINCIPE
University of Florida, Gainesville, FL, USA

WALID V. SOUSSOU
University of Southern California, Los Angeles, CA, USA

DAWN M. TAYLOR
Case Western Reserve University, Cleveland, OH, USA

and

PATRICK A. TRESCO
University of Utah, Salt Lake City, UT, USA

Springer

This document was sponsored by the National Science Foundation (NSF) and other agencies of the U.S. Government under an award from the NSF (ENG-0423742) to the World Technology Evaluation Center, Inc. The Government has certain rights in this material. Any opinions, findings, and conclusions or recommendations expressed in this material are those of the authors and do not necessarily reflect the views of the United States Goverment, the authors' parent institutions, or WTEC, Inc.

ISBN: 978-1-4020-8704-2 e-ISBN: 978-1-4020-8705-9

Library of Congress Control Number: 2008930994

Printed on acid-free paper.

9 8 7 6 5 4 3 2 1

springer.com

WTEC Panel on Brain-Computer Interfaces

Theodore W. Berger
University of Southern California, Los Angeles, CA, USA
John K. Chapin
State University of New York, Downstate Medical Center, Brooklyn, NY, USA
Greg A. Gerhardt
University of Kentucky, Lexington, KY, USA
Dennis J. McFarland
Wadsworth Center, Albany, NY, USA
José C. Principe
University of Florida, Gainesville, FL, USA
Walid V. Soussou
University of Southern California, Los Angeles, CA, USA
Dawn M. Taylor
Case Western Reserve University, Cleveland, OH, USA
Patrick A. Tresco
University of Utah, Salt Lake City, UT, USA

Acknowledgments

WTEC staff members wish to extend their gratitude and appreciation to all the panelists for their knowledge, enthusiasm, and dedication to this international benchmarking study of brain-computer interface R&D, and to all workshop presenters and site visit hosts for so generously sharing their time, expertise, and facilities with us. We also wish to thank Professor Jiping He of Arizona State University, who provided guidance and contacts regarding brain computer interface research laboratories in China, and Dr. Jason J. Burmeister of the University of Kentucky, who provided background material on sensor technologies for Chapter 2 of the report. For their sponsorship of this unique study, our sincere thanks go to the National Science Foundation, the Telemedicine and Advanced Technology Research Center of the U.S. Army Medical Research and Materiel Command, the National Institute of Neurological Disorders and Stroke of the National Institutes of Health, the National Space Biomedical Research Institute, the National Institute of Biomedical Imaging and Bioengineering of the National Institutes of Health, and the Margot Anderson Brain Restoration Foundation. Finally, our special thanks go to Dr. Semahat Demir for her vision and her unflagging support of the panel through all phases of this study.

R. D. Shelton
President, WTEC

Abstract

Brain-computer interface (BCI) research deals with establishing communication pathways between the brain and external devices. BCI systems can be broadly classified depending on the placement of the electrodes used to detect and measure neurons firing in the brain: in *invasive* systems, electrodes are inserted directly into the cortex; in *noninvasive* systems, they are placed on the scalp and use electroencephalography or electrocorticography to detect neuron activity. This WTEC study was designed to gather information on worldwide status and trends in BCI research and to disseminate it to government decisionmakers and the research community. The study reviewed and assessed the state of the art in sensor technology, the biotic–abiotic interface and biocompatibility, data analysis and modeling, hardware implementation, systems engineering, functional electrical stimulation, noninvasive communication systems, and cognitive and emotional neuroprostheses in academic research and industry.

The WTEC panel identified several major trends in current and evolving BCI research in North America, Europe, and Asia. First, BCI research throughout the world is extensive, with the magnitude of that research clearly on the rise. Second, BCI research is rapidly approaching a level of first-generation medical practice; moreover, BCI research is expected to rapidly accelerate in nonmedical arenas of commerce as well, particularly in the gaming, automotive, and robotics industries. Third, the focus of BCI research throughout the world is decidedly uneven, with invasive BCIs almost exclusively centered in North America, noninvasive BCI systems evolving primarily from European and Asian efforts, and the integration of BCIs and robotics systems championed by Asian research programs.

In terms of funding, BCI and brain-controlled robotics programs have been a hallmark of recent European research and technological development. The range and investment levels of multidisciplinary, multinational, multilaboratory programs in Europe appear to far exceed that of most university and government-funded BCI programs in the United States and Canada. Although several U.S. government programs are advancing neural prostheses and BCIs, private sources have yet to make a major impact on BCI research in North America generally. However, the U.S. Small Business Innovative Research grants (SBIRs) and Small Technology Transfer Research grants (STTRs) have been effective in promoting transition from basic research to precommercialized prototypes. In Asia, China is investing heavily in biological sciences and engineering in general, and the extent of investment in BCI and BCI-related research has grown particularly rapidly; still, the panel observed little coordination between various programs. Japanese universities, research institutes, and laboratories also are increasing their investment in BCI research. Japan is especially vigorous in pursuing nonmedical applications and exploiting its expertise in BCI-controlled robotics.

The WTEC panel concludes that there are abundant and fertile opportunities for worldwide collaborations in BCI research and allied fields.

WTEC Mission

WTEC provides assessments of international research and development in selected technologies under awards from the National Science Foundation (NSF), the Office of Naval Research (ONR), and other agencies. Formerly part of Loyola College, WTEC is now a division of the World Technology Research Center, a separate nonprofit research institute. Michael Reischman, Deputy Assistant Director for Engineering, is NSF Program Director for WTEC. Sponsors interested in international technology assessments and related studies can provide support for the program through NSF or directly through separate grants or GSA task orders to WTEC.

WTEC's mission is to inform U.S. scientists, engineers, and policymakers of global trends in science and technology. WTEC assessments cover basic research, advanced development, and applications. Panels of typically six technical experts conduct WTEC assessments. Panelists are leading authorities in their field, technically active, and knowledgeable about U.S. and foreign research programs. As part of the assessment process, panels visit and carry out extensive discussions with foreign scientists and engineers in their labs.

The WTEC staff helps select topics, recruits expert panelists, arranges study visits to foreign laboratories, organizes workshop presentations, and finally, edits and publishes the final reports.

WORLD TECHNOLOGY EVALUATION CENTER, INC. (WTEC)

R. D. Shelton, President
Michael DeHaemer, Executive Vice President
Geoffrey M. Holdridge, Vice President for Government Services
David Nelson, Vice President for Development
Ben Benokraitis, Assistant Vice President

Other WTEC Staff Members and Subcontractors Involved in this Study

Hassan Ali, Director of International Study Operations
Grant Lewison (Evaluametrics, Ltd.), Advance Contractor, Europe
Gerald Hane (Globalvation), Advance Contractor, Asia
Maria L. DeCastro, Manuscript Development and Support
Patricia M.H. Johnson, Director of Publications
Halyna Paikoush, Event Planner and Office Manager

Foreword

> We have come to know that our ability to survive and grow as a nation to a very large degree depends upon our scientific progress. Moreover, it is not enough simply to keep abreast of the rest of the world in scientific matters. We must maintain our leadership.[1]

President Harry Truman spoke those words in 1950, in the aftermath of World War II and in the midst of the Cold War. Indeed, the scientific and engineering leadership of the United States and its allies in the twentieth century played key roles in the successful outcomes of both World War II and the Cold War, sparing the world the twin horrors of fascism and totalitarian communism, and fueling the economic prosperity that followed. Today, as the United States and its allies once again find themselves at war, President Truman's words ring as true as they did a half-century ago. The goal set out in the Truman Administration of maintaining leadership in science has remained the policy of the U.S. Government to this day: Dr. John Marburger, the Director of the Office of Science and Technology (OSTP) in the Executive Office of the President, made remarks to that effect during his confirmation hearings in October 2001.[2]

The United States needs metrics for measuring its success in meeting this goal of maintaining leadership in science and technology. That is one of the reasons that the National Science Foundation (NSF) and many other agencies of the U.S. Government have supported the World Technology Evaluation Center (WTEC) and its predecessor programs for the past 20 years. While other programs have attempted to measure the international competitiveness of U.S. research by comparing funding amounts, publication statistics, or patent activity, WTEC has been the most significant public domain effort in the U.S. Government to use peer review to evaluate the status of U.S. efforts in comparison to those abroad. Since 1983, WTEC has conducted over 50 such assessments in a wide variety of fields from advanced computing, to nanoscience and technology, to biotechnology.

The results have been extremely useful to NSF and other agencies in evaluating ongoing research programs and in setting objectives for the future. WTEC studies also have been important in establishing new lines of communication and identifying opportunities for cooperation between U.S. researchers and their colleagues abroad, thus helping to accelerate the progress of science and technology generally within the international community. WTEC is an excellent example of cooperation and coordination among the many agencies of the U.S. Government that are involved in funding research and development: almost every WTEC study has

---

[1] Remarks by President Harry S. Truman on May 10, 1950, on the occasion of the signing of the law that founded the National Science Foundation. Public Papers of the Presidents 120: p. 338.

[2] http://www.ostp.gov/html/01_1012.html.

been supported by a coalition of agencies with interests related to the particular subject at hand.

As President Truman said over 50 years ago, our very survival depends upon continued leadership in science and technology. WTEC plays a key role in determining whether the United States is meeting that challenge, and in promoting that leadership.

Michael Reischman
Deputy Assistant Director for Engineering
National Science Foundation

Table of Contents

APPENDICES

List of Figures

List of Tables

Preface

This benchmarking panel study on brain-computer interfaces had broad sponsorship from the U.S. Government agencies and private organizations listed in in the Acknowledgments page of the report; it was organized by the World Technology Evaluation Center (WTEC). As the lead sponsoring program director for this study, I present this final report to the global brain-computer interface community on behalf of all the study participants and sponsors. This has been an informative, productive journey for all involved in the study. I would like to start by thanking those who contributed so much to this final product.

First, many thanks go to the panel chair, Ted Berger, and to all of the BCI panelists: John Chapin, Greg Gerhardt, Dennis McFarland, José Principe, Dawn Taylor, Patrick Tresco, and Walid Soussou (associate panelist). Next, our thanks go to the numerous eminent researchers from around the world whose input is a fundamental merit of this study. Gary Birch, John Donoghue, Daryl Kipke, Dan Moran, Richard A. Normann, David A. Putz, Andrew B. Schwartz, William Shain, and Krishna V. Shenoy presented at our North American BCI workshop on February 27, 2006. Twenty-seven leading institutions in Europe and Asia hosted panelists during site visits in May and October 2006. We are deeply grateful to all of those institutions and the many individuals who so generously shared their work and their insights with the panel.

My personal thanks go to Mike Reischman, Lynn Preston, and Bruce Hamilton of NSF for supporting this idea and for co-funding this study with me from the beginning. I also thank the following government colleagues for co-sponsoring this study: Ephraim Glinert (NSF/CISE), Joseph Pancrazio (NIH/NINDS), Kenneth Curley (TATRC), and Grace Peng (NIH/NIBIB). Two non governmental organi zations contributed funds to the study; I appreciate the support of Jeffrey Sutton of the National Space Biomedical Research Institute and Herman Edel of the Margot Anderson Brain Restoration Foundation. In addition to the contributions of the above-mentioned colleagues, I would like to recognize the efforts of Mike Roco (NSF), Nancy Shinowara (NIH/NICHD), and Bob Jaeger (NIDRR, now with NSF) for their technical input to me, the WTEC team, and the panelists, and for attending the planning meetings and workshops.

I acknowledge the WTEC team with special thanks to Mike DeHaemer (Executive Vice-President of WTEC), Hassan Ali (the manager for this study), and Duane Shelton (President of WTEC). Mike, Hassan, and Duane worked diligently from the initiation of the study. Grant Lewison (Evaluametrics, Ltd.) arranged the site visits in Europe, and Gerald Hane (Globalvation) arranged the site visits in Asia. Roan Horning provided computing and website support. Ben Benokraitis coordinated and reviewed the substantive work on the report. Maria DeCastro and Pat Johnson contributed editing support.

The study has been a great journey since my email to a few colleagues on November 10, 2004, in which I first proposed a study on Brain-Computer

Interfaces, and my initial meeting with WTEC representatives on January 3, 2005. Milestones along the way included meetings with sponsors in March and April 2005; the sponsors and chair meeting on October 14, 2005; the kickoff meeting with the BCI panelists and sponsors on December 2, 2005; the North American workshop on February 27, 2006; site visits to Europe in May–June 2006; the workshop "Review of International Research on Brain-Computer Interfaces" on July 21, 2006; site visits to Asia in October 2006; and the BCI international benchmarking teleconference (Asia-Japan) on December 14, 2006. This report is the final result of the myriad efforts of the study team, and the vision realized of a benchmarking study on brain-computer interface R&D.

BRAIN-COMPUTER INTERFACE SCIENCE

Brain-computer interfaces (BCIs) are defined as the science and technology of devices and systems responding to neural processes in the brain that generate motor movements and to cognitive processes (e.g., memory) that modify the motor movements. Advances in neuroscience, computational technology, component miniaturization, biocompatibility of materials, and sensor technology have led to a much improved feasibility of useful BCIs that engineers, neuroscientists, physical scientists, and behavioral and social scientists can develop as a large-scope team effort.

The WTEC BCI international assessment panel defined BCI technologies as either "invasive" (multielectrode arrays of tens to hundreds of electrodes implanted into cortical tissue from which "movement intent" is decoded), or "noninvasive" (multielectrode arrays emplaced on the surface of the skull to record changes in EEG state) in their control of computer cursors or other systems. The study results presented at the workshops on February 27 and July 21, 2006, indicated that the majority of BCI science in North America involves invasive technologies, and the majority of BCI science in Europe involves noninvasive technologies and also the development of biologically inspired robots. The panel presented findings that European efforts are more often integrated within a larger research scope, and European BCI systems involve a wider range of EEG-based applications. Overall, the panelists felt that European and Asian BCI work is highly competitive with that of the United States and that many opportunities exist for collaboration.

As indicated in this report, engineers around the world are working, in collaboration with neuroscientists, physical scientists, and social and behavioral scientists, to integrate and converge engineering tools and methods in the areas of sensors and signal processing, noninvasive and minimally invasive recording techniques from the brain and the peripheral nervous system, neural tissue engineering, neural imaging, nonlinear dynamics, chemical and biological transport, computational neuroscience and multiscale modeling, nano/micro technological neuroscience, control theory, systems integration, and robotics in order to permit control of movement where normal neural pathways do not exist. Transformational solutions being pursued are leading to better understanding of the central and peripheral nervous systems and pushing forward the frontier of scientific discovery.

The principal goal of BCI work is to enable people with neural pathways that have been damaged by amputation, trauma, or disease to better function and control their environment, through either reanimation of paralyzed limbs or control of robotic devices. BCI also extends to the fields of neurobiomimetics and complex hybrid neurobionic systems. BCI systems will have great societal impact, with growing interest on the part of industry to commercialize and market BCI systems for medical and nonmedical applications in the long term. The WTEC study identifies the following opportunities for multidisciplinary BCI teams to find transformational solutions:

- Studying multiple levels and multiple scales of neural functions and neural code
- Developing long-term biocompatibility between electronics and neural tissues
- Establishing bidirectional communication between biomimetic devices and the nervous system
- Developing hierarchically organized control systems for robotics and biomimetics
- Developing biologically inspired systems that will push the frontier for the development of autonomous intelligent systems ("conscious" self-adaptive systems)
- Engineering practical BCIs and even integrating BCIs with cyberinfrastructure.

RELATED ACTIVITIES AT NSF

In parallel to the WTEC BCI benchmarking study, NSF has sponsored several related neuroscience activities; some of the BCI panelists and I participated in those activities. The Steering Group workshop, "Brain Science as a Mutual Opportunity for the Physical Sciences, Mathematics, Computational Sciences and Engineering," took place in Arlington, VA, on August 21–22, 2006. It identified as broad areas of opportunity (1) instrumentation and measurement; (2) data analysis, statistical modeling, and informatics; (3) conceptual and theoretical approaches; and (4) brain like devices and systems. These four opportunity areas align with the WTEC panel's transformational solutions noted above.

A second workshop, "Brain Science at the Interface of Biological, Physical and Mathematical Sciences, Computer Science and Engineering: Analysis of New Opportunities," took place in Arlington, VA, March 5–6, 2007. The BCI-related opportunities and challenges that were identified at this workshop were:

1. Brain, mind, cognition, behavior, learning, development
2. Multiscale complexity; connectivity; nonlinear, nonstationary, stochastic control; stability; and adaptability
   (a) Neural coding and decoding (cognitive vs. neurophysiological)

3. Bioinspired systems
   (a) Abstracting from neuroscience principles to develop bioinspired systems
   (b) Replicating neural computation
   (c) Next generation of computing systems
4. Sensors, smart sensing, and bidirectional communication.

Research in neuroscience and cognition needs "bridging" of experimental and modeling work at the different scales of time (nanoseconds to years), of length (nanometers to meters), and of biology (atoms; molecules; molecular complexes; subcellular, cellular, multicellular elements; tissue, organs, organ systems, and organisms, up to entire populations). The *natural (biological) interfaces* of nervous systems have to be studied with multiscale (multilevel) approaches by interdisciplinary teams of life scientists, physical scientists, social scientists, behavioral scientists, mathematicians, and engineers who must work within a broad research framework. Engineers bring to these multidisciplinary teams workable methods and tools for analysis, recording, modeling, and implementation of new BCI technologies.

Bridging the sciences in the field of BCI from discovery to application or translation is a significant challenge. The Bioengineering Consortium (BECON, chaired by Dr. Michael Huerta, NIH/NIMH) formed a subcommittee called BECON Bridges on March 1, 2007, which Dr. Albert Lee (NIH/NIBIB) and I co-chaired. This subcommittee will determine the research areas in which the sciences needs to be bridged and what mechanisms can enable the bridging. BCI is one of those areas.

On July 27, 2007, the NSF Engineering Directorate released two Emerging Frontiers in Research and Innovation 2008 topics (EFRI-2008), one of which is BCI-related: "Cognitive Optimization and Prediction: From Neural Systems to Neurotechnology (COPN)."[1] The goal of COPN is to motivate engineers to reverse-engineer the prediction and optimization capabilities of the brain to facilitate usable design. While my NSF colleague Dr. Paul Werbos and I were developing COPN, the results of the WTEC BCI study were helpful.

Section IV of *National Science Foundation Investing in America's Future, Strategic Plan FY 2006–2011*[2] lists investment priorities for four strategic goals: Discovery, Learning, Research Infrastructure, and Stewardship. Under the Discovery strategic goal there are five topics listed (page 6 of the Strategic Plan), four of which are areas where BCI R&D can contribute.

---

[1] NSF. 2007. Emerging Frontiers in Research and Innovation, http://nsf.gov/publications/pub_summ.jsp?ods_key=nsf07579.

[2] NSF. 2006. The FY 2006–2011 strategic plan is available online at http://www.nsf.gov/pubs/2006/nsf0648/nsf0648.jsp.

SIGNIFICANCE OF BCI R&D TO THE U.S. ECONOMY AND SOCIETY

Based on the work of this panel and on the NSF discussions and activities noted above, it seems clear that BCI research and development activities can have an immediate and lasting impact on U.S. (and global) science and technology activities that far exceed their immediate, important, and exciting benefit to a relatively small number of citizens. The necessarily collaborative work towards BCI solutions depends on and at the same time advances work in many related high-tech fields. Thus, there is an inherently synergistic benefit to BCI work that operates on the cutting edge of many important fields of science and technology. At the same time, BCI work intersects with significant current trends in U.S. employment and in Federal support for science-based activities to enhance U.S. competitiveness relative to other nations.

BCI-Related Job and Educational Opportunities

According to the U.S. National Science Board,[3] occupational projections from the U.S. Bureau of Labor Statistics (BLS) predict that the employment in science and engineering occupations will increase faster then the overall growth rate for all occupations. In addition, the BLS Occupational Outlook Handbook, 2004–2005 edition, predicts that by 2012, top job growth will be in (1) healthcare and social assistance; and (2) biomedical, biotechnology, and bioengineering professions. Employment in biomedical engineering, biotechnology, and bioengineering is expected to increase by 21–35% by 2012. Thus, there are expected to be numerous promising career and job opportunities for biomedical engineers.

Education indicators sustain this outlook. The *IEEE Spectrum* survey results of February 2007, "Your Best Bet for the Future," identifies the top ten technology research and development fields that faculty would advise their students to pursue: the biomedical field is number one, and other fields in the top five, such as wireless/mobile (number 2) and nanotechnology (number 5), are relevant to biomedical R&D as well. More specifically, based on the American Society for Engineering Education six-year trend analysis (1999–2005),[4] BME, while still representing a small proportion of overall undergraduate and graduate degrees conferred, is one of the two fastest-growing disciplines at U.S. universities (the other is aerospace engineering). Of special note is the fact that BME is a field in which women

[3] National Science Board. 2004. Science and Engineering Indicators—2004. NSB-04-1. Arlington, VA: NSF.

[4] ASEE. 2007. 2006 profiles of engineering and engineering technology colleges. Washington, DC: ASEE. See also an online profiles sample at http://www.asee.org/publications/profiles/upload/2006ProfileEng.pdf.

represent a higher proportion than other engineering fields of tenure/tenure-track teaching faculty and degree recipients. All these indicators are promising for the pipeline and the diversity of engineers that will enter BME careers in academia, industry, government, or independent consultancy.

BCI and the Innovation and Competitiveness Debate

On August 9, 2007, President George W. Bush signed into law the "America Creating Opportunities to Meaningfully Promote Excellence in Technology, Education and Science (COMPETES) Act." America COMPETES authorizes research programs at the National Science Foundation (NSF), the National Institute of Standards and Technology (NIST) of the Department of Commerce, and the Department of Energy (DOE) Office of Science, with near-term doubling of funding. The bill also authorized $33.6 billion over fiscal years 2008 through 2010 for research and education programs across the Federal Government. The bill is intended to strengthen education and research in the United States related to science, technology, engineering, and mathematics (STEM). Many provisions of the legislation were developed based on recommendations made in two reports on competitiveness: *American Competitiveness Initiative: Leading the World in Innovation*[5] and *Rising Above the Gathering Storm: Energizing and Employing America for a Brighter Economic Future*.[6]

Other recent reports, articles, and statements have addressed the U.S. innovation and competitiveness debate. The American Competitiveness Initiative (ACI) recommends doubling funding over ten years on innovation-enabling research at three key Federal agencies (NSF, DOE, and NIST) that support high-leverage fields of of physical science, basic science, and engineering. ACI has three broad parts: (1) research in physical sciences and engineering (including 12 specific goals), (2) research and development (R&D) tax incentives, and (3) education and workforce. The report *Rising Above the Gathering Storm: Energizing and Employing America for a Brighter Economic Future* makes recommendations for K-12 education, research, higher education, and economic policy. The *Innovate America*[7] executive summary also makes recommendations under talent, investment, and infrastructure. BCI research is a strong contender as a field to promote U.S. technical leadership toward enhanced innovation and improved competitiveness, bringing attendant economic benefits.

---

[5] Office of Science and Technology Policy Domestic Policy Council. 2006 (February). Available online at http://www.ostp.gov/html/ACIBooklet.pdf.

[6] Committee on Science, Engineering, and Public Policy: National Academy of Sciences, National Academy of Engineering, and Institute of Medicine. 2007. Washington, DC: National Academies Press.

[7] Committee on Science, Engineering, and Public Policy: National Academy of Sciences, National Academy of Engineering, and Institute of Medicine. 2007. Washington, DC: National Academies Press.

CONCLUDING REMARKS

The WTEC BCI study presents the current status and future trends of BCI research in North America, Europe, and Asia. It will assist NSF and other U.S. Government agencies to perform strategic planning for future STEM programs and to accelerate discoveries and the progress of science and engineering. These are exciting times for life scientists, physical scientists, and engineers to work together in inter-disciplinary, innovation-enabling research fields. BCI is one of those fields that will enrich the innovation and competitiveness debate globally.

Semahat S. Demir, Ph.D.
Program Director
Biomedical Engineering Program
National Science Foundation
September 2007

Executive Summary

Theodore W. Berger

Brain-Computer Interface (BCI) research deals with establishing communication pathways between the brain and external devices. To provide program managers in U.S. research agencies as well as researchers in the field with a better understanding of the status and trends in BCI research abroad, in December 2005 the WTEC International Assessment of Brain-Computer Interface R&D was organized. Sponsors included

- National Science Foundation (NSF)
- Telemedicine and Advanced Technologies Research Center (TATRC) of the U.S. Army Medical Research and Materiel Command
- National Institute of Neurological Disorders and Stroke (NINDS) of the National Institutes of Health (NIH)
- National Space Biomedical Research Institute
- National Institute of Biomedical Imaging and Bioengineering (NIBIB) of NIH
- Margot Anderson Brain Restoration Foundation .

The study was designed to gather information on the worldwide status and trends in BCI research and to disseminate it to government decisionmakers and the research community. The study reviewed and assessed the state of the art in sensor technology, the biotic–abiotic interface and biocompatibility, data analysis and modeling, hardware implementation, systems engineering, functional electrical stimulation (FES), noninvasive communication systems, and cognitive and emotional neuroprostheses in academic research and industry. To provide a basis for comparison, the study began on February 27, 2006 with a workshop held at NSF entitled "Review of North American Research on Brain-Computer Interfaces." After convening this baseline workshop, a WTEC panel of U.S. experts visited seventeen sites in Europe and ten facilities in China and Japan involved in BCI research.

MAJOR TRENDS IN BCI RESEARCH

The WTEC panel identified several major trends that both characterize the present, and can be projected into the future, of Brain-Computer Interface Research in North America, Europe, and Asia. First, BCI research throughout the world is extensive, with the magnitude of that research clearly on the rise. BCI research is an unmistakable growth area—which because of the inherently interdisciplinary nature of BCIs, means growth in the interface between multiple key scientific areas, including biomedical engineering, neuroscience, computer science, electrical and computer engineering, materials science and nanotechnology, and neurology and neurosurgery. Thus, the panel sees future growth in BCIs as having a widespread influence in shaping the landscape of scientific research in general and radically altering the boundaries of interdisciplinary research in particular.

Second, BCI research is rapidly approaching a level of first-generation "medical practice"—clinical trials of invasive BCI technologies and significant home use of noninvasive, electroencephalography (EEG-based) BCIs. Because the threshold for substantial use of BCIs for medical applications is rapidly approaching, the panel predicts that BCIs soon will markedly influence the medical device industry. As a corollary, the panel sees that BCI research will rapidly accelerate in nonmedical arenas of commerce as well, particularly in the gaming, automotive, and robotics industries. Thus, the industrial influence of BCIs is certain to increase in the near future.

Third, the WTEC panel found that the focus of BCI research throughout the world was decidedly uneven, with invasive BCIs almost exclusively centered in North America, noninvasive BCI systems evolving primarily from European and Asian efforts, and the integration of BCIs and robotics systems championed by Asian research programs. Thus, the panel felt that there were abundant and fertile opportunities for worldwide collaborations that would allow the existing specializations in different regions of the globe to interact in a synergistic and productive manner. In this summary, we elaborate on these and other conclusions from the WTEC panel's study of Brain-Computer Interfaces (BCI) in North America, Europe, and Asia.

MAGNITUDE OF BCI RESEARCH

The magnitude of research and development of BCIs throughout the world will grow substantially, if not dramatically, in the next decades. There are multiple forces that are driving and will continue to drive this trend. One of the most fundamental forces accelerating BCI research is the continued advance in the science, engineering, and technology required for the realistic achievement of BCIs. The growth in neuroscience continues to be explosive, with new frontiers being reached every year in understanding principles of the central nervous system (CNS) structure and function and—importantly for BCI design—systems-level organization of the nervous system. Rapid advances in biomedical engineering and computer science are producing the methodologies required for predictive models of neural function that can interact with the brain in real time. The continuing achievements in microelectronics that allow ever-greater circuitry miniaturization together with increased speed and computational capacity are providing the next-generation hardware platforms for BCIs. This growing knowledge base and technological capability is creating the "bedrock" essential for developing BCI systems and powering ongoing advances in neural prostheses.

The strong recent and current investment in BCI research throughout the world virtually guarantees a continued high growth rate. BCI and brain-controlled robotics programs have been one of the hallmarks of the European Union's Sixth Framework Program (2002–2006) for Research and Technological Development. The large size and scope of these multidisciplinary, multinational, multilaboratory programs have been remarkable, with support levels far exceeding most BCI programs in the United States. Even if the scale of 7th Framework programs is reduced, the momentum of BCI research initiated by EU 6th Framework programs will not dampen for some time. Likewise, the panel was impressed by the formidable investment being made by China in biological sciences and engineering in general, and by the investment in BCI and BCI-related research in particular. Japanese universities and institutions also are unmistakably increasing their commitment to and investment in BCI research.

INVASIVE VERSUS NONINVASIVE BCI RESEARCH

It became clear to the panel during its study that there is a marked contrast in the worldwide distribution of "invasive" and "noninvasive" BCI research. Invasive systems interact with the brain directly, i.e., with electrodes that penetrate the brain or lay on the surface of the brain, while noninvasive systems interact with the brain indirectly by transmissions through the skull, e.g., electroencephalography (EEG), functional magnetic resonance imaging (fMRI), and magnetic sensor systems. The vast majority of invasive BCI research is currently being conducted in the United States. Virtually all BCI research in Europe is noninvasive, attributable in large part to constraints and intimidations imposed by animal rights organizations. BCI research in China appears to be almost exclusively noninvasive, though this reflects the relatively early stage of development of BCI research in that country. The massive modernization by China of its research programs in fundamental neuroscience and BCIs hopefully is leading to the emergence of a first-rate invasive BCI program. The panel felt that there is a strong need to maintain a worldwide balance between invasive and noninvasive approaches to BCI research and technology if the field of neural prostheses is to remain vigorous and viable. The panel was particularly impressed by the commitment in Europe and Japan to devote the substantial resources needed to explore the possibility of fMRI and magnetoencephalography (MEG) sensor technologies as the basis of noninvasive BCIs, despite the high cost of such technologies and the uncertain time span or probability of miniaturization to the appropriate scale for routine patient use.

NEED FOR MEDICAL BCI

One of the other forces driving the current acceleration in BCI research is societal demand for solutions to the problem of repairing the nervous system. An unassailable reality is that when the brain and spinal cord become damaged or diseased, they do not repair themselves. With the increasing size of the world population and particularly its increasing age, the number of future patients with such diagnoses as Parkinsonism and other tremor-related disorders and dementias including Alzheimer's disease, epilepsy, accident-induced spinal cord injuries, and peripheral neuropathies resulting from diabetes is likely to be staggering. The panel found that BCI researchers uniformly considered future health-related needs for BCIs to be a strongly motivating factor, with that motivation particularly great in populous countries like China.

In recognition of the current and future potential market for BCIs, the medical device industry has begun to accelerate development and market integration of BCI-related medical products. In the United States and Europe, evidence of medical industry collaborations with respect to BCI devices and systems is seen in an increasing number of startups and joint partnerships. As the bridge from research prototype to medical device strengthens, solutions are emerging to the specialized design requirements imposed by the CNS: sensor designs, mathematical models and their hardware implementations, and brain interface materials are increasingly becoming "biomimetic" and "neuromorphic" in nature. In addition, there are also power requirements and biocompatibility issues that are unique to the CNS.

SCOPE OF BCI RESEARCH: NONMEDICAL BCI

The need for medical applications of BCI research, i.e., repair of the nervous system, will remain the core driving force for BCIs at least in the near future. The panel also found evidence, however, that BCI research will increasingly widen to include nonmedical applications. This transition is already in progress in many European and Japanese BCI laboratories. Fundamental principles of BCIs were seen to generalize readily to brain control of video gaming and virtual reality environments. Intriguing extensions of BCIs to automotive industry problems were found in the form of measuring driver cognitive load. Multiple research programs included a focus on BCI-related principles for robotics control and comprehensive programs for integrating BCIs into everyday life to link the human sensorium more completely and interactively into the environment.

TRANSLATION/COMMERCIALIZATION OF BCI

The extent to which industry in Europe and Japan has embraced BCI-related research goals and the development of requisite technologies for BCIs is impressive. This high degree of industry commitment was perhaps most evidenced in Germany by institutional entities having the specific missions of actively promoting academic-industrial research interactions, garnering support for BCI research from industry sources, and transitioning the resulting BCI and BCI-related systems to industry for commercialization. Such entities house advanced technologies and equipment made available to startups with limited resources; research collaborations and partnerships could result in spinoffs that accelerate the entry of new BCIs and BCI technologies into the marketplace.

The EU 6th Framework research programs strongly encourage and to some degree require industrial involvement. Corporations involved in commercialization of BCI systems and/or BCI-related products are essentially able to participate in EU-sponsored research (with some restrictions) as a "collaborator" along with any other university or institute unit and are eligible to receive funds to conduct their respective component of the overall research project. Equally impressive was the degree to which BCI-related research issues were integrated into the agendas of major Japanese research institutes and corporations and the extent of government support of those private, and sometimes profit-making, entities. In general, the panel saw creative and highly flexible academic-industry collaborations that promoted the transition from laboratory-based to commercialized BCIs.

OPPORTUNITIES FOR WORLDWIDE COLLABORATIVE RESEARCH

Because of the rich, interdisciplinary nature of BCI-related research, the panel was able to readily identify multiple opportunities for worldwide collaborations. Foremost among these is a comprehensive effort to achieve a better understanding of the relation between noninvasive and invasive measures of cortical activity—EEG/MEG, local field potentials, and (population) single-unit activity. This issue was identified at multiple sites visited by the panel as one that is both fundamental to neuroscience and useful in the further development of BCIs. This problem also is complementary to the relative strengths of BCI research on the three continents.

Second, there is a plethora of new mathematical modeling and signal analysis methods being developed throughout the multiple countries involved in BCI research. Systematic evaluation of these methodologies and collaborative efforts to achieve synergy and avoid duplication would be beneficial to the forward movement of BCIs.

Third, there remain multiple electrode technologies used in North America, Europe, and Asia. Given the time required to develop and implement new electrode approaches and their associated electronics and signal processing protocols, dissemination of technological innovation and collaboration with respect to needed next-generation methods, e.g., "dry" EEG electrodes, could accelerate BCI research and development progress. Needed collaborations with respect to BCI-related microelectronics also were acknowledged. Several multinational collaborations and technology-sharing efforts that can attest to the beneficial effects of collaboration on BCI research include

- The joint DARPA Revolutionizing Prosthetics program (U.S.) and the robotics research program at the Polo Sant'Anna Valdera (Italy)
- U.S.-European use of the Watson Center BCI2000 system
- Multi Channel Systems and g.tec technologies.

The technologies developed within these collaborative programs are now used throughout the world in BCI research.

STUDY HIGHLIGHTS: BCI R&D IN NORTH AMERICA AND EUROPE

Science of BCIs

- The majority of BCI science in NA (North America) involves "invasive" technologies, i.e., recordings from arrays of electrodes implanted into the brain.
- The majority of BCI science in Europe involves "noninvasive" technologies, i.e., recordings from arrays of electrodes mounted onto the surface of the skull.
- Other fundamental differences between U.S. and European BCI efforts:
  - European efforts are more often integrated within a larger research scope of developing "hybrid bionic systems."
  - European BCI systems involve a wider range of EEG-based applications.
  - The panel saw many opportunities for synergy and collaboration with European BCI investigators.
  - Overall, the panel felt that, in terms of quality and sophistication, European BCI efforts are highly competitive with those of the United States.

Interdisciplinary/Programmatic Structure for BCI Research

- Programs are defined on a decade-long time scale.
- High risk is "comfortably" inherent in programmatic definitions.
- Fundamental science is considered an equal to practical outcomes.

- In general, the panel found a strong European commitment to long-term, visionary, high-risk, interdisciplinary research, in other words, the foundation required for successful development of BCIs.
- U.S. counterparts include DARPA initiatives, NSF ERC programs, and NINDS Neural Prosthetics.
- The scale of multi-investigator projects possible under EU programs exceeds that found in the United States; multidisciplinary teams necessary for BCI research are more readily created in the EU system.

Funding for BCI Research

- Consistent with the large, multidisciplinary BCI teams found in Europe, the scale of European BCI research funding is substantial.
- Only NSF Engineering Research Centers (e.g., Biomimetic Microelectronic Systems Center at USC) and the largest DARPA programs (e.g., Revolutionizing Prosthetics) compete with EU programs.
- In part, this reflects the consistent investment by European countries in fundamental science and technology, in addition to investing in the engineering and applications aspects of BCI:
  - Tübingen, Germany: research-dedicated fMRI and MEG systems for non-invasive BCI
  - Freiburg, Germany: large-scale research program in nonlinear dynamics of brain function
  - Lausanne, Switzerland: world's most advanced electrophysiological and modeling analysis of cortical circuitry.

Translation/Commercialization of BCI Research

- The European system has created specific mechanisms and institutions for cooperative activity between academia and industry; there is a high level of transitioning BCI research.
- The European system is more effective than U.S. systems in integrating Industrial and academic efforts; there is substantial support from industry for BCI research.

Extension of BCI Research to Patient Populations

- There are several compelling examples of integrated research, development, and clinical applications in both Europe and the United States:

  - University of Aalborg, University of Tübingen, La Sapienza University
  - Wadsworth Center, Case Western Reserve University.
- Collaborations between the United States and Europe on "best practices" in clinical applications of BCIs would be beneficial.

Educational/Training Programs in BCI

- Surprisingly little attention is paid to developing formal, BCI-specific training programs at the undergraduate, graduate, or postdoctoral levels.
- The United States clearly has more comprehensive, well-developed educational/training programs in BCI, with greater sensitivity to recruiting underrepresented minorities.
- New programs for interdisciplinary training are under development in Europe at Aalborg University and Scuola Superiore Sant'Anna.

STUDY HIGHLIGHTS: BCI R&D IN ASIA

China

Overall Scope and Magnitude of BCI Research in China

- Although BCI research in China only started within the last ten years, it is already substantial in its scope and impressive in its accomplishments.
- BCI algorithm development already leads the field.
- Current BCI research is focused on low-cost, low-technology solutions—a reflection of socioeconomic demand, i.e., large population and relatively low economic status.
- Extension to clinical settings and commercialization of BCIs are barely begun.
- Future BCI research will incorporate "systems-level" solutions evolving from fundamental, invasive studies of brain function.

Future Growth of BCI Research in China

- Growth rate is now high and will remain high into the future.
- BCI research will benefit from broad, large-scale investment in biological/medical sciences, engineering/microelectronics, and mathematics/computer sciences.
- Evidence exists for targeted, high-priority investment in BCI/biomedical engineering.
- New facilities of world-class caliber for BCI/biomedical engineering:

  - Tsinghua University: new biomedical engineering building/facilities
  - East China Normal University: new state-of-the-art multisite electrophysiological facilities; new genetic mouse-breeding facilities
  - Shanghai Jiao-Tong University: new campus; new multidisciplinary facilities for biomedical engineering, microelectronics, computing

- Strong, high-level academic/government support exists.
- Associations between different disciplines, critical for the development of BCIs, are already forming.
- Strong commitments to education and large student/faculty population exist.
- Invasive BCI programs are just now emerging, but commitment is clear and investment has begun.

Relations with Industry/Commercialization

- BCI research is in its beginning stages in China, but it is too early for significant industrial involvement or commercialization.
- Nevertheless, there are multiple patents, and researchers are conscious of commercialization.

Funding and Funding Mechanisms

- The primary funding source for BCI research in China is the government.
- Funding entities include the Chinese Ministry of Science and Technology, "NNSF China" (National Natural Science Foundation of China), and the China High-Tech Research and Development Program.

Training Programs and Educational Mechanisms

- Little attention is now paid to developing BCI-specific training programs at any level: undergraduate, graduate, or postdoctoral.
- Because of the early stage of development of BCI programs in China, efforts are focused on forming foundational departments and programs (e.g., biomedical engineering); as a consequence, traditional disciplines have precedence.

Japan

Overall Scope and Magnitude of BCI Research in Japan

- BCI research in Japan should be evaluated within a context very different than that of China; critical factors for Japan are:

  - Mature neuroscience and engineering research environments
  - World-leading robotics programs (output of motor BCI systems)
  - Integrated academic-industrial research agendas/partnerships.

- Like China, Japan also is "discovering" BCI research (in terms of BCI-directed research currently representing a relatively small percentage of its total current research effort), but Japan appears to conduct BCI research in the following ways:
  - As an extension of the challenge of understanding the brain
  - As an extension of its now well-developed robotics programs (BCI-controlled robotics platforms).
- BCI research in Japan is currently almost exclusively noninvasive, despite the many experimentally-based Japanese neuroscience programs. This results from the following:
  - A deliberate decision motivated by estimates of the ultimate user base (users other than those requiring nervous system repairs)
  - High-level technologies within Japanese research and industrial entities for noninvasive BCI research, e.g., combined fMRI, MEG, NIRS.
- Japan has a "broader" perspective on BCIs than most other countries:
  - BCIs are not just for medical applications and nervous system repair
  - BCIs are integrated into everyday life of "normal" individuals (e.g., enhancing desired movements, enhanced cognitive function)
  - Commercial issues with respect to both medical and nonmedical applications of BCIs are already being considered
  - Ethical issues are already elevated to a significant level of importance

Future Growth of BCI Research in Japan

- Future growth will *increase* from the present at a relatively low rate.
- Driving forces for future growth include
  - Commercial value of nonmedical applications
  - Increasing size of aging population: need for "assistive" BCI applications
  - Increased need for "smart" security/safety sensor-actuator systems

Relations with Industry/Commercialization

- BCI research already is becoming well integrated with large-scale industry:
  - Nippon Telegraph and Telephone (NTT)
  - Advanced Technology Research Institute (ATR).

- Growth of industrial involvement should increase in future years.
- There is the issue of need to balance supporting BCI growth within "agile," small-sized companies against supporting BCI growth within the less dynamic, but better-funded large-sized companies.

Funding and Funding Mechanisms

- BCI research is primarily sponsored by the government.
- Counter to recent trends in the United States, Japan continues to "bridge the gap" between academic and industrial research with funding from industry.

Training Programs and Educational Mechanisms

- Relatively little attention is paid to specialized training programs for BCIs. This probably reflects funding levels that are sufficiently broad-based that specialized training programs are unnecessary.

CONCLUSIONS

Magnitude of BCI Research

- In general terms, the magnitude of BCI research throughout the world will grow substantially, if not dramatically, in future years.
- There are multiple driving forces:
  - Continued advances in underlying science and technology
  - Increasing demand for solutions to repair the nervous system
  - Increase in the aging population world-wide; need for solutions to age-related, neurodegenerative disorders, and for "assistive" BCI technologies
  - Commercial demand for nonmedical BCIs.

Scope of BCI Research

- The need for nervous system repair will remain the core driving force for BCIs.
- BCI research will increasingly widen to include nonmedical BCIs because of commercial demand, e.g., video games, automobile industry.
- There is a long-standing need for "intelligent" robotics.

Invasive Versus Noninvasive BCI Research

- The majority of invasive BCI research is now being conducted in the United States; this is likely to remain the case for decades into the future.
- European BCI research will be limited to the noninvasive domain for the foreseeable future as a result of the strong influence of animal rights advocates.
- China's BCI research programs will increasingly become more balanced in terms of invasive and noninvasive technologies as China's BCI programs grow:
  - Noninvasive BCIs will be in high demand because of the large population and limited healthcare funding.
  - Invasive BCIs will become increasingly attractive because of strong growth in fundamental neuroscience/engineering and the lack of animal rights movements.
- Japan's research programs will continue to focus on brain-robotics BCIs and how to utilize high-tech, noninvasive methodologies as the basis for BCIs.

Opportunities for Worldwide Collaborative Research

- The relationship between EEG/MEG, local field potentials, and (population) single-unit activity measures of cortical activity remains an issue that is both fundamental to neuroscience and useful in the context of developing BCIs. Cooperation in this research area could stimulate and maintain U.S.-European-Asian collaborations.
- There remain multiple electrode technologies throughout the world for recording and stimulating neural tissue.
  - Systematic evaluation of these technologies, with respect to defined needs/ conditions, would be extremely helpful
  - Development of new technologies is essential (e.g., "dry" EEG electrodes, small-feature-size micro/nanoscale electrodes)
- The issue of biocompatibility between micromachined devices and brain tissue, particularly within the context of recording-stimulation functionality maintained for implant periods greater than one year, remains a high priority.
- There is a need to identify spatiotemporal patterns of population, ensemble unit firing.
  - Multiple theoretical/modeling approaches have been proposed and utilized as part of BCI projects throughout the world
  - Systematic evaluation of these methods—and development of new approaches—is sorely needed

- Solutions addressing the issue of hardware implementations of BCI models remain opportunistic and not approached in a rigorously defined manner. Still to be explored methodically are
  - Analog vs. digital vs. hybrid design advantages
  - Integration of low-power design constraints
  - Potential synergies between the designs for medical and nonmedical applications.

CHAPTER 1

Introduction

Theodore W. Berger

BACKGROUND AND SCOPE

The impetus behind research into the establishment of communications pathways between the brain and external devices, or brain-computer interfaces (BCI), can be traced back to studies conducted in the 1970s postulating algorithms that correlated the firing patterns of motor cortex neurons with specific muscular responses. In the intervening decades, advances in computer and sensor technologies, component miniaturization, and materials biocompatibility, as well as our ever-improving understanding of the human central nervous system (CNS), have served to accelerate research into the development of truly effective BCI systems.

Today, BCI systems can be broadly classified into two categories, depending on the placement of the electrodes used to detect and measure neurons firing in the brain. In *invasive* systems, electrodes are inserted directly into brain tissue. In *non-invasive* systems, electrodes are placed on the scalp and use electroencephalography (EEG) or electrocorticography (ECoG) to detect neuron activity. Other sensing methods employed in BCI systems in an auxiliary capacity include magnetoencephalography (MEG), thermography, functional magnetic resonance imagery (fMRI) interpretation, and analysis of near infrared spectrum (NIRS) activity.

Currently, governments, universities, and private industry around the world are engaged in a wide variety of research projects related to various aspects of BCI. As just one measure of the increase in interest, the number of BCI-related scientific papers published in technical journals and at conferences has doubled every year since 2002.

To provide program managers in U.S. research agencies as well as researchers in the field a better understanding of the status and trends in BCI research abroad, in December 2005 the National Science Foundation (NSF), the Army Telemedicine and Advanced Technologies Research Center (TATRC), the National Institute of Biomedical Imaging and Bioengineering (NIBIB) and the National Institute of Neurological Disorders and Stroke (NINDS) of the National Institutes of Health

T.W. Berger et al., Brain-Computer Interfaces, 1–6.

(NIH), the National Space Biomedical Research Institute, and the Margot Anderson Brain Restoration Foundation sponsored the *WTEC International Assessment of Brain-Computer Interfaces*. The study was designed to gather information on the worldwide status and trends in BCI research and to disseminate it to government decision makers and the research community. The study participants reviewed and assessed the state of the art in sensor technology, interface and compatibility, data analysis and modeling, hardware implementation, systems engineering, and functional electrical stimulation (FES) in academic research and industry.

Questions of interest to the sponsoring agencies to be addressed by the study included the following:

- What is the state of science worldwide, including investigators and funding profiles?
- What are the gaps, holes, and needs? What are the "grand challenges," and are they being addressed?
- What kinds of clinical studies have been initiated?

As BCI research continues to accelerate into the foreseeable future, this study will help researchers to collaborate and exchange scientific data more effectively and to direct more focused research into research areas that offer promising results.

METHODOLOGY

Once the agency sponsors established the scope of the assessment, WTEC recruited a panel of U.S. experts chaired by Theodore W. Berger, Professor of Biomedical Engineering and Neurosciences, David Packard Professor of Engineering, and Director of the Center for Neural Engineering at the University of Southern California (see Table 1.1). The assessment was initiated by a kickoff meeting on December 12, 2005 at the NSF headquarters in Arlington, Virginia. Participants discussed the scope of the project and the need for a North American baseline workshop, candidate sites in Europe and Asia for panel visits, the overall project schedule, and assignments for the final report.

Table 1.1 Panel Members

#	Panelist	Affiliation
1	Theodore W. Berger (Panel Chair)	University of Southern California
2	John K. Chapin	SUNY Downstate Medical Center
3	Greg A. Gerhardt	University of Kentucky
4	Dennis J. McFarland	Wadsworth Center
5	José C. Principe	University of Florida
6	Dawn M. Taylor	Case Western Reserve University
7	Patrick A. Tresco	University of Utah

The panelists, sponsors, and WTEC convened a North American Baseline Workshop on February 27, 2006, at NSF to report on noninvasive and minimally invasive BCI using EEG and ECoG; sensors, signal processing, and biocompatibility in invasive BCI; systems integration and modeling; and translation and comercialization issues. Table 1.2 lists the speakers and the titles of their presentations.

The international assessment phase of the WTEC study commenced in late May 2006 with two weeks of visits to the 17 European sites shown in Table 1.3. That trip concluded with an outstanding meeting in Frankfurt, Germany, on June 3, 2006, in which the panelists reviewed and compared their site visits in Europe. A second round of site visits to ten facilities in China and Japan took place during the last week of October 2006, as shown in Table 1.4. During its visit to China, the

Table 1.2 Speakers and Presentations at the North American Baseline Workshop

Name	Affiliation	Presentation Title
Theodore Berger	University of Southern California	WTEC International Assessment of Brain-Computer Interface Research
Gary Birch	Neil Squire Foundation	Asynchronous BCI and Brain Interface Research
Dan Moran	Washington University	Electrocorticographic (ECoG) Control of Brain-Computer Interfaces
Dennis McFarland	Wadsworth Center	Commentary: Summary of EEG/ECoG
Daryl Kipke	University of Michigan	Implantable Microscale Neural Interface Devices for BCI Systems
Richard Normann	University of Utah	Applications of Penetrating Microelectrodes in Nervous System Disorders
William Shain	Wadsworth Center	Understanding Biological Responses to Inserted Neural Prosthetic Devices: Building a Foundation to Promote Improved Tissue Integration and Device Performance
Patrick Tresco	University of Utah	Commentary
Greg Gerhardt	University of Kentucky	Commentary
Krishna Shenoy	Stanford University	Decoding Movement Plans for Use in Neural Prosthetic Devices
Andy Schwartz	University of Pittsburgh	Useful Signals from Motor Cortex
Dawn Taylor	Case Western Reserve University	Commentary
José Principe	University of Florida	Commentary
John Donoghue	Brown University	Neuromotor Prosthesis/Direct Brain Interfaces
David Putz	Ad-Tech Medical Instrument Corporation	The Path from Research & Development to FDA Approval to Commercialization
John Chapin	SUNY Downstate Medical Center	Commentary
Greg Gerhardt	University of Kentucky	Commentary

Table 1.3 Sites Visited in Europe

#	Country	Site	#	Country	Site
1	Austria	Graz University of Technology	10	Germany	Berlin Brain-Computer Interface (BBCI)
2	Austria	Guger Technologies OEG (g.tec)	11	Germany	Multi Channel Systems (MCS)
3	Belgium	European Union—Research Directorate General	12	Germany	University of Freiburg
4	Denmark	Aalborg University	13	Germany	University of Tübingen
5	England	University of Oxford	14	Italy	Polo Sant'Anna Valdera
6	France	CEA (Atomic Energy Commission)	15	Italy	The Santa Lucia Foundation
7	France	Physiology of Perception and Action Laboratory (CNRS/College de France)	16	Scotland	University of Edinburgh
8	Germany	Max Planck Institute for Biochemistry	17	Switzerland	Swiss Federal Institute of Technology
9	Germany	Natural and Medical Sciences Institute and Retina Implant (NMI)			

Table 1.4 Sites Visited in Asia

#	Country	Site	#	Country	Site
1	China	Huazhong University of Sciences and Technology	6	China	Wuhan University
2	China	Shanghai Institute of Brain Functional Genomics	7	Japan	RIKEN Brain Science Institute
3	China	Tsinghua University, Department of Electrical Engineering	8	Japan	Advanced Telecommunications Research Institute
4	China	Tsinghua University Institute of Microelectronics	9	Japan	NTT Communication Science Laboratories
5	China	Shanghai Jiao Tong University	10	Japan	Waseda University

WTEC panel was privileged to attend a symposium on BCIs sponsored by Shanghai Jiao-Tong University's Institute of Laser Medicine and Biophotonics, at which approximately 75–100 faculty and students heard presentations from a dozen faculty members whose laboratories are actively developing BCIs.

WTEC hosts in both Europe and Asia demonstrated a wide range of BCI research and systems in various stages of development in laboratory settings. This included computer-based animal and human testing of invasive and noninvasive systems; research and experimentation protocols; experimentation aimed at improving signal and pattern recognition; and hardware and software development. The panelists noted that the degree of collaboration between the biological and engineering sciences varied widely among the institutes visited.

Following the conclusion of the European round of site visits but prior to the visits to China and Japan, the panel reconvened for a final workshop at NSF on July 21, 2006, to present its findings and conclusions. Presentations focused on the following topics:

- Sensor technologies
- Biotic–abiotic interfaces
- Modeling, architectures, and signal processing
- Robotics and prosthetics
- FES and rehabilitation applications
- Communication devices
- Cognitive and emotional prostheses
- Organizational and translational issues.

OVERVIEW OF THE REPORT

In Chapters 2 and 3, Drs. Gerhardt and Tresco review some of the major technical issues involved in recording electrophysiological activity from multi-site arrays. These involve micro-electrode materials and manufacturing procedures (Chapter 2), and biocompatibility and integration of the electrodes with brain tissue (Chapter 3). Chapter 4 reviews signal processing and modeling methodology commonly used in BCI research to date, with future issues and directions for future research indicated as well (Drs. Principe and McFarland). In Chapter 5, John Chapin reviews how various BCI systems promise to help people overcome paralysis caused by damage to the brain, spinal cord, spinal nerves, or muscles. Although cell biology research may ultimately yield definitive cures for paralysis, at least into the near future the restoration of motor function will likely depend on continued progress in electronic and computer technologies. In Chapter 6 Dawn Taylor summarizes recent progress in FES for a variety of lifesaving and motor-control applications. She reminds us that BCI-derived options must be considered within the broader context of techniques and technologies that are (or will soon be) available to users. In Chapter 7, Dennis McFarland discusses how recent advances in EEG-based BCI communications systems promise mobility and control to people who have experienced loss of voluntary and/or involuntary muscle control. The twin challenges of limited bandwidth and system complexity must be overcome if today's proof-of-principle systems are to become tomorrow's successful applications. In Chapter 8, Walid Soussou and Theodore Berger present developments in cognitive and emotional prostheses to address cognitive impairments such as memory loss, mood or personality alterations, behavioral changes, and emotional dysfunction. Finally, in Chapter 9, Theodore Berger reviews issues of funding for research organizations, translation-commercialization, and education-training.

Appendix A contains biographies of the delegation members, and Appendixes B and C include detailed reports for each of the sites visited during the international assessment; a glossary is provided in Appendix D.

Additional information, documentation, and photographs for all phases of the WTEC International Assessment of Brain-Computer Interfaces are available on the WTEC website at http://www.wtec.org/bci/. In particular, a list of foreign and domestic BCI-related research programs, professional organizations, and conferences is provided at http://www.wtec.org/bci/BCI_Research_Programs.htm.

CHAPTER 2

Sensor Technology

Greg A. Gerhardt and Patrick A. Tresco

INTRODUCTION

This chapter deals with an overview of sensors used in the collection of data for Brain-Computer Interface (BCI) technology. For the purposes of this chapter, we divide sensor technologies into two basic categories. First, we discuss "invasive" technologies, which entail brain surgery procedures for implantation involving primarily multielectrode recordings from arrays of microelectrodes implanted directly into the brain to measure action potentials from single cells. This is a major growth area for sensor technologies and will be the major focus of this chapter. However, we caution that most of this technology is under development in animal models and is not yet approved for human use. In addition, measurements from subdural or epidural strips of electrode arrays used to record cortical potentials somewhat analogous to EEG-type recordings on the surface of the skull will be discussed, as this is currently the greatest application for use of these invasive electrodes in humans for (primarily) epilepsy surgery. However, this could help increase the growth of other BCI applications. Second, we discuss "noninvasive" technologies, which primarily involve multielectrode EEG recording arrays of "wet" silver (Ag) or gold (Au) conducting paste electrodes that are placed on the surface of the skull to record EEG activity. These electrodes are commercially available from a number of sources, but surprisingly, there has been limited growth in this area. We caution that "noninvasive" electrodes have largely been used acutely and may be more invasive to the scalp when used in future, more chronic, applications of BCI technology by humans at home or work. Additional tech-nology development in this area will be briefly discussed.

We do not discuss other types of recording electrodes such as EMG electrodes and associated electrodes, which are covered in other sources. In addition, we do not discuss deep-brain stimulation (DBS) technology, which is used extensively in patients with movement disorders (Kossof et al., 2004). This area, however, should be monitored as the chronic implantation of the stimulating electrodes for DBS is a

T.W. Berger et al., Brain-Computer Interfaces, 7–29.

clinical forum for development of long-lasting brain electrode technologies and a test bed for development of brain-compatible BCI devices (see Chapter 3).

Electrodes are enabling technologies to allow information from the brain to be encoded by computer algorithms to provide input and control of BCI devices. Without these devices we cannot transfer information from the brain that can be used to control BCI instrumentation. As such, it is too often assumed that the technologies surrounding sensors for BCI are fully worked out and that there is little room for improvement. In reality, there is a tremendous potential for growth of these devices and need for new types of both invasive and noninvasive electrode technologies to further pursue BCI applications. The major challenges are discussed at the end of this chapter.

The purpose of the present chapter is to review the current sensor technologies used for invasive and noninvasive BCI approaches throughout North America, Europe, and Asia. We have visited and/or interacted with key laboratories with expertise in these areas. Although not completely comprehensive, this chapter gives an overview of the major sensor technologies that are being developed for potential BCI applications.

We are pleased to acknowledge the extensive assistance of Jason J. Burmeister, our colleague at the University of Kentucky, for helping us prepare this chapter.

BCI SENSOR WORLD OVERVIEW

Most BCI science in North America involves "invasive" sensor technologies, i.e., multielectrode recordings from arrays of microelectrodes implanted directly into the brain. This is the greatest area of growth in sensor technology.

Most BCI science in Europe involves "noninvasive" sensor technologies, i.e., using multielectrode recordings from arrays of EEG electrodes mounted onto the surface of the skull. This sensor technology has experienced a very limited growth and requires substantial improvement. Certain BCI sites in Europe are capable of providing sensor technologies that could aid in the advancement of "invasive" sensor technologies; however, this is not their current plan.

Even with respect to noninvasive technologies, many European sites collaborate with, or utilize paradigms that were developed in the United States, such as at the Wadsworth Center in Albany NY.

In Asia, there is a clear emphasis on less expensive EEG BCI approaches. Reasons include the large population in China and the need for low-cost, noninvasive BCI technology for improved public healthcare there. Japan is also focused on noninvasive EEG-based BCI technologies. There is rapid economic growth and science spending in China and Japan that will propel all BCI technology development forward. In addition, there are clear indications that facilities are available and there is interest in invasive BCI technology in China. Overall, the panel believes Asia has the manufacturing facilities and infrastructure to drive development of new invasive BCI technology development that could rival or exceed U.S. efforts in five to ten years.

MAJOR TYPES OF SENSORS FOR BCI TECHNOLOGY

History of Direct Implantable Electrodes

The history of implanting electrode arrays in the CNS (see Chapter 3 for historical references and additional papers) dates back to the early work of Hess in the 1930s with initial implants in felines. This set the stage for investigators in the 1950s, such as Heath and Olds (Heath et al., 1953; Olds et al., 1971; Baumeister, 2006), to use implantable electrodes primarily for electrical stimulation of the brain, but also for recording. In the late 1950s, Fischer and colleagues were the first to use a variety of different metal-type electrodes and single-wire electrodes and also started to investigate any pathology resulting from the effects of wire electrodes (see Chapter 3). However, the more modern adaptation of implantable electrodes occurred in the 1970s. Selman and Bach in the early 1970s started using coated microwires for electrophysiological recordings, and in the early 1980s Chapin and Woodward (1986) reported the development of 50 μm tungsten microwire arrays for multiple single-unit recordings. Basically, this type of technology is used today by many laboratories for the more routine multiple single-unit recordings and many applications of BCI in animals. However, some of the problems of multiwire arrays relate to precise control of the electrode recording sites and issues surrounding the viability of individual wires.

Between 1970 and 1975, Wise and Angell (Wise et al., 1970; Wise and Angell, 1975) introduced the concept of using integrated chip (IC) technology to develop microelectrodes. Over the next years, numerous papers were published, and in the 1980s the seminal work of BeMent and coworkers (BeMent et al., 1986; Drake et al., 1988) was the first development of a multisite microelectrode arrays from silicon. A few years later, in the early 1990s, the first silicon-based monolithic multishank electrode array was developed, which is now used by numerous laboratories and is even used for human BCI applications by Donoghue and coworkers (Hochberg et al., 2006). In general, microelectrodes can provide a means to electrically stimulate and record both electrophysiological activity and chemical activity of neurons in the brain and spinal cord (Hochberg et al., 2006; Burmeister and Gerhardt, 2006). There have been many reports too numerous to cite for this chapter of the design and use of microelectrodes for electrophysiological recordings (Anderson et al., 1989; Burmeister and Gerhardt, 2006; Cheung, 2007). In addition, in part we have discussed some of this technology in a recent chapter (Burmeister and Gerhardt, 2006).

Wire-Type Microelectrodes

Currently, the workhorse electrode for recording multiple single-unit action potential activity from the brains of animals is through the use of what are termed microwire array bundles. These generally involve the use of 13–200 μm-diameter, Teflon®-coated tungsten (W) or iridium (Ir) wires arranged in bundles of 16–64 or even hundreds of wires. Some of the longest BCI-type recordings for 1.5 years have been carried out with these types of electrodes (see also Chapter 3).

Most wire-type microelectrodes are constructed by sealing a metal (tungsten, gold, platinum, iridium, platinum-iridium, stainless steel) wire in an insulating material. The metal wires from the brain and the connections between the recording wires are insulated using Teflon or plastics. The microelectrode surface area is determined by cutting the exposed wire to a desired length. Typical wire electrodes range in diameter from 13–200 μm, with an exposed length of up to 1 mm. Wire electrodes are widely used for recordings in rats, monkeys, cats, and more recently, mice (see Table 3.1 in Chapter 3, Burmeister and Gerhardt, 2006; Ludvig, 2001; Chapin and Nicolelis, 2001; Chapin, 2004; Chiganos et al., 2006; Lin et al., 2006). Figure 2.1 shows an example of a high-density array and integrated microdrive for recordings from as many as 128 wires from freely moving mice (Lin et al., 2006). In addition, this microwire bundle incorporates a microdrive device so that the microwire electrodes can be repositioned for optimum performance during the recordings. Additional information about wire electrodes can be found in other sources (Burmeister and Gerhardt, 2006).

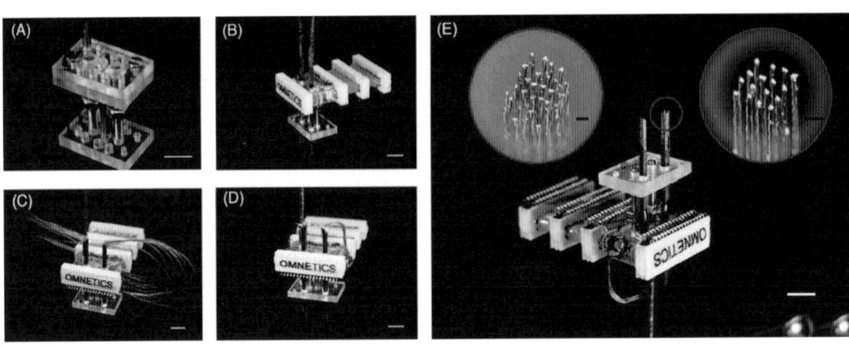

Figure 2.1. Construction of a high-density ensemble recording microdrive for mice. (a) is the base foundation for the microdrive; (b) indicates four 36-pin connector arrays positioned at the base of the microdrive in parallel (each bundle of 32 pieces—for stereotetrodes—or 16 pieces (for tetrodes) of polyimide tubing was glued to an independently movable screw nut on the microdrive base); (c) is a microdrive on the assembly stage (the free ends of electrode wires are wrapped around to adjacent connect pins); (d) is a fully assembled, adjustable 128-electrode microdrive; (e) indicates that 128 channels can be formatted with either tetrodes (right inset) or stereotetrodes (left inset) on each bundle. The tip of the two electrode bundles was shaped at a certain angle (10°–20°) to fit the contour of the dorsal CA1 cell layer. Black scale bars in red circles of (e) are 100 μm. White scale bars in (a–d) are 3 mm (Lin et al., 2006; © The Society for Neuroscience).

Traditional wire-type microelectrodes are still in wide use for several reasons. First, they can be purchased from several vendors or constructed from commercially available materials (Sugiyama et al., 1994; Williams et al., 1999; Rennaker et al., 2005; Lin et al., 2006; Burmeister and Gerhardt, 2006). Second, very small microelectrodes can be constructed (Lin et al., 2006; Burmeister and Gerhardt, 2006). Third, they are established in the field. However, traditional wire microelectrodes have disadvantages. Because they are handmade, large variability between individual microelectrodes with inconsistent geometries can result. Surface area variability resulting in altered response characteristics can be caused by irregularities in the cut tip and the junction between the metal and the insulating material. Because of the needed supplies and materials as well as the art of their production, many labs have difficulty assembling reproducible microelectrodes.

Mass-Fabricated Microelectrodes

Photolithographic methods employed in the microcircuit industry are used for the mass fabrication of microelectrodes (Burmeister and Gerhardt, 2006; Cheung, 2007). Recording surfaces as small as 5–10 μm can be routinely produced now, and in the future, surfaces as small as 0.1–4 μm will be developed using photolithography methods (Smith et al., 2004). This rivals or exceeds some of the smallest traditional microelectrode tips for intracellular recordings. However, less expensive screen-printing methods can be used to fashion features as small as 50–100 μm if very small microelectrode features are not required. In addition, multiple designs of microelectrodes can be patterned simultaneously on the same substrate, allowing for large numbers of microelectrodes to be simultaneously fabricated, reducing production costs. Also, micromachining procedures may be used to construct microelectrodes with multiple recording sites in well-defined spatial arrangements that may be used to record from layered brain structures. The microelectrodes can be designed to conform to brain structures. Improved quality of microelectrodes may be achieved by allowing experts in the semiconductor industry to fabricate the microelectrodes, thereby avoiding the inherent costs of setting up in-house microfabrication facilities (e.g., Thin Film Technologies, Inc.).

There are four basic layers to most microelectrodes constructed using thin-film techniques. The substrate is the first layer, which often is composed of silicon, ceramic, silicon, silica/glass, or polyimide. An insulating layer such as silicon nitride often covers the substrate when a silicon substrate is used. An adhesion layer of titanium or chromium may be applied to the substrate to allow the active metal to adhere to the substrate surface if needed. Photolithography or screen printing is used to lay out the microelectrode recording sites, connecting lines, and bonding pads using the desired noble metals such as Au, Pt, or Ir. An insulating layer such as polyimide, silicon nitride, or alumina is applied to the connecting lines (Burmeister and Gerhardt, 2006). After application of the insulating layer, only the recording sites and bonding pads are exposed. Microelectrodes constructed using eight or more photomasks with very specialized layers have been reported (Anderson et al., 1989; Bai et al., 2000; Burmeister and Gerhardt, 2006; Najafi et al., 1990). Numerous microelectrodes can be formed on a single substrate at the same time using this approach. The final shape of the microelectrodes is achieved by chemical etching, laser cutting, or diamond saw procedures. Finally, the bonding pads of the individual microelectrodes are wire-bonded to a larger printed circuit board (PCB) holder or "paddle" that is more easily handled and connected to recording equipment.

Silicon-Based Microelectrodes

Silicon was the first substrate to be used to construct multisite, semiconductor-based microelectrodes, and there have been many reports of such microelectrodes for brain recordings and brain tissue stimulation (Anderson et al., 1989; Schmidt et al., 1993; Kovacs et al., 1994; Della Santina et al., 1997; Bai et al., 2000; Najafi et al., 1990; Yoon et al., 2000; Vetter et al., 2004; Kipke et al., 2003; Burmeister and Gerhardt, 2006). The option of using chemical etching is one of the desirable properties of silicon as a substrate. Individual microelectrodes can be formed from a single substrate simultaneously without the need for laser machining or sawing. Small features such as channels in the substrate can be constructed. Very thin microelectrodes may be fashioned by etching to reduce the substrate thickness. Substrates as thin as 6–15 μm have been reported (BeMent et al., 1986; Drake et al., 1988; Hetke et al., 1994; Burmeister and Gerhardt, 2006). However, a very thin silicon substrate is flexible and fragile. Flexibility is both desirable and a liability. Once implanted, flexible microelectrodes have the ability to move with the tissue and possibly minimize damage. However, one must caution that long, thin, flexible silicon electrodes can be difficult to implant. An insulating layer between the metal and the silicon substrate may be necessary to reduce electrical crosstalk between adjacent recording sites because silicon is a semiconductor (Moxon et al., 2004; BeMent et al., 1986; Drake et al., 1988; Hetke et al., 1994; Ensell et al., 2000; Burmeister and Gerhardt, 2006).

The semiconductor properties of silicon can be altered by doping. Also, silicon is very compatible with onboard circuitry. Silicon has many features that have made it widely used as the foundation for forming microelectrode arrays. Photographs of some silicon-based microelectrodes constructed at the Center for Neural Communication Technology at the University of Michigan, which is the home to some of the greatest contributions to BCI microelectrode technology, are shown in Figure 2.2 (Anderson et al., 1989, Bai et al., 2000, Najafi et al., 1990, BeMent et al., 1986, Drake et al., 1988, Hetke et al., 1994). These represent many of the current designs that have been used for BCI applications in rats and nonhuman

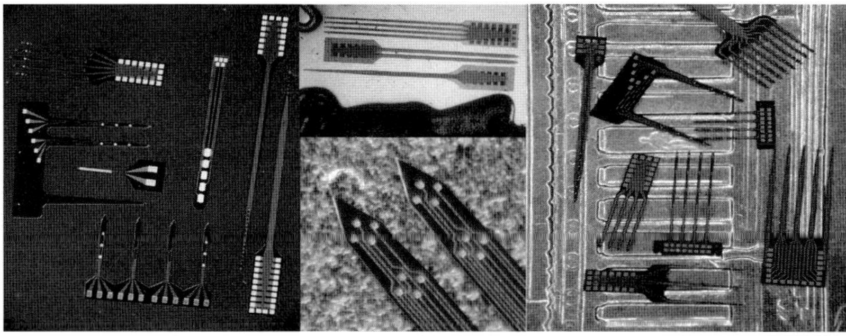

Figure 2.2. Photomicrograph of silicon-based microelectrode arrays constructed at the University of Michigan. Michigan probe photos were provided by David Anderson at the University of Michigan Center for Neural Communication Technology, an NIH/NCRR Resource Center. Used with permission, from *Encyclopedia of Sensors* (Burmeister and Gerhardt, 2006).

primates. In addition, this grouping of microelectrodes shows some of the versatile designs afforded by this approach. The option of chemical etching procedures is one of the greatest advantages silicon has as a substrate material. The micro-electrode thickness as well as shape can be altered using etching. Isotropic etchant (10% hydrofluoric acid, 90% nitric acid) is used for thinning of the substrate. An etch of ethylene-diamine-pyrocatechol water (EDP) is used to separate the individual microelectrodes from the silicon substrate (Burmeister and Gerhardt, 2006). A layer of silicon nitride patterned onto the silicon wafer can be used to define the intended microelectrode shape. Silicon nitride stops the etchant from reacting with the substrate. Alternatively, the etchant may also be stopped by selectively doping the substrate with boron (Bai et al., 2000; Najafi et al., 1990, Ensell et al., 2000).

A promising silicon-based electrode array design has been developed by the VSAMUEL consortium (European Union, grant IST-1999–10073 termed ACREO [ACREO AB, Sweden]) on microelectrode arrays (Jensen et al., 2006; Yoshida et al., 2001). These electrodes have one to eight recording shafts, are very versatile and flexible, and appear to have very promising insertion mechanics (Jensen et al., 2006). These also represent the major microelectrode manufacturing capabilities in the European Union, which strongly competes with the technologies being developed in the United States and Asia. Figure 2.3 shows representative designs.

Novel devices can be integrated onto the sensors using silicon-based microelectrodes. Holes have been etched into the substrate to aid in securing the microelectrode into brain tissue and to perhaps better integrate the electrode into the brain extracellular space (Kovacs et al., 1992, 1994; Della Santina et al., 1997; Burmeister and Gerhardt, 2006). Multiple flow channels for the delivery of chemicals/drugs, while performing electrophysiological recordings, have been etched into the silicon probe substrate (see Figure 2.4) (Chen et al., 1997; Rathnasingham et al., 2004; Burmeister and Gerhardt, 2006).

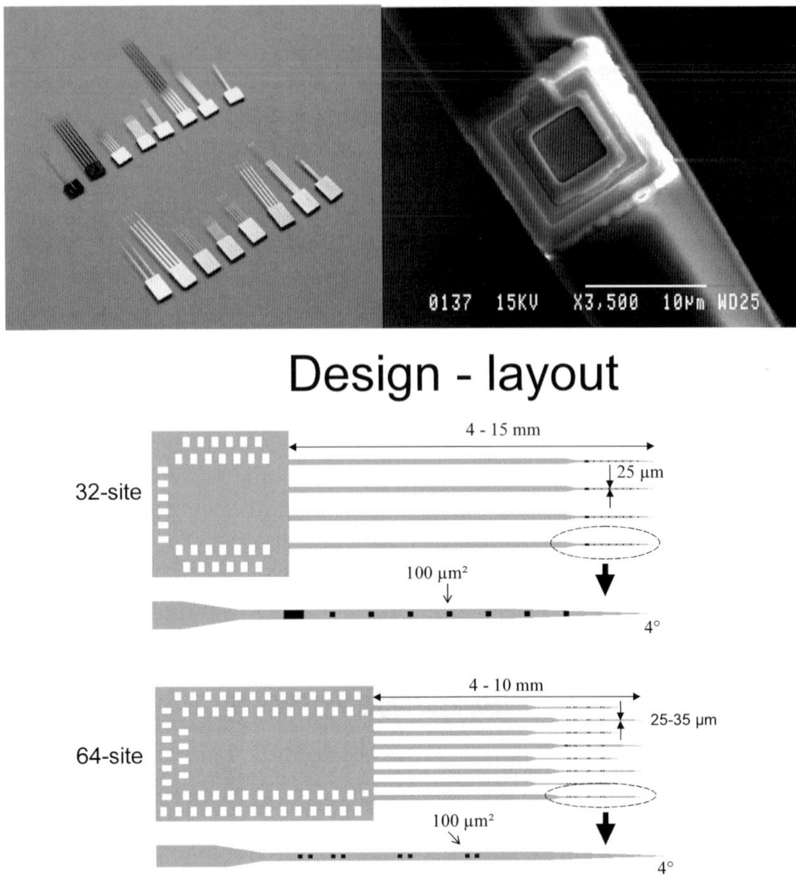

Figure 2.3. (Top-left) examples of silicon-based ACREO microelectrode arrays; (top-right) micrograph of an individual ACREO microelectrode recording site; (bottom) schematic of the ACREO microelectrode arrays (Photographs courtesy of ACREO AB, Sweden).

Integrated Ag/AgCl reference electrodes have been included on microelectrode arrays (Burmeister and Gerhardt, 2006; Pancrazio et al., 1998). Microdrives have been integrated into the microelectrode design for *in situ* adjustments after implantation (Burmeister and Gerhardt, 2006). An integrated polysilicon microheating device has been constructed (Chen and Wise, 1997). On-electrode amplification and signal processing may be achieved by including VLSI chips on the silicon substrate (see Figure 2.5 with integrated amplification) (Patterson et al., 2004; Bai and Wise, 2001; Pancrazio et al., 1998; Csicsvari et al., 2003). Silicon-based microelectrodes allow "hybrid" microelectrode designs to be manufactured.

Electrophysiological arrays with 100 recording sites have been developed to provide an interface for prosthetics, which is the foundation for the seminal work of Norman, Donoghue, and coworkers (Nordhausen et al., 1996; Hochberg et al.,

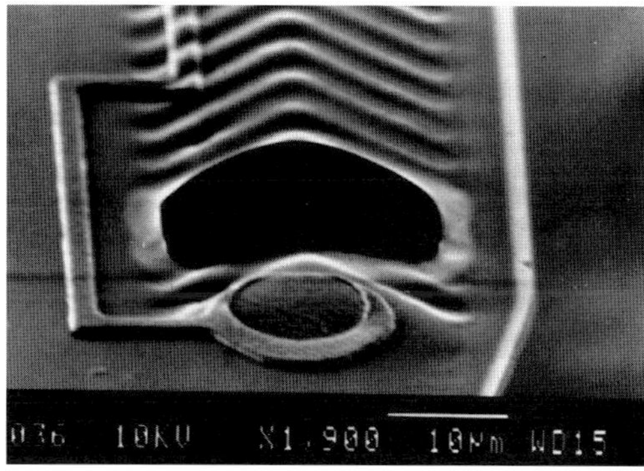

Figure 2.4. SEM of a microchannel on a silicon-based microelectrode for delivery of chemicals into CNS tissue. (Michigan probe photos provided by David Anderson at the University of Michigan Center for Neural Communication Technology, an NIH/NCRR Resource Center; reprinted with permission from *Encyclopedia of Sensors* [Burmeister and Gerhardt, 2006]).

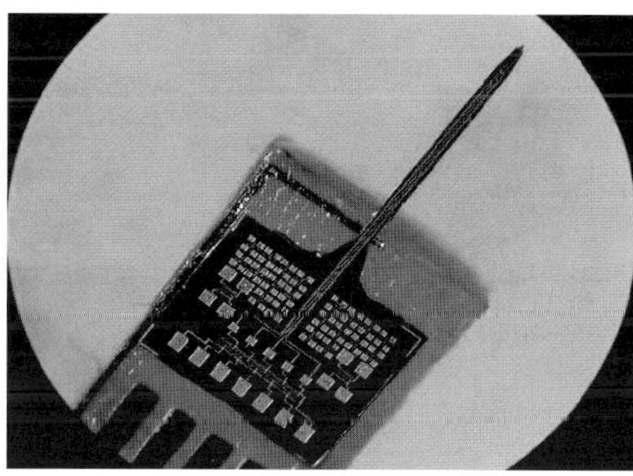

Figure 2.5. Photomicrograph of a silicon-based microelectrode for electrophysiological recordings with on-chip amplification is shown (photograph provided by Sung June Kim of Inter-University Semiconductor Research Center at Seoul National University, Korea; reprinted with permission from *Encyclopedia of Sensors* [Burmeister and Gerhardt, 2006]).

2006; Warren et al., 2001; Schmidt et al., 1993; Branner et al., 2004; Burmeister and Gerhardt, 2006). These designs are currently being used in humans and represent the first BCI microelectrode arrays that have been sterilized and used in both nonhuman and human primate trials. Individual microelectrode "shafts" extend 1.5 mm from the 10 × 10 mm planer substrate. The shaft tips are metalized with Pt over doped silicon for conduction down the shaft. The conducting doped silicon is insulated using glass and silicon nitride. Figure 2.6 shows a SEM of one of the "Utah" electrodes. Similar three-dimensional microelectrode arrays can be constructed by combining many planar silicon multishank microprobes (Hoogerwerf and Wise, 1994; Bai et al., 2000; Burmeister and Gerhardt, 2006), as shown in Figure 2.7. For brain-slice recordings, planar microelectrode arrays have been used to map neuronal communication (Borkholder et al., 1997; Burmeister and Gerhardt, 2006).

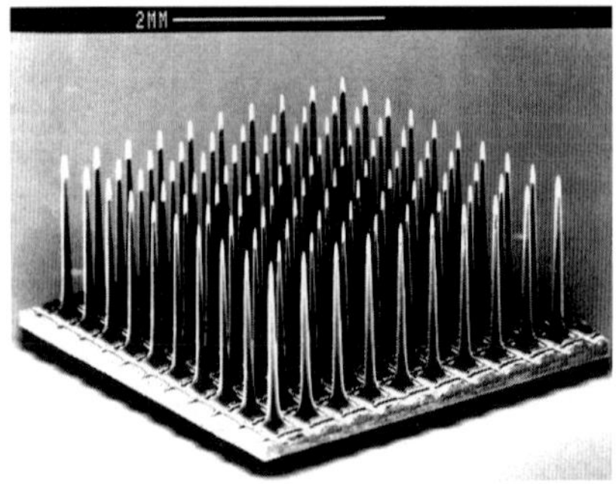

Figure 2.6. SEM of Utah Electrode Array (UEA) for visual prosthetics. The array consists of 100 individual microelectrode "shafts" that extend 1.5 mm from the 10 × 10 mm planar substrate (SEM provided by Richard A. Normann, Department of Bioengineering, University of Utah, Salt Lake City; reprinted with permission from *Encyclopedia of Sensors* [Burmeister and Gerhardt, 2006]).

Figure 2.7. Photomicrograph of a multishank probe formed using several silicon-based microelectrodes. There are multiple recording sites on each shaft for recordings at different brain depths. (Michigan probe photos provided by the University of Michigan Center for Neural Communication Technology, an NIH/NCRR Resource Center; reprinted with permission from *Encyclopedia of Sensors* [Burmeister and Gerhardt, 2006]).

Ceramic-Based Microelectrodes

The insulator ceramic (alumina, Al_2O_3) has been used as a substrate to reduce crosstalk between adjacent connecting lines (Burmeister and Gerhardt, 2001, 2006; Burmeister et al., 2000). Ceramic is mechanically strong, allowing for development of microelectrodes that can access much deeper brain structures (up to 5–6 cm vs. 2–4 mm for silicon). Precise placement of the microelectrode in tissue without flexing or breaking can be achieved. Multisite microelectrodes on ceramic substrates for use in animal models have been constructed (Moxon et al., 2004; Burmeister et al., 2000).

Individual microelectrodes must be mechanically cut from the wafer because the ceramic is not compatible with standard etching procedures. Laser machining is the most flexible way to cut the microelectrodes from the bulk wafers enabling formation of complex shapes. However, due to the stepping of the laser, laser machining can produce rough edges that can cause potential problems with microelectrode insertion into tissues. Much smoother microelectrode edges may be formed using a diamond saw, which polishes as it cuts; thus unnecessary tissue damage may be avoided. Minimal CNS tissue damage is required to study the biology of the intact brain. When using a diamond saw it is more difficult to form complex shapes because saws generally cut in straight lines. Figure 2.8a is a photograph of a complex microelectrode shape cut by laser machining. Figure 2.8b is a simple ceramic substrate microelectrode shape formed by a computer-controlled diamond saw. Figure 2.8c is a magnification of this microelectrode's smooth edges. The use of excimer lasers may provide smoother edges than conventional laser machining. Thinner microelectrodes may be achieved by polishing the ceramic substrate (Moxon et al., 2004).

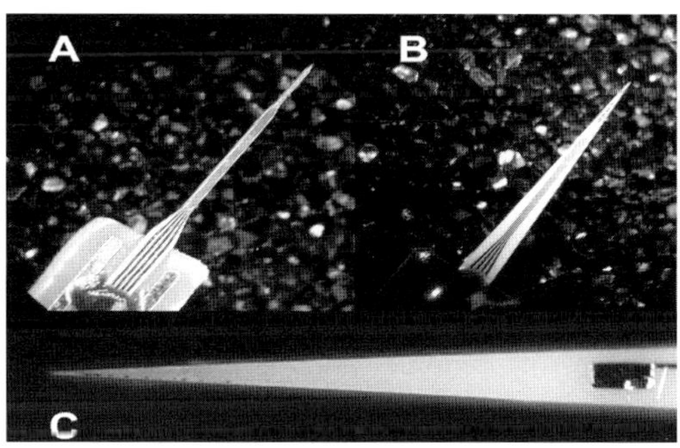

Figure 2.8. (a) Photograph of a complex ceramic substrate-based microelectrode shape cut by laser machining; (b) a less complex microelectrode shape formed by a computer-controlled diamond saw; (c) a magnification of the microelectrode's much smoother edge (Reprinted with permission from *Encyclopedia of Sensors* [Burmeister and Gerhardt, 2006]).

Figure 2.9 shows microelectrodes assembled on ceramic substrates that have been polished to make them between 38 to 51 μm thick with a tip width of 60 μm. The alumina insulating layer is applied using ion-beam-assisted deposition. These 20 × 80 μm platinum recording sites with 200 μm spacing have been used to record single-neuron action potentials *in vivo* for up to 24 weeks.

Numerous four- and five-site Pt microelectrodes on ceramic substrates have been developed. The versatility of the lithographic methods can be seen in Figure 2.10. In general, recording sites are either grouped in side-by-side pairs or in a linear arrangement. Two recent designs configure the microelectrodes in a linear arrangement similar to the previously reported 50 × 50 μm microelectrodes (Burmeister et al., 2000). The new designs have larger Pt recording sites of 50 × 100 and

50 × 150 μm in order to investigate whether larger recording sites can record better single-unit activity or lower detection limits for chemical recordings.

Two other new designs have two sets of microelectrodes arranged side-by-side: 25 × 100 and 25 × 300 μm. Recording-site dimensions vary from 10 × 10 μm to 25 × 300 μm, depending on the application. Other designs (dimensions in μm) include 10 × 10 serial (200 spacing), 20 × 20 serial (200 spacing), 50 × 50 serial (200 spacing), 25 × 100 pairs (15 spacing), 50 × 100 serial (200 spacing), 50 × 150 serial (200 spacing), 25 × 300 pairs (15 spacing), 25 × 300 pairs (30 spacing), 50 × 50 serial (400 spacing), 15 × 300 "eliminator," and 15 × 300 "T-eliminator." This also shows the versatility of such microelectrode fabrication approaches. Although the ceramic-base, multisite microelectrodes were originally intended to be disposable (one-time use), a cleaning procedure has been developed to allow for multiple uses due to the durability of the materials *in vivo* (Burmeister et al., 2002).

Figure 2.11 shows several designs of 8-site "conformal" microelectrodes that are under development for different brain region recordings in rats and monkeys. The individual electrodes may be chosen based on the brain region(s) and type of recordings of interest. For instance, two or more recording sites placed toward the tip of the microelectrode are useful in studying thin layers of cells such as the Purkinje cells in the cerebellum or pyramidal cells in the hippocampus. Multiple measures can be accomplished in the brain region of interest by providing a large concentration of recording sites at the tip. By spreading out the recording sites over a larger vertical distance, layered and/or larger brain structures such as the hippocampus, cortex, and striatum may be studied. Various species of animals may require different sizes and features of the microelectrode. In addition, the recording site density of the ceramic-based microelectrodes can be increased by

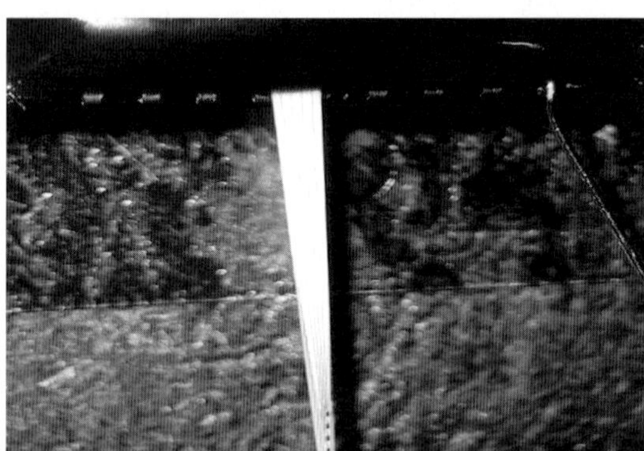

Figure 2.9. Photomicrograph of a ceramic-based microelectrode constructed on a thinner substrate with an alumina insulating layer. Alumina is applied using ion-beam-assisted deposition. The substrate thickness is between 38 to 51 μm with a tip width of 60 μm. The 20 × 80 μm platinum recording sites have been used to chronically record single-neuron action potentials *in vivo* for up to 8 weeks (Figure provided by Karen A. Moxon, Drexel University; reprinted with permission from *Encyclopedia of Sensors* [Burmeister and Gerhardt, 2006]).

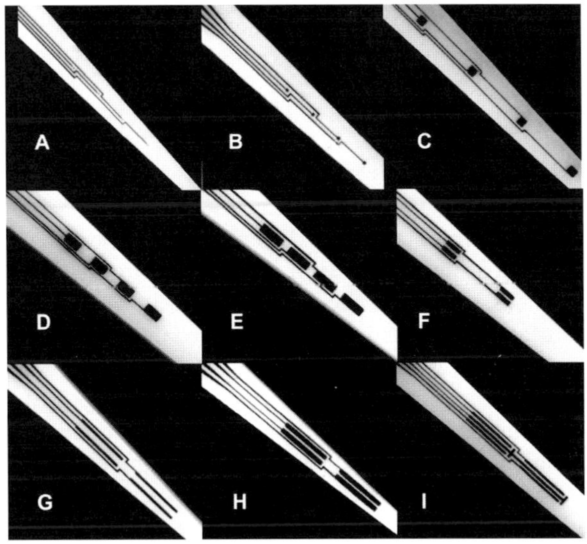

Figure 2.10. Photomicrographs of several ceramic-based multisite microelectrode designs. (a) 100 μm^2 serial—10 × 10 μm recording sites; (b) 400 μm^2 serial—20 × 20 μm recording sites; (c) 2,500 μm^2 serial—50 × 50 μm recording sites with 400 μm center-to-center spacing; (d) 5,000 μm^2 serial—100 × 50 μm recording sites; (e) 7,500 μm^2 serial—150 × 50 μm recording sites; (f) 2,500 μm^2 pairs—100 × 25 μm recording sites; (g) 4,500 μm^2 pairs 300 × 15 μm recording sites, 30 μm spacing; (h) 7,625 μm^2 pairs—305 × 25 μm recording sites; (i) 4,500 μm^2 eliminator—300 × 15 μm recording sites (Photographs are courtesy of Mr. Peter Huettl of the Center for Microelectrode Technologies University of Kentucky, Lexington, Kentucky; reprinted with permission from *Encyclopedia of Sensors* [Burmeister and Gerhardt, 2006]).

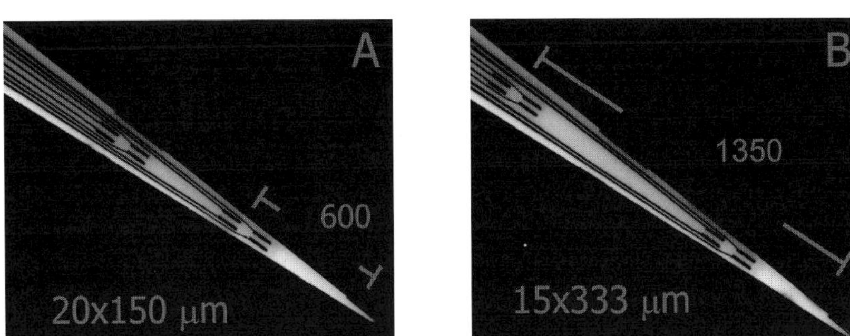

Figure 2.11. Layouts of ceramic-based "conformal" microelectrodes with eight recording sites. Parts (a) and (b) each have four pairs of 20 × 150 μm recording sites separated by 1,350 μm and 600 μm, respectively (Photographs courtesy of Mr. Peter Huettl at the Center for Microelectrode Technologies, University of Kentucky, Lexington, Kentucky).

forming sites on the front and back of the substrate. Finally, several recording sites in the array may be used to electrically stimulate, and the others can be used for electrophysiological or neurochemical recordings.

Polyimide-Based Microelectrodes

Polyimide films, trade name Kapton® (DuPont, Circleville, OH), have been used as a substrate as well as the top insulator for microelectrodes used for intracortical implantation Besides polyimide, the polyimide precursor Parylene (DuPont) can be spun onto surfaces as a liquid then polymerized at high temperatures (200°C). Microelectrodes less than 20 μm thick have been constructed (Rousche et al., 2001). Polyimide as a substrate is very structurally flexible. Figure 2.12 shows a photomicrograph of a three-dimensional multishank microelectrode designed for intracortical implantation. Although the flexibility of polyimide can make implantation difficult, a flexible microelectrode may in certain cases contribute to less tissue damage. Guide incisions in the neural tissue are often needed to prevent the micro-electrode shaft from buckling upon microelectrode implantation (Rousche et al., 2001). Polyimide microelectrodes have even been driven through tissue using surgical suture (Gonzalez and Rodriguez, 1997). The substrate may be folded to provide some rigidity (Takahashi et al., 2003).

As with other substrates, perforations or holes in the polyimide have been used to help secure the microelectrodes in place (Gonzalez and Rodriguez, 1997). Multiple layers can be used to construct useful microelectrodes. Wells may be constructed by simply leaving an open via in a polyimide layer (Rousche et al., 2001).

Figure 2.12. Photograph of a polyimide-based microelectrode array for intracortical implantation. The semitransparent polyimide substrate can be folded to achieve multishank arrays. The metal connecting lines are visible (Photograph provided by Daryl Kipke of the University of Michigan Center for Neural Communication Technology; reprinted by permission from *Encyclopedia of Sensors* [Burmeister and Gerhardt, 2006]).

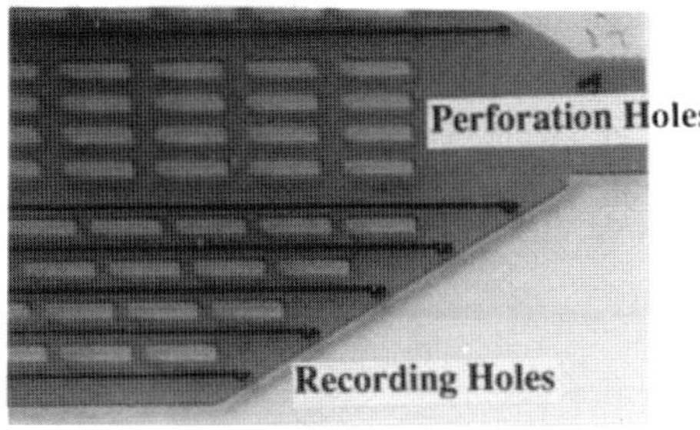

Figure 2.13. Magnification of several recording sites on a polyimide-based microelectrode with perforation holes to help secure the microelectrode in tissue (Used with permission from Elsevier Publishing; adapted from Gonzalez and Rodriguez. 1997; also in *Encyclopedia of Sensors* [Burmeister and Gerhardt, 2006]).

Connectors

Connecting microelectrodes to recording equipment is a major problem for microelectrode fabrication. Often, the microelectrode is secured to a PCB holder or "paddle." The recording sites are electrically connected to the holder by wire bonding from the pads on the microelectrode to pads on the connector. Metal lines (usually Au or Pt) run the length of the holder to pins, or some other type of connecting device. These may be connected to electronic equipment using dual-inline-pin (DIP) sockets or zero-insertion-force (ZIF) sockets.

Another approach to attach microelectrodes to recording equipment combines flexible polyimide ribbon and silicon ribbon cables (Hetke et al., 1994; Bragin et al., 2000; Akin et al., 1999; Kipke et al., 2003). The same photolithographic techniques and basic processes used to construct the silicon microelectrode probes are used to fabricate miniature, flexible, multi-lead silicon ribbon cables consisting of a long, thin, silicon substrate that supports multiple dielectrically encapsulated leads. The ends of the cable are thicker with exposed metal pads for bonding the cable either to a microelectrode or to a connector. The main cable itself can be electrically shielded with an outer barrier layer (typically Au or polysilicon) over the upper dielectrics. This layer makes contact to the silicon substrate so that the leads are electrically shielded as well as sealed, effectively making the cable a multilead "coaxial" structure. Because ribbon cables can be integrated into the microelectrode itself, the need for bonding, soldering, or encapsulation between the microelectrode and the interconnect system is eliminated. Ribbon cables as thin as 4–5 μm have been reported. Flexibility is maintained in all dimensions providing functionality for periods of at least one year (Hetke et al., 1994).

ECoG Strip Electrodes

A growing area of study involves the use of electrocorticographic (ECoG) recordings for BCI (Felton et al., 2007; Marzullo et al., 2005; Leuthardt et al., 2004). This technology grew out of clinical EEG recordings through the work of Jasper and Penfield in the 1930s through the 1950s. The technology has been primarily used by surgeons to record from cortical areas in patients with drug refractory epilepsy to determine the best surgical targets for transaction. We do not review this extensive area as applied to epilepsy surgery. Rather, we discuss the electrodes that are available for such recordings in humans as these electrodes, although invasive, may possess many of the features that make them ideal for BCI applications. First, the safety of the technology, at least acutely, has been tested in thousands of human subjects. Second, ECoG has higher spatial resolution than EEG (tenths of millimeters vs. centimeters) and newer electrode designs (see Figure 2.14) possess spatial resolution closer to that of direct penetrating electrode recordings. Third, the signals recorded from the surface of the brain exhibit higher amplitudes with broader band widths. Fourth, patients undergoing epilepsy surgery constitute a large test bed for investigating BCI technology that is starting to be investigated in the United States and Europe. Finally, such proven technologies may have better long-term stability *in vivo*, but this is still to be determined. One of the largest manufacturers of ECoG electrodes for human recordings is Ad-Tech Medical Instrument Corporation (Racine, WI). It designs and manu-factures about 70% of the sterilized ECoG electrodes used throughout the world. Ad-Tech is an FDA- and ISO13485-registered manufacturer of high-quality medical devices. Ad-Tech, which successfully distributes its electrodes in more than 40 countries, has been active in the design, development, manufacture, and marketing of intracranial monitoring strip-type, grid-type, depth-type, and other related

Figure 2.14. Subdural ECoG microgrid for epidural recordings (Reprinted with permission from D. Moran).

electrodes for more than 22 years. These electrodes are used primarily by comprehensive epilepsy centers and major institutions/medical centers that provide brain mapping in their neurological programs. These electrodes are made of implant silicone or polyurethane with microconductors attached to stainless steel or platinum contacts (usually 7 or 10 mm disks) that populate the dielectric area.

Figure 2.15 shows numerous Ad-Tech ECoG strip electrodes ranging in size from 4 to 64 recording sites. Proprietary connectors/cables attach these electrodes to commercial monitoring equipment. More than 100 medical journal papers have been written on the use of Ad-Tech's products for the treatment of epilepsy and other neurological disorders and diseases (Kossoff et al., 2004; Pan et al., 2005; Ad-Tech [http://www.adtechmedical.com/articles.htm]).

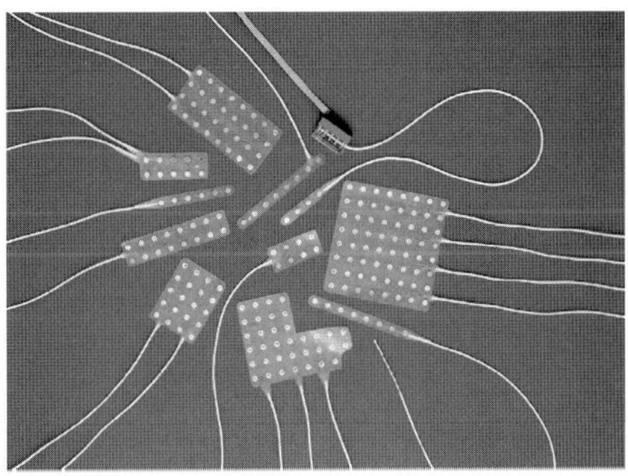

Figure 2.15. Four-to-64-site ECoG recording strip electrodes (Reprinted with permission from Ad Tech Medical Instruments).

Noninvasive EEG Sensors for BCI

Nearly all BCI studies using noninvasive sensors involve the use of Ag or Au disk electrodes with conducting paste that are affixed to the skull using some type of head cap configuration to facilitate the application of the EEG electrodes. Limited progress has been made in improving these devices over the last two decades to rapidly and comfortably affix them to the skull of a BCI user. Head caps have been developed that aid in the measurement and placement of 64 to 256 EEG electrodes using the "International 10–20 grid system." Suppliers of head caps and electrodes are numerous and include g.tec (Guger Technologies OEG), Grass Technologies, BioSemi, and others. For a variety of BCI technologies, g.tec is a source of one of the best head caps used in the field involving wet electrode recordings, as shown in Figure. 2.16. Its unique head cap for EEG electrodes design allows for some of the best signal-to-noise achievable in the business from wet electrode technology. In particular, the electrode cap design requires extra time for attachment of electrodes but achieves excellent signal-to-noise characteristics. This highly versatile design can be employed with other g.tec products and amplifiers, as well as other suppliers of such instrumentation.

A promising improvement is the 128- and 256-channel active "pin-type" Ag electrodes and head cap design distributed by BioSemi (Amsterdam, The Netherlands). This company's active electrode designs have potentially improved signal-to-noise capabilities without the need for Faraday-cage shielding for BCI recordings (see Figure 2.17). In addition, there are promising "dry-type" electrode configurations that have been under development using carbon nanotube electrodes and other dry-type sensor designs (Ruffini et al., 2006; Fonseca et al., 2007).

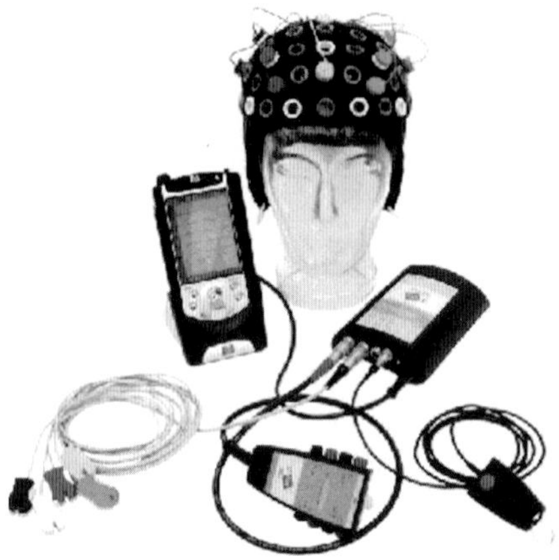

Figure 2.16. A g.tec head cap system for EEG recordings (Reprinted with permission from g.tec).

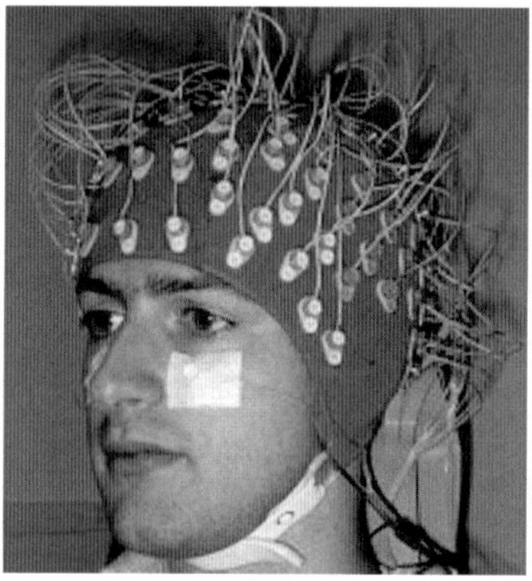

Figure 2.17. BioSemi 128-channel active EEG system (Courtesy of BioSemi).

The process of fitting individuals with EEG electrodes with head caps, however, is time consuming, requires testing of individual electrodes for their impedance, and results in a system that is not comfortable or practical for routine BCI use. There is a need for development of "dry electrodes," which could be used without the preparation required for the current designs. In addition, active electrode designs (such as sold by BioSemi) are needed to improve signal-to-noise ratios of such recordings in practical, real-world applications.

MAJOR CHALLENGES FOR PRODUCING BCI SENSORS

There are major questions that need to be addressed for the development of both noninvasive and invasive sensors that can be used for practical, real-world applications of BCI technology. These are as follows:

- How long do current sensors really last?
- How do we make dry EEG electrodes that allow for ease of application and use?
- How do we develop sensors that last for 5–20 years?
- How do we develop a systematic and scientific approach to developing "biologically-based," implantable microelectrodes and surface electrodes?

Perhaps the largest challenge in the area of implantable electrodes for BCI is the development of electrode arrays that will function for 5–20 years *in vivo*. By far the longest recordings from the CNS of individual unit activity with respect to the context of BCI technology have been achieved by the use of microwire arrays. In fact, more than one-and-a-half years of recording using microwire arrays in nonhuman primates was reported in 2003 (Nicolelis et al., 2003). Unfortunately, this has not been reliably achieved by methodology involving the silicon, ceramic, or polyimide-based multielectrode arrays that have many advantages for future recordings involving BCI technology. Dry EEG electrodes with improved signal-to-noise ratio and ease of use are also needed for noninvasive BCI applications.

In the context of multielectrode arrays, one of the groups that have achieved the greatest amount of success and the greatest following of investigators resides at the University of Michigan. In fact, the greatest number of silicon-based microelectrodes implanted in a nonhuman primate has been achieved at the University of Michigan. Here, Drs. Schwartz and Kipke have been able to record, for more than a year, 60 functional, silicon, microelectrode channels that were implanted in an awake monkey, resulting in more than 90 high-quality recording spikes. This is ground-breaking work that demonstrated the ability of the BCI to control a mechanical limb through recordings of the individual unit activity involving multiple single-unit array electrodes of the silicon type. These studies and the seminal work of Dr. John Donoghue and co-workers (Hochberg et al., 2006; Song et al., 2005) will help shape the development of reliable, long-lasting, tissue-compatible BCI sensors in the years to come (see Chapter 3).

SUMMARY AND CONCLUSIONS

The majority of BCI science in Europe involves "noninvasive" sensor technologies, i.e., multielectrode recordings from arrays of EEG electrodes mounted onto the surface of the skull. This sensor technology has experienced limited growth and needs substantial improvement. Even with respect to noninvasive technologies, many European sites collaborate with, or utilize paradigms that were developed in the United States (Wadsworth Center, Albany, NY).

In Asia, there is clear emphasis on inexpensive, EEG-BCI approaches as the population is large and there is a need for low-cost, noninvasive BCI technology for improved health care in China. In addition, Japan is also focused on noninvasive, EEG-based BCI technologies. However, there is rapid economic growth and science spending in China and Japan that will propel BCI technology in Asia. In addition, there are clear indications that interest and facilities are available to pursue invasive, BCI-sensor technology in China. Asia has manufacturing facilities and infrastructure to drive development of new, invasive, BCI-sensor development that could rival or exceed the efforts in the United States in five to ten years.

REFERENCES

Akin, T., B. Ziaie, S.A. Nikles, and K. Najafi. 1999. A modular micromachined high-density connector system for biomedical applications. *IEEE Trans. Biomed. Eng.* 46(4):471–480.

Anderson, D.J., K. Najafi, S.J. Tanghe, D.A. Evans, K.L. Levy, J.F. Hetke, X.L. Xue, J.J. Zappia, and K.D. Wise. 1989. Batch-fabricated thin-film electrodes for stimulation of the central auditory system. *IEEE Trans. Biomed. Eng.* 36:693–704.

Bai, Q., and K.D. Wise. 2001. Single-unit neural recording with active microelectrode arrays. *IEEE Trans. Biomed. Eng.* 48:911–920.

Bai, Q., K.D. Wise, and D.J. Anderson. 2000. A high-yield micro assembly structure for three-dimensional microelectrode. *IEEE Trans. Biomed. Eng.* 47:281.

Baumeister, A.A. 2006. Serendipity and the cerebral localization of pleasure. *J. His. Neurosci.* 15(2):92–98.

BeMent, S.L., K.D. Wise, D.J. Anderson, K. Najafi, and K.L. Drake. 1986. Solid-state electrodes for multichannel multiplexed intracortical neuronal recording. *IEEE Trans. Biomed. Eng.* 33:230.

Borkholder, D.A., J. Bao, N.I. Maluf, E.R. Perl, and G.T. Kovacs. 1997. Microelectrode arrays for stimulation of neural slice preparations. *J. Neurosci. Meth.* 77(1):61–66.

Bragin, A., J. Hetke, C.L. Wilson, D.J. Anderson, J. Engel Jr, and G. Buzsaki. 2000. Multiple site silicon based probes for chronic recordings in freely moving rats: implantation, recording, and histological verification. *J. Neurosci. Meth.* 98(11):77–82.

Branner, A., R.B. Stein, E. Fernandez, Y. Aoyagi, and R.A. Normann. 2004. Long-term stimulation and recording with a penetrating microelectrode array in cat sciatic nerve. *IEEE Trans. Biomed. Eng.* 51(1):146–157.

Burmeister, J.J., and G.A. Gerhardt. 2001. Self-referencing ceramic based multisite microelectrodes for the detection and elimination of interferences from the measurement of l-glutamate and other analytes. *Anal. Chem.* 73:1037–1042.

Burmeister, J.J., and G.A. Gerhardt. 2006. Neurochemical arrays. In *Encyclopedia of sensors*, Vol. 6, Eds. C. Grimes, E. Dickey, and M.V. Pishko. Stevenson Ranch, CA: American Scientific Publishers, 525.

Burmeister, J.J., K. Moxon, and G.A. Gerhardt. 2000. Ceramic-based multisite microelectrodes for electrochemical recordings. *Anal. Chem.* 72:187–192.

Burmeister, J.J., F. Pomerleau, M. Palmer, B.K. Day, P. Huettl, and G.A. Gerhardt. 2002. Improved ceramic-based multisite microelectrode for rapid measurements of 1-glutamate in the CNS. *J. Neuro. Meth.* 119:163–171.

Chapin J.K., and D.J. Woodward 1986. Distribution of somatic sensory and active-movement neuronal discharge properties in the MI-SI cortical border area in the rat. *Exp. Neurol.* 91(3):502–523.

Chapin, J.K., M.A.L. Nicolelis, 2001. Brain Control of Sensorimotor Prostheses. In *Neural prostheses for restoration of sensory and motor function*. Eds. Chapin, J.K., and K.A. Moxon. Boca Raton, FL: CRC Press, 235–261.

Chapin, J.K. 2004. Using multi-neuron population recordings for neural prosthetics. *Nat. Neurosci.* 7:452.

Chen, J., and K.D. Wise. 1997. A silicon probe with integrated microheaters for thermal marking and monitoring of neural tissue. *IEEE Trans. Biomed. Eng.* 44:770–774.

Chen, J., K.D. Wise, J.F. Hetke, and S.C. Bledsoe Jr. 1997. A multichannel neural probe for selective chemical delivery at the cellular level. *IEEE Trans. Biomed. Eng.* 44:760.

Cheung, K.C. 2007. Implantable microscale neural interfaces. *Biomed. Microdevices* Jan. 25 (epub), PMID: 17252207.

Chiganos, T.C., W. Jensen, and P.J. Rousche. 2006. Electrophysiological response dynamics during focal cortical infarction. *J. Neural Eng.* 3(4):15–22.

Csicsvari, J., D.A. Henze, B. Jamieson, K.D. Harris, A. Sirota, P. Bartho, K.D. Wise, and G. Buzsaki. 2003. Massively parallel recording of unit and local field potentials with silicon-based electrodes. *J. Neurophysiol.* 90:1314–1323.

Della Santina, C.C., G.T. Kovacs, and E.R. Lewis. 1997. Multi-unit recording from regenerated bullfrog eighth nerve using implantable silicon-substrate microelectrodes. *J. Neurosci. Meth.* 72(1):71–86.

Drake, K.L., K.D. Wise, J. Farraye, D.J. Anderson, and S.L. BeMent. 1988. Performance of planar multisite microprobes in recording extracellular activity. *IEEE Trans. Biomed. Eng.* 35:719.

Ensell, G., D.J. Banks, P.R. Richards, W. Balachandran, and D.J. Ewins. 2000. Silicon-based microelectrodes for neurophysiology, micromachined from silicon-on-insulator wafers. *Med. Biol. Eng. Comput.* 38(2):175–179.

Felton E.A., J.A. Wilson, J.C. Williams, P.C. Garell. 2007. Electrocorticographically controlled brain-computer interfaces using motor and sensory imagery in patients with temporary subdural electrode implants. Report of four cases. *J. Neurosurg.* 106(3):495–500.

Fonseca C., J.P., Silva Cunha R.E. Martins, V.M. Ferreira, J.P. Marques de Sa, M.A. Barbosa, and A. Martins da Silva. 2007. A novel dry active electrode for EEG recording. *IEEE Trans. Biomed. Eng.* 54(1):162–165.

Gonzalez, C., and M. Rodriguez. 1997. A flexible perforated microelectrode array probe for nerve and muscle tissues. *J. Neurosci. Meth.* 72:189–195.

Heath R.G., S.M. Peacock, Jr., and W. Miller, Jr. 1953. Induced paroxysmal electrical activity in man recorded simultaneously through subcortical and scalp electrodes. *Trans. Am. Neurol. Assoc.* 3:247–250.

Hetke, J.F., J.L. Lund, K. Najafi, K.D. Wise, and D.J. Anderson. 1994. Silicon ribbon cables for chronically-implantable microelectrode arrays. *IEEE Trans. Biomed. Eng.* 41:314.

Hochberg, L.R., Serruya, M.D, Friehs, G.M, Mukand, J.A., Saleh, M, Caplan, A.H., Branner, A., Chen, D., Penn, R.D., and Donoghue, J.P. 2006. Neuronal ensemble control of prosthetic devices by a human with tetraplegia. *Nature* 442:164–171.

Hoogerwerf, A.C. and K.D. Wise. 1994. A three-dimensional microelectrode array for chronic neural recording. *IEEE Trans. Biomed. Eng.* 41:1136–1146.

Jensen, W., K. Yoshida, and U.G. Hofmann. 2006. *In vivo* implant mechanics of flexible, silicon-based ACREO microelectrode arrays in rat cerebral cortex. *IEEE Trans. Biomed. Eng.* 53(5): 934–940.

Kipke, D.R., R.J. Vetter, J.C. Williams, and J.F. Hetke. 2003. Silicon-substrate intracortical microelectrode arrays for long-term recording of neuronal spike activity in cerebral cortex. *IEEE Trans. Neural Syst. Rehab. Eng.* 11(2):151–155.

Kossoff, E.H., E.K. Ritzl, J.M. Politsky, A.M. Murro, J.R. Smith, R.B. Duckrow, D.D. Spencer, and G.K. Bergey. 2004. Effect of an external responsive neurostimulator on seizures and electrographic discharges during subdural electrode monitoring. *Epilepsia* 45(12):1560–1567.

Kovacs, G.T., C.W. Storment, and J.M. Rosen, 1992. Regeneration microelectrode array for peripheral nerve recording and stimulation. *IEEE Trans. Biomed. Eng.* 39:893.

Kovacs, G.T., C.W. Storment, M. Halks-Miller, C.R. Belczynski Jr, C.C. Della Santina, E.R. Lewis, and N.I. Maluf. 1994. Silicon-substrate microelectrode arrays for parallel recording of neural activity in peripheral and cranial nerves. *IEEE Trans. Biomed. Eng.* 41:567.

Leuthardt, E.C., G. Schalk, J.R. Wolpaw, J.G. Ojemann, and D.W. Moran. 2004. A brain-computer interface using electrocorticographic signals in humans. *J. Neural Eng.* 1:63–71.
Lin, L., G. Chen, K. Xie, K.A. Zaia, S. Zhang, and J.Z. Tsien. 2006. Large-scale neural ensemble recording in the brains of freely-behaving mice. *J. Neurosci. Meth.* 155(1):28–38.
Ludvig, N. 2001. Drug deliveries into the microenvironment of electrophysiologically monitored neurons in the brain of behaving rats and monkeys. In *Neural prostheses for restoration of sensory and motor function*. Eds. Chapin, J.K., and K.A. Moxon. Boca Raton, FL: CRC Press, 263–283.
Marzullo T.C., J.R. Dudley, C.R. Miller, L. Trejo, D.R. Kipke. 2005. Spikes, local field potentials, and electrocorticogram characterization during motor learning in rats for brain-machine interface tasks. *Conf Proc IEEE Eng. Med. Biol. Soc.* 1:429–431.
Moxon, K.A., S.C. Leiser, G.A. Gerhardt, K.A. Barbee, and J.K. Chapin. 2004. Solid-state electrodes for multichannel multiplexed intracortical neuronal recording. *IEEE Trans. Biomed. Eng.* 51:647.
Najafi, K., J. Ji, and K.D. Wise. 1990. Scaling limitations of silicon multichannel recording probes. *IEEE Trans. Biomed. Eng.* 37:1.
Nicolelis, M.A., D. Dimitrov, J.M. Carmena, R. Crist, G. Lehew, J.D. Kralik, and S.P. Wise. 2003. Chronic, multisite, multielectrode recordings in macaque monkeys. *Proc. Natl. Acad. Sci. USA*, 100(19):11041–11046.
Nordhausen, C.T., E.M. Maynard, and R.A. Normann. 1996. Single-unit recording capabilities of a 100-microelectrode array. *Brain Res.* 726:129.
Olds, J., W.S. Allan, and E. Briese. 1971. Differentiation of hypothalamic drive and reward centers. *Am. J. Physiol.* 221(1):368–375.
Pan, J.W., J.H. Kim, A.A. Cohen-Gadol, C. Pan, D.D. Spencer, and H.P. Hetherington. 2005. Regional energetic dysfunction in hippocampal epilepsy. *Acta Neurol. Scand.* 111(4):218–224.
Pancrazio, J.J., P.P. Bey Jr., A. Loloee, S. Manne, H.C. Chao, L.L. Howard, W.M. Gosney, D.A. Borkholder, G.T. Kovacs, P. Manos, D.S. Cuttino, and D.A. Stenger. 1998. Description and demonstration of a CMOS amplifier-based system with measurement and stimulation capability for bioelectrical signal transduction. *Biosens. Bioelectron.* 13(9):971–979.
Patterson, W.R., Y.K. Song, C.W. Bull, I. Ozden, A.P. Deangellis, C. Lay, J.L. McKay, A.V. Nurmikko, J.D. Donoghue, and B.W. Connors. 2004. A microelectrode/microelectronic hybrid device for brain implantable neuroprosthesis applications. *IEEE Trans. Biomed. Eng.* 51:1845.
Rathnasingham, R., D.R. Kipke, S.C. Bledsoe Jr, and J.D. McLaren. 2004. Characterization of implantable microfabricated fluid delivery devices. *IEEE Trans. Biomed. Eng.* 51:138–145.
Rennaker, R.L., A.M. Ruyle, S.E. Street, and A.M. Sloan. 2005. An economical multichannel cortical electrode array for extended periods of recording during behavior *J. Neurosci. Meth.* 142(1):97–105.
Rousche, P.J., D.S. Pellinen, D.P. Pivin Jr, J.C. Williams, R.J. Vetter and D.R. Kipke. 2001. Flexible polyimide-based intracortical electrode arrays with bioactive capability. *IEEE Trans. Biomed. Eng.* 48:361–371.
Ruffini, G., S. Dunne, E. Farres, P.C.P. Watts, E. Mendoza, S.R.P. Silva, C. Grau, J. Marco-Pallares, L. Fuentemilla, and B. Vandecasteele. 2006. ENOBIO: First tests of a dry electrophysiology electrode using carbon nanotubes. In *Proceedings of the 28th IEEE EMBS Annual International Conference*, New York City, Aug 30–Sept 3, 2006. Piscataway, NJ: IEEE Engineering in Medicine and Biology Society (EMBS), 1826–1829.
Schmidt, S., K. Horch, and R. Normann. 1993. Biocompatibility of silicon-based electrode arrays implanted in feline cortical tissue. *J. Biomed. Mater. Res.* 27(11):1393–1399.
Smith, S.L., J.W. Judy, and T.S. Otis. 2004. An ultra small array of electrodes for stimulating multiple inputs into a single neuron. *J. Neurosci. M.* 133:109–114.
Song, Y.K., W.R. Patterson, C.W. Bull, J. Beals, N. Hwang, A.P. Deangelis, C. Lay, J.L. McKay, A.V. Nurmikko, M.R. Fellows, J.D. Simeral, J.P. Donoghue, and B.W. Connors. 2005.

Development of a chipscale integrated microelectrode/microelectronic device for brain implantable neuroengineering applications. *IEEE Trans. Neural Syst. Rehab. Eng.* 13:220.

Sugiyama, K., W.K. Dong, and E.H. Chudler. 1994. A simplified method for manufacturing glass-insulated metal microelectrodes. *J. Neurosci. Meth.* 53:73.

Takahashi, H., T. Ejiri, M. Nakao, N. Nakamura, K. Kaga, and T. Herve. 2003. Microelectrode array on folding polyimide ribbon for epidural mapping of functional evoked potentials. *IEEE Trans. Biomed. Eng.* 50(4):510–516.

Vetter, R.J., J.C. Williams, J.F. Hetke, E.A. Nunamaker, and D.R. Kipke. 2004. Chronic neural recording using silicon-substrate microelectrode arrays implanted in cerebral cortex. *IEEE Trans. Biomed. Eng.* 51:896.

Warren, D.J., E. Fernandez, and R.A. Normann. 2001. High-resolution, two-dimensional spatial mapping of cat striate cortex using a 100-microelectrode array. *Neuroscience* 105:19.

Wise K.D., and J.B. Angell. 1975. A low-capacitance multielectrode probe for use in extracellular neurophysiology. *IEEE Trans. Biomed. Eng.* 22(3):212–219.

Wise K.D., J.B. Angell and A. Starr 1970. An integrated-circuit approach to extracellular microelectrodes. *IEEE Trans. Biomed. Eng.* 17(3):238–247.

Williams, J.C., R.L. Rennaker, and D.R. Kipke. 1999. Long-term neural recording characteristics of wire microelectrode arrays implanted in cerebral cortex. *Brain Res. Brain Res. Protoc.* 4:303–313.

Wilson, J.A., E.A. Felton, P.C. Garell, G. Schalk, and J.C. Williams. 2006. An ECoG-based brain-computer interface using multimodal control. *IEEE Trans. Neural Syst. Rehab. Eng.* 14(2):246–250.

Yoon, T.H., E.J. Hwang, D.Y. Shin, S.I. Park, S.J. Oh, S.C. Jung, H.C. Shin, and S.J. Kim. 2000. A micromachined silicon depth probe for multichannel neural recording. *IEEE Trans. Biomed. Eng.* 47:1082.

Yoshida, K., W. Jensen, P. Norlin, M. Kindlundh, and U.G. Hofmann. 2001. Characterization of silicon microelectrodes from the EU VSAMUEL project. *Biomediinische Technik* 446.

CHAPTER 3

The Biotic–Abiotic Interface

Patrick A. Tresco and Greg A. Gerhardt

INTRODUCTION

Brain-computer interfaces (BCI), or brain-machine interfaces (BMI), are systems designed to aid humans with central nervous system disabilities, including disabilities in movement, communication, and independent control of one's environment (Donoghue, 2002; Friehs et al., 2004; Lebedev and Nicolelis, 2006; Schwartz et al., 2006). Although these same approaches have the potential to augment normal function, as currently envisioned this new class of biomedical devices is being developed to help those with disabilities. As such, these devices may be useful for patients suffering from a variety of conditions including spinal cord injury, musculodegenerative diseases, stroke, amyotrophic lateral sclerosis, or other neurological or neuromuscular diseases. The intent of these devices and their associated components is to provide or supplement motor or sensory function that has been lost. The theoretical basis for such devices lies in our ability to detect neural signals and translate volitional commands into control signals for external devices including computers, robotics, or other machines. The acquisition of neural signals has traditionally occurred in the cerebral cortex, and the recording of these signals from implanted electrodes has a fairly extensive history.

Although several forms of technology are being developed, this chapter will focus exclusively on our present knowledge of the foreign body response to invasive technologies. Generally speaking, such devices are small by present biomedical-devices standards and are implanted in the cortex, the most superficial aspect of the mammalian brain. As currently designed, they penetrate a few millimeters depending on the target region and species. They contain multiple recording or stimulating sites located on one or more penetrating shafts that consist of conducting ceramics, metals, or polymers and have at least one insulating material (see the figures in Chapter 2 that illustrate some of the hardware under development). At present, such devices are tethered to insulated wires that exit the skull and lead to external amplifiers and other devices that can be substantial in size and are not

T.W. Berger et al., Brain-Computer Interfaces, 31–45.

very portable. Due to the nature of their design, the recording devices are frequently referred to as "penetrating electrode arrays."

To date, CNS recording devices have taught us much about the functional organization and neurophysiological underpinnings of the mammalian cortex and other brain regions, and appear, based on evidence to date, to become increasingly utilized in a variety of healthcare applications that will improve the quality of life of those affected with CNS-related disabilities (Lebedev and Nicolelis, 2006; Schwartz et al., 2006). The future economic impact of the technology appears equally significant. Coupled with other emerging technologies, we find ourselves on the doorstep of understanding one of the most elaborate and complex systems ever studied the human brain. What has been shown is that neural signals can be recorded from different brain regions by a number of different technologies in a variety of species for periods of time extending from months to well over a year. With direct relevance to BCI or BMI technology, neural activity can be interpreted and used to control a computer or robot or prosthetic device.

With the feasibility and proof of principle firmly established for various BCI and BMI applications, some of the focus has shifted to understanding how to maintain consistent, long-term operation of the implanted devices. Even though current designs usually perform as intended in short-term studies and applications, the major limitation of our current state of technology is inconsistent performance in chronic or long-term applications, which limits clinical implementation of this promising technology (Polikov et al., 2005; Schwartz et al., 2006).

Although the brain-tissue response to implanted electrode arrays is believed to be a major contributing factor, the precise mechanisms that cause inconsistent recording performance are unknown. Thus, until the mechanisms that underlie loss of function are understood, we are unlikely to develop rational strategies to improve their usefulness.

This chapter first discusses what is perceived to be the problem and establishes a foundation of common terminology. It then provides evidence that invasive electrodes can function over extended time frames as a proof of concept that this new class of biomedical device as currently envisioned can be biocompatible. It then reviews the current state of understanding of what happens following implantation of electrodes into the mammalian brain, trying where possible to identify gaps in our knowledge in an attempt to shed light on what still needs to be done. It concludes with a discussion of various strategies are under development to modulate the biotic–abiotic interface in an attempt to achieve better integration into brain tissue and achieve superior device performance. We caution that most of the technology focused on augmenting device performance, as promising as it may appear, is still under development, and for the most part has not been replicated sufficiently to understand its ultimate impact on advancing the field.

By all indications, a full understanding of what needs to be done to consistently interface various hardware with the variety of potential neural targets is still far off. A major obstacle at present is in understanding the science responsible for loss of function. This understanding is unlikely to occur in the near term without enhanced and targeted funding to increase the number of investigators working in the field. The scientific breadth and depth needs to advance sufficiently, as has occurred with cochlear implants, so that the challenge shifts from lack of scientific knowledge to engineering.

BCI ABIOTIC-BIOTIC INTERFACE WORLD OVERVIEW

It is generally held by the scientific and engineering communities that maintaining a stable, long-term interface between an implanted recording electrode and adjacent neural circuitry is one of the major challenges that limit the widespread clinical implementation of BCI/BMI-based therapies.

The majority of science in North America is focused on describing the events that accompany the implantation of multielectrode recording arrays into brain tissue over time, with a particular emphasis on describing the temporal and spatial nature of the events that take place at the biotic–abiotic interface and assessing their potential to affect device function. Although the broad brush strokes are in place, significant detail is lacking, thus limiting the development of rational strategies to enhance consistency of long-term performance.

In Europe, China and Japan, there appears to be little direct work in the BCI/BMI-related domain that is focused specifically on understanding the biological underpinnings of invasive sensor biocompatibility. Here the emphasis of the research has been directed more toward the development of noninvasive technology and to adding intelligence to the robotic or external components of such devices. It was apparent from our visits in Europe and Asia that numerous groups are planning or developing implantable neural interfaces or are developing technology that has the potential to significantly improve the performance of invasive sensors for BCI and BMI applications. Clearly, the emphasis on such technology is increasing. The future potential capability of this community to improve or displace existing technology is significant.

A number of laboratories, mostly in North America, have explicitly acknowledged focused efforts at developing strategies to manipulate the tissue response in an effort to improve the long-term function of BCI/BMI and related invasive technologies. While some results appear promising, it is still too early to know which, if any, of these will ultimately improve the consistency of long-term recording performance of penetrating electrodes. Alternatively, perhaps practitioners need to radically change the approach as it is currently envisioned.

After an analysis of the peer-reviewed literature, two workshops, and visits abroad, we conclude that major gaps still remain in our understanding of the science behind the loss of function that occurs over time with the use of penetrating recording electrode arrays. It appears that this is unlikely to change significantly at the present levels of funding. Investments in science and technology are increasing abroad. We suggest that targeted funding also be provided in the United States to increase our knowledge of the underlying science of CNS implant biocompatibility in order to maintain a leadership position in this sector, with corresponding general benefits to U.S. healthcare and the economy.

The Major Challenge: Consistent, Long-Term, Functional Integration

The key to the long-term operation of penetrating recording electrodes is in consistently and reproducibly maintaining connectivity with the system of interest. It is impossible to implant anything as large as a penetrating electrode into brain tissue without causing some damage at the site of implantation. Therefore, the goal is to achieve a response that allows the device to function as intended without causing unacceptable harm to the patient. The term used to describe this condition is biocompatible; notice that the definition is conditional. For a BCI or BMI device to be biocompatible, by definition it must be functional; that is, it must be capable of recording the activity of neurons, and the information sensed must serve some intended function. The definition does not require zero response, and it does not necessarily require that every electrode site record activity that is maintained over its lifetime. To be biocompatible, the tissue response to the implanted electrode, or risk, must be offset by the benefit of the device; that is, it has to remain functional. Ambiguity is often derived from the conditional nature of the definition. As the definition implies, at one point in time a device can be biocompatible, and a little later it may not be. Notice also that by definition, materials cannot be considered biocompatible unless they serve some measurable function. The ambiguity of terms makes a critical reading of the literature somewhat confusing, especially for students and members of constituencies outside the field who do not understand the nuance of the term. The ideal goal is full or seamless integration with nervous system tissue, or the achievement of a functional symbiosis between the biotic and abiotic interface that maintains device function over the lifetime of the patient. At present, we are far from the ideal.

Proof of Principle

For many BCI and BMI applications, a sufficient number of recording sites in an implanted array must be located exceedingly close to actively depolarizing neuronal cell bodies. Moreover, these neuronal cell bodies must remain viable independently and maintain their integrated activity and connectivity with the rest of the central nervous system. Practitioners in the field will tell you that recording sites need to be placed within a few hundred micrometers to as little as 50 μm away from the neuronal cell bodies in order to sense single-unit activity, and slightly farther away to record local-field potentials. The literature also supports this view (Buzsaki and Kandel, 1998; Henze et al., 2000; Rall, 1962; Mountcastle, 1957). One of the major challenges is to determine how to achieve a higher level of consistency than is possible with the current state of the art.

Despite this seemingly difficult specification hurdle, the technology of recording devices has progressively advanced from the benchtop into the clinic (Hochberg et al., 2006; Schwartz et al., 2006). The earliest identifiable publications that describe the idea behind the approach may be credited to Schmidt (Schmidt, 1980; Schmidt et al., 1976). Since then, a variety of investigators have illustrated the potential of using recording devices to facilitate motor function. The earliest study demonstrating the use of a brain-computer interface in humans used a neurotrophic cone electrode implanted into the cortex of three patients who reportedly gained the ability to move a cursor on the computer screen through volitional commands recorded by indwelling electrodes (Kennedy et al., 2000). Since then, a multishank, silicon-microelectrode array was implanted into several paralyzed patients who demonstrated a substantial gain of function for volitionally moving a computer cursor (Hochberg et al., 2006).

Despite the successful experimental work in humans, the bulk of the proof of principle for BCIs and BMIs has been derived from studies in nonhuman primates (Musallam et al., 2004; Santhanam et al., 2006; Serruya et al., 2002; Taylor et al., 2002; Wessberg et al., 2000), as well as the contribution from numerous groups that have developed hardware or have used the hardware to advance our knowledge of neuroscience (Table 3.1).

As it stands, a number of groups have reported the ability to record signals for periods ranging from months to several years. Collectively, the publication record shows that the implementation of such technology for BCI and BMI applications is clearly possible. Furthermore, it can be achieved using a variety of designs, including glass microelectrodes, ceramic-based sensors, microwires insulated with a variety of materials, and doped silicon. These designs may be constructed as planar arrays or as multipoint tip electrodes, indicating that economic opportunity

Table 3.1 Longevity of Recording Performance in the CNS

Year	First Author	Species Implanted	Electrode Type	Functional Period of Signal Recording (day)
1976	Schmidt	Monkey	Parylene-coated iridium wires	223 138
1977	Loeb	Monkey	Parylene-coated iridium wires	136
1984	Legendy	Cat	Parylene-coated platinum-iridium wires	9–25
1988	Schmidt	Monkey	Parylene-coated iridium wires	1,144
1989	Kennedy	Rat	Neurotrophic cone electrode (glass)	201[a]
1992	Kennedy	Monkey	Neurotrophic cone electrode (glass)	~450
1993	Carter	Cat	Michigan electrode (silicon)	~30
1994	Hetke	Guinea Pig	Michigan electrode (silicon)	~330

(Continued)

Table 3.1 (Cont.)

1998	Rousche	Cat	Utah intracortical electrode array (UEA; silicon)	~390[a]
1999	Williams	Guinea Pig	Polyimide-insulated tungsten wires	283[a]
				81
				101
				151
				35
				55
				51
				54
1999	Liu	Cat	Iridium wires	242
2000	Kennedy	Human	Neurotrophic cone electrode (glass)	426[a]
2003	Cui	Guinea Pig	Polypyrrole-coated Michigan electrodes (silicon)	14
2003	Kipke	Rat	Michigan electrode (silicon; 4-shank)	382
2003	Nicolelis	Monkey	Teflon-coated stainless steel microwires	~540
2004	Moxon	Rat	Ceramic-based microelectrodes	~91
2004	Kennedy	Human (40 years old)	Neurotrophic cone electrode (glass)	>636
2004	Vetter	Rat	Michigan electrode (silicon)	127
2005	Johnson	Rat	Michigan electrode (silicon)	>131 (when voltage biasing occurred)
2005	Rennaker	Rat	Tungsten microwires	21 (manual insertion)
				42 (mechanical insertion)
2005	Suner	Monkey	Bionic (Cyberkinetics; silicon multishank array)	569[a]
				870 (no data provided)
				425 (no data provided)
				92 (no data provided)
				1,264 (no data provided)
2006	McCreery	Cat	Parylene-coated iridium wires	220
				343
				320
				302
				293
2006	Ludwig	Rat	Michigan electrode (silicon)	42[a]
2006	Hochberg	Human (25 years old)	Bionic (Cyberkinetics; silicon multishank array)	~300[a]
2006	Hochberg	Human (55 years old)	Bionic (Cyberkinetics; silicon multishank array)	~330
2006	Liu	Cat	Iridium microwires	1,061

[a]Denotes electrode was still functioning at time of publication

exists. Moreover, the record supports the notion that it is indeed possible to have a long-lasting biocompatible recording electrode implanted in the mammalian cortex.

It is clear from discussions with practitioners that this kind of performance is not achieved routinely and represents a smaller subset of the total cases. We believe that it is safe to say that this type of performance is not the norm even though this point is hard to make from a review of the archival literature. Animals implanted with nonfunctioning electrodes are typically not used for experiments, and hence a rich source of failure analysis is not readily available. Nonetheless, anecdotal information informally deliberated at conferences among participating scientists acknowledges the challenge. Indeed, one can find discussions in the peer-reviewed literature that draw attention to performance problems with chronic recording electrode arrays where typically the number of functional recording sites and the quality of signals observed diminish over time (Burns et al., 1974; Liu et al., 1999; Ludwig et al., 2006; Nicolelis et al., 2003; Rousche and Normann, 1998; Schmidt et al., 1976; Williams et al., 1999).

The State of the Science

The available evidence emerging in numerous fields indicates that the biological processes that accompany the implantation of such devices into brain tissue involve the integration of different cellular and molecular events. Indeed, a mechanistic understanding of the type that allows manipulation of the biocompatibility of implanted electrode arrays is still beyond our grasp.

Studies from numerous groups performed on a variety of devices in a number of species have begun to sort out the details, and the broad dynamics of the process have been uncovered (Biran et al., 2005; Polikov et al., 2005). Collectively, the research has revealed certain patterns of response regardless of the size or the type of device or of the materials employed. Many investigators believe that the cellular response to the implanted electrode contributes significantly to inconsistent performance, and this belief may be traced back to several pioneering studies. These seminal studies showed that the number of electrode sites capable of recording well-defined single units decreased with time following implantation (Burns et al., 1974; Schmidt et al., 1976). The investigators postulated that the foreign body response, particularly astrocyte encapsulation that forms around the implanted electrode, may be responsible for the performance problems. Despite a lack of direct experimental evidence to support the hypothesis, many practitioners believe that astrocytic encapsulation is one of the major contributing factors to decreasing device performance. On the other hand, it is possible that a portion of inconsistent recording performance may have nothing to do with the tissue response and may be a normal attribute of an exceedingly plastic network-based physiology.

Studies conducted over the last half century evaluated the cortical brain tissue response to indwelling electrodes, both passive and active. The earliest studies showed that the foreign body response to indwelling electrodes involved reactive encapsulation of astrocytes and fibroblasts as well as activation of leucocytes, macrophages, and microglia, which was accompanied by neuronal degeneration (Bickford et al., 1957; Collias and Manuelidis, 1957). Collectively, these early studies established that the biotic–abiotic interface was well defined and composed of astrocytes surrounding the implantation tract. The implantation tract was well established by one month and remained stable through six months, the longest time frame studied. Macrophages, which are generally not observed in the normal cortical parenchyma, were observed at all time points at the brain tissue device interface. Similar observations have been reported with the latest generation of implants (Biran et al., 2005).

The earliest reports to describe the cellular nature of the interface indicated a reduction in synapses adjacent to the gliotic sheath surrounding implanted electrodes, whereas normal synaptic density was found just outside of this region (Collias and Manuelidis, 1957; Schultz and Willey, 1976). Astrocytes were observed to span a region 50–100 μm away from the edge of the electrode, and meningeal cells were observed in the gliotic sheath that may have migrated from the overlying meninges. These early studies, which used insulated metal wires as electrode arrays, showed that foreign-body, giant cells were always present adjacent to the implanted electrode.

Other researchers built upon this work by examining the brain tissue response to a variety of materials using approaches that attempted to preserve the interface (Babb and Kupfer, 1984; Dymond et al., 1970; Robinson and Johnson, 1961; Schultz and Willey, 1976; Stensaas and Stensaas, 1976, 1978). These studies taught us that the tissue response surrounding the implant could be quite variable, which may have reflected differences in the dorsal ventral architecture of the cortical columns, differences in the physical and chemical attributes of the implants, and differences in implantation techniques. At the end of the day, we learned that a wide variety of materials in such devices appeared safe whereas others were not.

Although descriptive studies refined our understanding of the range of usable materials, other pioneering work began to examine the usefulness of semiconductor technology for fabricating high-count, neural interfaces (see Chapter 2 for details). This led to a series of targeted funding initiatives by the National Institute of Neurological Disorders and Stroke (NINDS) of the National Institutes of Health (NIH) that culminated in the formation of the Neural Prosthesis Program. For over 30 years, NINDS has supported grants and contracts in numerous areas within the neural prosthesis field, including functional neuromuscular stimulation, deep-brain stimulation, multielectrode cuffs for nerve interfaces, cortical microelectrode arrays, biocompatibility of neural interfaces, implantable neural stimulators, and brain/computer interfaces. One of the noteworthy results of this funding was recognizing the importance of the biological understanding driving innovation in the field and a shift from emphasizing the materials component of the biomaterials to a shared emphasis on the biology and the materials aspects of the technology.

Increased funding opportunities and awareness of the challenges pulled researchers from other areas into the field and increased the collective knowledge of the brain-tissue response to implanted electrodes. At its initiation, the Neural Prosthesis Program was funded primarily through contracts; however, the program now makes use of both grant and contract mechanisms to enable progress in the field. The transition from contracts has been facilitated by the increasingly widespread recognition of the importance of data-driven research. Program funding, along with the organization of workshops, conferences, and symposia, has been an effective driving force to attract researchers from allied fields. As a result, major changes have occurred in the way designers and fabricators envision neural interfaces.

One of the first papers to describe the tissue response to the newer generation of microelectrodes described gliosis and neuronal loss in the recording zone surrounding implanted electrodes (Edell et al., 1992). This study was one of the first to report increased gliosis near the tips of the implanted electrodes, which has encouraged others to model the biomechanics of implant design and generate hypotheses regarding the relationship between tethering forces and gliosis (Lee et al., 2005a; Subbaroyan et al., 2005); however, these models have not been completely validated experimentally.

The discovery of cell-specific antigens led to the increasing use of immunofluorescent histology to describe the spatial arrangement of specific cell types involved in the response. What we have learned to date is that the response to the newest generation of implanted microelectrode arrays resembles what has been reported in the past for simple insulated wires and other biomaterials. The major observation is the presence of encapsulating hypertrophic astrocytes that appear regardless of device type or design or whether the device is free-floating or tethered (Biran et al., 2005; Hoogerwerf and Wise, 1994; Schmidt et al., 1993; Szarowski et al., 2003; Turner et al., 1999). These observations motivated hypotheses that astrogliotic encapsulation contributes to the failure of such devices to maintain connectivity with adjacent neurons. The reasoning is that the reaction increases the distance between the recording site and nearby neurons. In addition, astrocytes can form a syncytium owing to their expression of junctional complexes (Lee et al., 2005b; Nagy and Rash, 2000). Although this has not been experimentally shown as a mechanism of function loss, it may affect electrode impedance. The thought is that astrocytes increase extracellular tortuosity in the surrounding tissue, which increases the path length for diffusion of solutes enhancing impedance at electrode sites or moves viable neurons out of the recording zone through hypertrophy and the overexpression of matrix (Polikov et al., 2005; Sykova, 2005).

Even though this hypothesis seems reasonable and may account for some of the loss of function, it is important to note that such responses happen irrespective of whether an electrode is functional or not. To the best of our knowledge, no study has established that astrocytic encapsulation is incompatible with device function or is the primary cause for the inconsistent device performance that challenges the clinical implementation of such technology. At least one study attempted to understand the impact of surrounding silicon planar electrode arrays with cells involved in the foreign-body response and found that such cells indeed increase impedance at 10 kHz but not to a level that would be expected to impede recording *in vivo* (Merrill and Tresco, 2005). In addition, no study has shown that astrocytic encapsulation changes with a time constant, which might explain the inconsistency in performance that occurs beyond the initial month-long period over which it becomes established. Clearly other mechanisms are also at play.

Another prominent observation is persistent macrophage activation at the surface of the device and in the tissue immediately surrounding the implant (Biran et al., 2005; Schmidt et al., 1993; Szarowski et al., 2003). These observations occur irrespective of the indwelling time or type of recording array and suggest that such devices are a persistent source of inflammatory stimuli. As mentioned earlier, macrophages at the biotic–abiotic interface are not seen in the newer generation of devices. Instead, macrophages have been observed to accompany the foreign body response to the earliest implanted stimulating and recording electrodes. It is a general observation seen with all types of currently implanted devices.

Since macrophages can be a source of neurotoxic cytokines, are known to be toxic to oligodendrocytes, and inhibit progenitor division, they may impede healing or replenishment of damaged cells resulting from low-grade persistent inflammation and may contribute to inconsistent device performance over time (Hendriks et al., 2005). These last areas have been unexplored with respect to their potential to contribute to electrophysiological disturbances of recording electrodes.

In addition to the persistence of inflammatory cells, studies have observed significant reductions in nerve fiber density and neuronal cell bodies in the tissue immediately surrounding implanted electrodes (Biran et al., 2005; Edell et al., 1992). Persistent up-regulation of inflammatory cells and neurodegeneration does not accompany stab wound injuries in brain tissue. Therefore, loss of neurons is not caused by the initial mechanical trauma of electrode implantation but is associated with the foreign body response, possibly due to secondary cell loss associated with neuroinflammatory events. This has been observed near more natural foreign bodies occurring in MS, HIV infection, and Alzheimer's disease. Removal of key neurons has the potential to inactivate specific circuitry within the cortical column leading to electrophysiological deficits. Obviously, loss of neurons in the recording zone may also contribute to loss of function of such devices. However, to the best of our knowledge there are no studies that have examined whether neuron viability in the recording zone declines with indwelling time. Moreover, no studies have examined the relationship between neuronal loss and recording inconsistency over time. In addition, we currently have no knowledge of which of the many types of neurons in the cortical column may be affected by the foreign-body response. Clearly, much work remains to be done.

In summary, it appears that a number of factors may contribute to inconsistent performance of invasive recording electrodes. Although glial encapsulation clearly can be a problem, the cellular and molecular aspects of neuroinflammation may also be important contributing factors. The science emerging in the areas of neuro-inflammation may be particularly important in understanding electrical instability in chronic recording devices. For the most part, this newer body of work has reinforced the foundation of knowledge established by earlier studies using light and electron microscopy. Unfortunately, it has not yet provided the specific insights needed to drive improvements in device function.

STRATEGIES UNDER DEVELOPMENT TO IMPROVE ELECTRODE PERFORMANCE

A number of labs, mostly in North America, have explicitly acknowledged focused efforts to develop strategies to improve the long-term function of BCI/BMI and related invasive technologies. The strategies can be grouped into a number of different categories including pharmacological approaches, micro/nanoscale surface science, new materials, novel hardware design, insertion technology and adjustable depth electrodes and wireless technology. We briefly describe some of these developments below.

The mechanical mismatch between electrodes and surrounding brain tissue has been hypothesized as one of the major factors that determine biocompatibility of indwelling recording electrodes (Lee et al., 2005; Subbaroyan et al., 2005). Several groups have used finite element models to show that current designs are associated with increased strain fields at the biotic–abiotic interface. They propose that the strain fields will exacerbate the brain-tissue response given the movements that likely occur with normal respiration and changes in blood pressure during the cardiac cycle (Lee et al., 2005; Subbaroyan et al., 2005). Indeed, a recent paper has shown that the general tissue response to tethered microelectrode arrays is significantly greater with respect to glial encapsulation, macrophage activation, and loss of adjacent neurons when compared to the same electrode implanted as a free-floating implant. The paper suggests that wireless floating designs may be associated with less tissue reactivity (Biran et al., 2007).

Along these lines, it has been argued that making electrodes out of softer polymeric materials may also reduce the associated brain tissue response (Rousche et al., 2001; Subbaroyan et al., 2006; Yuen and Agnew, 1995). Likewise, a recent report from the Kipke group of the University of Michigan introduced a novel open architecture electrode design that places the recording sites on a thin supporting member. This design removes the function from the most reactive main shaft of the electrode similar to the tip electrode designs of microwire arrays and the Utah electrode array (Seymour and Kipke, 2006). A preliminary report suggests that this design reduces cellular encapsulation and is associated with less neuron loss; however, to the best of our knowledge, a fully functional recording electrode of this design has not been demonstrated.

To create a softer interface, the Bellamkonda group of Georgia Tech has examined the use of layer-by-layer electrostatic deposition of polyelectrolytes and laminin (He and Bellamkonda, 2005). They reported that such coatings reduce astrogliosis after four weeks of implantation compared to uncoated controls (He et al., 2006). Also, the Martin group of the University of Michigan has been developing alginate coatings for electrodes with sustained-release capabilities delivering anti-inflammatory agents (Kim and Martin, 2006). Also, the Martin group has been developing high-surface-area, fuzzy, conducting polymers for application at the recording sites, some of which incorporate growth factors (Cui et al., 2001, 2003). Similarly, the Bellamkonda group is developing strategies to immobilize endogenous anti-inflammatory agents like α-melanocyte-stimulating hormone, a neuromodulatory peptide that appears promising as an approach to reducing inflammation around the electrode (Zhong and Bellamkonda, 2005).

The community is concerned with minimizing the trauma associated with device implantation especially with regard to vasculature damage, which is believed to be an important contributor to downstream events. Minimizing trauma may also improve biocompatibility (Bjornsson et al., 2006). Toward this end, investigators are developing novel means of mechanically controlling insertion technology (Rennaker et al., 2005) as well as developing adjustable-depth electrodes that may be moved after implantation to achieve more consistent recording (Kralik et al., 2001; Musallam et al., 2007).

Despite the promise of these strategies, it is still too early to know which, if any, will ultimately improve the consistency of long-term recording performance of penetrating electrodes as most of the developmental work has not been performed on fully functional electrodes. Whereas in some cases the tissue response has been shown to be improved, it is not yet clear whether recording function will be improved in the same way. Therefore, it is still too early to know whether such approaches will meet the challenge posed by the foreign body response. It is possible that the field needs to radically change its approach as it is currently envisioned. This represents a significant opportunity for the next generation of scientists and engineers.

SUMMARY AND CONCLUSIONS

After an analysis of the peer-reviewed literature, two workshops, and visits abroad, we conclude that major gaps still remain in our understanding of the science behind the loss of function that occurs over time with the use of penetrating recording electrode arrays. It appears that this is unlikely to change significantly at the present levels of funding. Therefore, we suggest that targeted funding be provided in the United States to increase our knowledge of the underlying science of CNS implant biocompatibility in order to maintain a leadership position in this sector, as well as to accelerate the technological improvements that will be necessary for this technology to contribute to improving U.S. healthcare and economic well-being.

REFERENCES

Babb, T.L. and W. Kupfer. 1984. Phagocytic and metabolic reactions to chronically implanted metal brain electrodes, *Exp. Neurol.* 86(2):171–182.

Bickford, R.G., G. Fischer, and G.P. Sayre. 1957. Histologic changes in the cat's brain after introduction of metallic and plastic coated wire used in electro-encephalography. *Mayo Clin. Proc.* 32(1):14–21.

Biran, R., D.C. Martin, and P.A. Tresco. 2007. The brain tissue response to implanted silicon microelectrode arrays is increased when the device is tethered to the skull. *J. Biomed. Mater. Res. Part A* 82A(1):169–178.

Biran, R., D.C. Martin, and P.A. Tresco. 2005. Neuronal cell loss accompanies the brain tissue response to chronically implanted silicon microelectrode arrays. *Exp. Neurol.* 195(1):115–126.

Bjornsson, C.S., S.J. Oh, Y.A. Al-Kofahi, Y.J. Lim, K.L. Smith, J.N. Turner, S. De, B. Roysam, W. Shain, and S.J. Kim. 2006. Effects of insertion conditions on tissue strain and vascular damage during neuroprosthetic device insertion. *J. Neural Eng.* 3(3):196–207.

Burns, B.D., J.P. Stean, and A.C. Webb. 1974. Recording for several days from single cortical neurons in completely unrestrained cats. *Electroencephalogr. Clin. Neurophysiol.* 36(3): 314–318.

Buzsaki, G., and A. Kandel. 1998. Somadendritic backpropagation of action potentials in cortical pyramidal cells of the awake rat. *J. Neurophysiol.* 79(3):1587–1591.

Collias, J.C., and E.E. Manuelidis. 1957. Histopathological changes produced by implanted electrodes in cat brains; Comparison with histopathological changes in human and experimental puncture wounds. *J. Neurosurg.* 14(3):302–328.

Cui, X., V.A. Lee, Y. Raphael, J.A. Wiler, J.F. Hetke, D.J. Anderson, and D.C. Martin. 2001. Surface modification of neural recording electrodes with conducting polymer/biomolecule blends. *J. Biomed. Mater. Res.* 56(2):261–272.

Cui, X., J. Wiler, M. Dzaman, R.A. Altschuler, and D.C. Martin. 2003. *In vivo* studies of polypyrrole/peptide coated neural probes. *Biomaterials* 24(5):777–787.

Donoghue, J.P. 2002. Connecting cortex to machines: Recent advances in brain interfaces. *Nature Neurosci.* 5:1085–1088.

Dymond, A.M., L.E. Kaechele, J.M. Jurist, and P.H. Crandall. 1970. Brain tissue reaction to some chronically implanted metals. *J. Neurosurg.* 33(5):574–580.

Edell, D.J., V.V. Toi, V.M. McNeil, and L.D. Clark. 1992. Factors influencing the biocompatibility of insertable silicon microshafts in cerebral cortex. *IEEE Trans. Biomed. Eng.* 39(6):635–643.

Friehs, G.M., V.A. Zerris, C.L. Ojakangas, M.R. Fellows, and J.P. Donoghue. 2004. Brain-machine and brain-computer interfaces. *Stroke* 35(11 Suppl 1):2702–2705.

He, W. and R.V. Bellamkonda. 2005. Nanoscale neurointegrative coatings for neural implants. *Biomaterials* 26(16):2983–2990.

He, W., G.C. McConnell, and R.V. Bellamkonda. 2006. Nanoscale laminin coating modulates cortical scarring response around implanted silicon microelectrode arrays. *J. Neural Eng.* 3(4):316–326.

Hendriks, J.J., C.E. Teunissen, H.E. de Vries, and C.D. Dijkstra. 2005. Macrophages and neurodegeneration. *Brain Res. Brain Res. Rev.* 48(2):185–195.

Henze, D.A., Z. Borhegyi, J. Csicsvari, A. Mamiya, K.D. Harris, and G. Buzsaki. 2000. Intracellular features predicted by extracellular recordings in the hippocampus *in vivo*. *J. Neurophysiol.* 84(1):390–400.

Hochberg, L.R., M.D. Serruya, G.M. Friehs, J.A. Mukand, M. Saleh, A.H. Caplan, A. Branner, D. Chen, R.D. Penn, and J.P. Donoghue. 2006. Neuronal ensemble control of prosthetic devices by a human with tetraplegia. *Nature* 442(7099):164–171.

Hoogerwerf, A.C. and K.D. Wise. 1994. A three-dimensional microelectrode array for chronic neural recording. *IEEE Trans. Biomed. Eng.* 41(12):1136–1146.

Kennedy, P.R., R.A. Bakay, M.M. Moore, K. Adams, and J. Goldwaithe. 2000. Direct control of a computer from the human central nervous system. *IEEE Trans. Rehab. Eng.* 8(2):198–202.

Kim, D.H. and D.C. Martin. 2006. Sustained release of dexamethasone from hydrophilic matrices using PLGA nanoparticles for neural drug delivery. *Biomaterials* 27(15):3031–3037.

Kralik, J.D., D.F. Dimitrov, D.J. Krupa, D.B. Katz, D. Cohen, and M.A. Nicolelis. 2001. Techniques for long-term multisite neuronal ensemble recordings in behaving animals. *Methods* 25(2):121–150.

Lebedev, M.A. and M.A. Nicolelis. 2006. Brain-machine interfaces: past, present and future, *Trends Neurosci.* 29(9):536–546.

Lee, H., R.V. Bellamkonda, W. Sun, and M.E. Levenston. 2005a. Biomechanical analysis of silicon microelectrode-induced strain in the brain. *J. Neural Eng.* 2(4):81–89.

Lee, I.H., E. Lindqvist, O. Kiehn, J. Widenfalk, and L. Olson. 2005b. Glial and neuronal connexin expression patterns in the rat spinal cord during development and following injury. *J. Comp. Neurol.* 489(1):1–10.

Liu, X., D.B. McCreery, R.R. Carter, L.A. Bullara, T.G. Yuen, and W.F. Agnew. 1999. Stability of the interface between neural tissue and chronically implanted intracortical microelectrodes. *IEEE Trans. Rehab. Eng.* 7(3):315–326.

Ludwig, K.A., J.D. Uram, J. Yang, D.C. Martin, and D.R. Kipke. 2006. Chronic neural recordings using silicon microelectrode arrays electrochemically deposited with a poly(3,4-ethylenedioxythiophene) (PEDOT) film. *J. Neural Eng.* 3(1):59–70.

Merrill, D.R. and P.A. Tresco. 2005. Impedance characterization of microarray recording electrodes in vitro. *IEEE Trans. Biomed. Eng.* 52(11):1960–1965.

Mountcastle, V.B., P.W. Davies, and A.L. Berman. 1957. Response properties of neurons of cat's somatic cortex to peripheral stiumuli. *J. Neurophysiol.* 20(4):374–407.

Musallam, S., M.J. Bak, P.R. Troyk, and R.A. Andersen. 2007. A floating metal microelectrode array for chronic implantation. *J. Neurosci. Meth.* 160(1):122–127.

Musallam, S., B.D. Corneil, B. Greger, H. Scherberger, and R.A. Andersen. 2004. Cognitive control signals for neural prosthetics. *Science* 305(5681):258–262.

Nagy, J.I. and J.E. Rash. 2000. Connexins and gap junctions of astrocytes and oligodendrocytes in the CNS. *Brain Res. Brain Res. Rev.* 32(1):29–44.

Nicolelis, M.A., D. Dimitrov, J.M. Carmena, R. Crist, G. Lehew, J.D. Kralik, and S.P. Wise. 2003. Chronic, multisite, multielectrode recordings in macaque monkeys. *Proc. Natl. Acad. Sci. USA* 100(19):11041–11046.

Polikov, V.S., P.A. Tresco, and W.M. Reichert. 2005. Response of brain tissue to chronically implanted neural electrodes. *J. Neurosci. Meth.* 148(1):1–18.

Rall, W. 1962. Electrophysiology of a dendritic neuron model. *Biophys. J.* 2(2pt2):145–167.

Rennaker, R.L., S. Street, A.M. Ruyle, and A.M. Sloan. 2005. A comparison of chronic multichannel cortical implantation techniques: manual versus mechanical insertion. *J. Neurosci. Meth.* 142(2):169–176.

Robinson, F.R. and H.T. Johnson. 1961. Histopatholgical studies of tissue reactions to various metals implanted in cat brains. *ASD Technical Report* 61-397, Wright-Patterson Air Force Base, Ohio, 1–16.

Rousche, P.J. and R.A. Normann. 1998. Chronic recording capability of the Utah intracortical electrode array in cat sensory cortex. *J. Neurosci. Meth.* 82(1):1–15.

Rousche, P.J., D.S. Pellinen, D.P. Pivin, Jr., J.C. Williams, R.J. Vetter, and D.R. Kipke. 2001. Flexible polyimide-based intracortical electrode arrays with bioactive capability. *IEEE Trans. Biomed. Eng.* 48(3):361–371.

Santhanam, G., S.I. Ryu, B.M. Yu, A. Afshar, and K.V. Shenoy. 2006. A high-performance brain-computer interface. *Nature* 442(7099):195–198.

Schmidt, E.M. 1980. Single neuron recording from motor cortex as a possible source of signals for control of external devices. *Ann. Biomed. Eng.* 8(4–6):339–349.

Schmidt, E.M., M.J. Bak, and J.S. McIntosh. 1976. Long-term chronic recording from cortical neurons. *Exp. Neurol.* 52(3):496–506.

Schmidt, S., K. Horch, and R. Normann. 1993. Biocompatibility of silicon-based electrode arrays implanted in feline cortical tissue. *J. Biomed. Mater. Res.* 27(11):1393–1399.

Schultz, R.L and T.J. Willey. 1976. The ultrastructure of the sheath around chronically implanted electrodes in brain. *J. Neurocytol.* 5(6):621–642.

Schwartz, A.B., X.T. Cui, D.J. Weber, and D.W. Moran. 2006. Brain-controlled interfaces: movement restoration with neural prosthetics. *Neuron* 52(1):205–220.

Serruya, M.D., N.G. Hatsopoulos, L. Paninski, M.R. Fellows, and J.P. Donoghue. 2002. Instant neural control of a movement signal. *Nature* 416(6877):141–142.

Seymour, J. and D.R. Kipke. 2006. Ultra-fine structures on neural probes reduce cellular encapsulation. *Soc. Neurosci. Annu. Meet.* 354:16.

Stensaas, S.S. and L.J. Stensaas. 1978. Histopathological evaluation of materials implanted in the cerebral cortex. *Acta Neuropathol.* (Berl) 41(2):145–155.

Stensaas, S.S. and L.J. Stensaas. 1976. The reaction of the cerebral cortex to chronically implanted plastic needles, *Acta Neuropathol.* (Berl) 35(3):187–203.

Subbaroyan, J., D.C. Martin, and D.R. Kipke. 2005. A finite-element model of the mechanical effects of implantable microelectrodes in the cerebral cortex. *J. Neural Eng.* 2(4):103–113.

Subbaroyan, J., T.D.K. Yoshida, and D.R. Kipke. 2006. Chronic tissue response evoked by variably flexible intracortical polymer implant systems. Atlanta, GA: *Society for Neuroscience Abstracts.*

Sykova, E. 2005. Glia and volume transmission during physiological and pathological states, *J. Neural Transm.* 112(1):137–147.

Szarowski, D.H., M.D. Andersen, S. Retterer, A.J. Spence, M. Isaacson, H.G. Craighead, J.N. Turner, and W. Shain. 2003. Brain responses to micro-machined silicon devices. *Brain Res.* 983(1–2).23–35.

Taylor, D.M., S.I. Tillery, and A.B. Schwartz. 2002. Direct cortical control of 3D neuroprosthetic devices. *Science* 296(5574):1829–1832.

Turner, J.N., W. Shain, D.H. Szarowski, M. Andersen, S. Martins, M. Isaacson, and H. Craighead. 1999. Cerebral astrocyte response to micromachined silicon implants. *Exp. Neurol.* 156(1): 33–49.

Wessberg, J., C.R. Stambaugh, J.D. Kralik, P.D. Beck, M. Laubach, J.K. Chapin, J. Kim, S.J. Biggs, M.A. Srinivasan, and M.A. Nicolelis. 2000. Real-time prediction of hand trajectory by ensembles of cortical neurons in primates. *Nature* 408(6810):361–365.

Williams, J.C., R.L. Rennaker, and D.R. Kipke. 1999. Long-term neural recording characteristics of wire microelectrode arrays implanted in cerebral cortex. *Brain Res. Brain Res. Protoc.* 4(3):303–313.

Yuen, T.G. and W.F. Agnew. 1995. Histological evaluation of polyesterimide-insulated gold wires in brain. *Biomaterials* 16(12):951–956.

Zhong, Y. and R.V. Bellamkonda. 2005. Controlled release of antiinflammatory agent alpha-MSH from neural implants. *J. Control Rel.* 106(3):309–318.

CHAPTER 4

BMI/BCI Modeling and Signal Processing

José C. Principe and Dennis J. McFarland

INTRODUCTION

Due to the large differences in the biophysical characteristics between multi-microelectrode array and EEG/ECoG recordings, this report treats them separately. The signal processing modalities required to spatially resolve, extract features, and interpret intent depend on the selection of recording modality, which includes electroencephalography (EEG), electrocorticography (ECoG), local field potentials (LFPs), and single-neuron action potential recordings (single units).

Brain-machine interfaces (BMIs) are significantly different from BCIs, although they are posed to solve the same problem: translating a subject's intent into robotic commands. BCIs work with the macroscopic brain activity (mostly EEG) that is known to correlate with behavior, but in a diffuse and unspecific way. Capitalizing on the available knowledge of EEG research and on machine learning techniques, BCIs already have achieved successes and are ready to be used in patients (see the site reports in Appendixes B and C). But their applicability to the full gambit of functions needed in the unrestricted interaction of a subject with the environment is limited. On the other hand, BMIs probe the brain at many different levels of abstraction (microscopic as in spikes and mesoscopic as in local field potentials, or LFPs), and therefore offer potentially better performance at the cost of being much more demanding in terms of the brain functional organization at these different levels, i.e., the neuron code, the neural assembly, and its cytoarchitecture. Perhaps BCIs will be merged someday with BMIs when the macroscopic information is incorporated. Due to this finer integration of the BMI with brain signals, deeper neurophysiology knowledge is required to conduct this type of research than for the BCI counterpart. Larger multidisciplinary teams need to be assembled to advance research in this area. All the successful groups in the United States have a combination of expertise ranging from basic neuroscience (electrophysiology), computational neuroscience, signal processing and modeling, and advanced electronic and computer systems. And all of this may still be insufficient!

T.W. Berger et al., Brain-Computer Interfaces, 47–64.

MULTIMICROELECTRODE ARRAY TECHNIQUES

Brain-machine interfaces here denote systems that use spike-train data. Two basic classes of methods are used with multimicroelectrode array data: one class uses the spike-train data directly; one class uses an estimate of the instantaneous spike rate through binning, i.e., counting the number of spikes in a time interval (the bin). There is a preponderance of binned models in BMI research.

Binned Data Models

The BMI experimental paradigm lends itself nicely to statistical signal processing methodologies used to derive optimal models from data. Indeed, in the BMI setting the researcher synchronously has available both the input to the BMI (the spike trains) and the desired response (hand position). The problem can be then framed in terms of "decoding," by which spike occurrences of individual neurons are translated into hand positions. Since the data are collected by multielectrode arrays, and each electrode can potentially sense several neurons, spike sorting is commonly utilized to identify individual neurons. Accurate spike detection and sorting is a critical step for high-performance decoding models since its role is to identify the neuromodulation related to movement intent and separate execution from the background noise.

Accurate and automated spike detection and sorting remains an ongoing research topic (Fee et al., 1996; Lewicki, 1998; Wood et al., 2004). Provided that the spike features have been accurately extracted, the neuronal spike firings become the decoding model input. It is possible to translate the decoding problem into a system identification framework, where a parametric linear or nonlinear system is trained directly from the collected data to achieve outputs close to the hand positions (Figure 4.1). Model building has been extensively studied in control theory and signal processing, so there is a wealth of methods that can be utilized (Soderstrom and Stoica, 1989).

The only difficulty for a straightforward application of system identification is that these models have been derived for continuously varying signals, but in BMIs, the inputs are spike trains. Binning the spikes firing of each neuron over an appropriate time window (~50 to ~100 ms) has been widely used to smooth the spike trains and provide a time scale closer to behavior (Nicolelis, 2001).

Perhaps the first account of a BMI (Chapin et al., 1999) used a linear model trained by least squares with one second (i.e., the current and nine past bins) of previous spike data per neuron as a memory buffer to predict the lever press in a rat model. It turns out that least squares with past inputs is equivalent to the Wiener solution for a finite impulse-response filter. Therefore, the first BMI was trained with a Wiener filter (Haykin, 2002). The bulk of the results reported in the literature

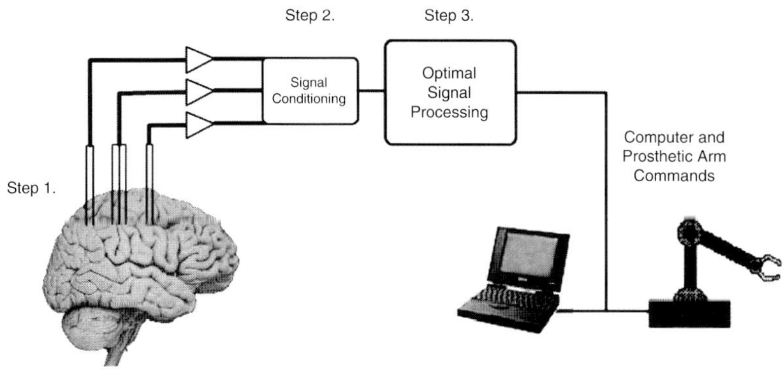

Figure 4.1. System identification framework.

use this simple but powerful linear solution (Wessberg et al., 2000; Serruya et al., 2002; Carmena et al., 2003; Shenoy et al., 2003).

Of course, there are many aspects that need the attention of the researcher to fine-tune the results. First, these models tend to have many parameters (number of time bins × number of channels × number of outputs). Each parameter is a degree of freedom in the solution, and as is well known, the model must be properly regularized for good generalization. Regularized least squares, subspace projections using partial least squares, or special memory structures, have been used to reduce the number of free parameters (Kim et al., 2006). Neural selection has also been attempted using sensitivity analysis and variable selection procedures (Sanchez et al., 2003). Second, the tradeoffs between timing resolution (bin size) and memory depth needed to optimally solve the system identification problem have not been directly addressed (Wu et al., 2005). Third, the Wiener solution assumes stationnarity, i.e., it assumes that the spike-train statistics do not change over time. This is an unrealistic assumption that can be counteracted using sample-by-sample adaptation—the famous least mean square (LMS) algorithm—to track the optimal solution through time, or by partitioning the solution space into a set of local linear models that switch automatically among themselves (Kim et al., 2003).

The third assumption in the Wiener filter is that the input-output map is linear, which may not be the case. Dynamic neural networks (a time-delay neural network or recurrent neural networks [Chapin et al., 1999; Gao et al., 2003; Sanchez et al., 2002]) can be utilized to enhance the simple linear solution at the expense of longer and more careful training algorithms. More recently, methods based on Support Vector Machine regression have also been reported (Wang et al., 2004). All these enhancements improve the original solution in a marginal but statistically significant way. The reported performance ranges from 60% to 90%, but since each research group is using its own data and the performance changes from task to task, animal to animal, and even from day to day, it is very difficult to compare results.

A second aspect that complicates the comparison of decoding techniques is the simplicity of the movement tasks studied. The next generation of decoding models will have to assess performance in multipostural reaching, grasping, and holding under the influence of force. Here the possible advantages of nonlinear models may be highlighted. We refer to Kim et al. (2006) for a comparison of models on the same data set, but the best way to test performance of different algorithms is to create an open competition as the BCI community is doing so successfully. All in all, the viability of decoding the intent of motion contained in multidimensional spike trains collected from the motor cortex has been established by numerous research groups and should be considered an established scientific fact.

A second class of nonlinear models for BMIs was derived from system theory, using the concept of Volterra series expansions in Hilbert spaces. A Volterra series is similar to a Taylor series expansion in functional spaces, where the terms are convolution integrals of products of impulse responses called kernels. The Volterra models are universal approximators, but they require the estimation of many parameters. Therefore, when applied in practice, one of the design difficulties is how to minimize the parameters, either by limiting the expansion to the second or third term, or by using polynomial approximations, such as Laguerre polynomials (Marmarelis, 1993), or preferably, Kautz functions, to decrease the number of free parameters for the impulse responses. Although originally developed for continuous amplitude signals, Volterra series have been proposed to model neuronal receptive fields—the reverse-correlation or white-noise approach (Marmarelis and Marmarelis, 1978). More recently, they have been applied to cognitive BMIs to model the CA3 region of the hippocampus (Song et al., 2002). Cognitive BMIs attempt to restore lost cognitive functions such as memory (Berger et al., 2001).

The other class of models that has been used for BMIs uses a generative approach instead of the input-output modeling described above (Moran and Schwartz, 1999; Wu et al., 2002; Taylor et al., 2002). Generative models can benefit from a Bayesian formulation and they can offer a general approach to estimate the biological response from the multichannel spike-train input. The probabilistic approach analyzes and infers the biological response as a state variable of the neural dynamical system from a sequence of noisy observations of the neural activity. In the Bayesian models it is necessary to build an observation measurement model, which relates the measurement of the noisy neural activity to the states, the tuning curve (Dayan and Abbott, 2001). The probabilistic state space formulation and the updating of information are rooted on the Bayesian approach of incorporating information from measurements. A recursive algorithm is used to construct the posterior probability density function of the biological response at each time, which embodies all available statistical information and in principle yields the solution to the decoding problem. By estimating the expectation of the posterior density (or by maximum likelihood estimation), the movement estimate can be recovered probabilistically from the multichannel neural recordings.

The Kalman filter is the best known of these models; several groups (Wu et al., 2002; Sanchez et al., 2002a, b) have applied it to BMI. One of the strong assumptions of the Kalman filter is that the (neural activity) time series was generated by a linear system, which means the tuning function is only a linear filter. The other strong assumption is the Gaussian distribution of the posterior density of the kinematic stimulus, given the neural spiking activities at every time step, which reduces all the richness of the interactions to second-order information (mean and covariance). These two assumptions may be too restrictive for BMI applications, considering the large dimensionality of the input and the much smaller dimension of the state vector.

The particle filter framework lifts all the restrictions of the Kalman filter (linearity and Gaussian assumption) but complicates substantially the computational algorithms. As its counterpart, the particle filter also uses sequential estimation of the posterior at each step. But due to the generality of the model, there is no closed-form solution and therefore the posterior has to be estimated by probing. To help create an estimate of the posterior density, a set of samples using Monte Carlo sampling drawn from a properly determined density that is estimated at each step is sent through the system with the present parameters. The peak of this posterior (or another central moment) is considered as the state estimate. Particle filters have also been applied to BMIs (Brockwell et al., 2004), where the tuning function was assumed as an exponential operation on linear filtered velocities (Schwartz, 2001; T. Matsumoto, Waseda University site visit).

Spike-Train Methods (Point Process)

The spike-data methods have been applied primarily to understand how neurons encode information (Simoncelli et al., 2004). In motor-control BMIs, the problem is actually the reverse, where a process called *decoding* (Paninski et al., 2004) identifies how a spike train in motor cortex can explain the movement of a limb. However, the primary methodologies are still inspired by the *encoding* methods. For example, the population vector method of Georgopoulos and colleagues (1986) is a generative model of the spike activity based on the tuning curve concept (preferential firing for a given hand position/speed) that has been extensively utilized in encoding methods. In BMIs, the population vector technique has been championed by Schwartz and collaborators (2001).

All the encoding methods effectively model the probability density function (PDF) of the spike firings. PDF estimation is a difficult problem that is seldom attempted because it requires lots of data and stationary conditions. An alternative methodology that effectively bypasses this requirement is the use of maximum likelihood methods assuming a specific PDF. In neuroscience, the Poisson distribution assumption is very common because it has been validated in numerous experimental setups, but it cannot account for multimodal firing histograms that are often found in neurons. The Poisson model has been improved with a time varying mean to yield what is called the inhomogeneous Poisson model. Unfortunately, the extension of these methods to multineuron spike trains is still based on the assumption of spike independence, which does not apply when neurons are part of neural assemblies.

A general point process adaptive filtering paradigm was recently proposed by Brown et al. (2001) to probabilistically reconstruct the hand position from the discrete observation of the neural firing. This algorithm modeled the neural spike train as an inhomogeneous Poisson process feeding a kinematic model through a nonlinear tuning function. This approach also embodies the conceptual Bayesian filtering algorithm: predicting the posterior density by a linear state update equation and revising it with the next observation measurement. The point process filter analogs of the Kalman filter, recursive least squares, and steepest descent algorithms were derived and recently compared in the decoding of tuning parameters and states from the ensemble neural spiking activity (Eden et al., 2004). The point process analog of the Kalman filter performs the best because it provides an adjustable step size to update the state, which is estimated from the covariance information. However, this method still assumes that the posterior density of the state vector, given the discrete observation, is always Gaussian, which is certainly not the case. A Monte Carlo sequential estimation algorithm on the point process was recently proposed to infer the kinematic information directly from the neural spike train (Wang et al., 2006). Given the neural spike train, the posterior density of the kinematic stimulus was estimated at each time step without the Gaussian assumption. The preliminary simulations showed a better velocity reconstruction from exponentially tuned neural spike trains.

BMI methods that derive the kinematic information from the neural activity using a Bayesian formulation require preknowledge of the neuron receptive properties, and an essential stationary assumption is used when the receptive field is built from a block of data, which may not account for changes in response of the neural ensemble from open-to-closed loop experiments (Tillery et al., 2003). Moreover, the good initialization of all the parameters in the algorithm can directly affect the results of the prediction of the subject's movements in BMI. This is because all the probabilistic approaches are based on the Bayesian formulation to construct the posterior density at each time step from the prior density of the kinematic state, which is the posterior density of the previous time step.

Relations between Spike Trains and Local Field Potentials

One intriguing area of BMI research that is growing in importance is quantifying the relationships between local field potentials (LFPs) and spike trains (Donoghue et al., 1998; Donchin et al., 2001). The basic idea is to complement the information contained in the spike trains with that contained in LFPs. There are two good reasons. First, the spike trains are very specific but they are also local to the neuron. Due to the physical limitation of the number of electrodes in microelectrode arrays, the number of neurons sensed will always be a minute fraction of motor cortex neurons. Second, neuron firings exist at a time scale of milliseconds, whereas behavior exists at the time scale of seconds. Linking the two time scales requires a model-driven rather than a data-driven approach, which so far is based mostly on averaging (e.g., binning). LFPs can come to the rescue because they represent aggregate (mesoscopic) activity of millions of neurons. LFPs are defined here as the low frequency activity (0.5–1 kHz) collected by the high-impedance tip of the microelectrode. Obviously, LFPs are not as specific as spike activity, but they provide the "state" of millions of neurons and they have the time scale of behavior. Moreover, LFPs can perhaps be related to electrocorticography (ECoG) and from there to electroencephalograms (EEG), which has the added advantage of being noninvasive. Understanding the relation between spike trains and LFPs becomes central to this endeavor.

The association between spike firings and LFPs has been investigated in a stimulus-related manner. Researchers have described the temporal structure in LFPs and spikes where negative deflections in LFPs were proposed to reflect excitatory, spike-causing inputs to neurons in the neighborhood of the electrode (Arieli et al., 1995). An appropriate feature-detection method proposes to explore correlation between the amplitude-modulated (AM) components of the movement-evoked (local field) potentials (mEPs) and single-unit activities recorded at the same electrode across all movement trials as stimuli (Mehring et al., 2003; Wang et al., 2006). The correlation between pairs of perievent time histograms (PETH) and mEPs at the same electrode showed high correlation coefficients for some neurons, which indicates that the extracellular dendritic potentials contain an indication of the level of neuronal output. A critical demonstration of this relationship was the process of averaging the LFP and single-unit activity across the lever-press trials that reduced the noise contamination caused by the brain spontaneous activity. However, to be useful for BMIs, this technique must be improved to a real-time analysis, and a multiscale analysis framework should be developed to effectively utilize the joint information.

We had the opportunity to visit one of the leading groups working on LFPs for BMIs: Albert Ludwigs University and University Hospital in Freiburg, Germany. They showed that the integration of LFPs with spikes improved the overall determination of the movement direction (Mehring et al., 2003). They also showed that the statistical pattern recognition techniques based on penalized discriminant analysis and Support Vector Machines are significantly better than the population vector of the Gaussian Mixture model for this task. This group also illustrated the use of a frequency band decomposition of single-trial LFPs, averaged over the scalp, to decode movement direction in center-out movement tasks. The frequency bands that better code the information are less than 4, 6 to 13 Hz, and 63 to 200 Hz. This group is also investigating the use of ECoG grids in epileptic patients to infer movement direction (Rickert et al., 2005). They showed that the ECoG electrodes collected synchronously with movement over M1 and PM carry substantial information about the movement direction. They demonstrate 80% of correct identification of movement direction, and 0.7 cc between predicted and hand velocity. These results are exciting since they indicate the possibility of using the ECoG over the motor cortex to infer intention of movement. Since the relationship between the ECoG and the EEG is well understood, it may be possible to extend these results to the scalp with the use of inverse models.

EEG/ECoG RECORDINGS

Signal processing with EEG and ECoG recordings is similar in many respects. In both cases the signal is a field resulting from the activity of large numbers of neurons. In both cases, signals can be detected only if many of these neurons are synchronized. Asynchronous activity should cancel so that the resulting signal reflects only that portion of neuronal activity that is synchronous. BCI signal processing can be thought of as involving two phases, feature extraction and translation (Wolpaw et al., 2002; Leuthardt et al., 2004; Sanchez et al., 2006). EEG and ECoG signals contain transient, time-domain signals phase-locked to events such as the P300 (Donchin et al., 2000) and motor potentials (Bashashati et al., 2006). These field potentials also contain frequency-domain signals such as the mu rhythm (Pfurtscheller et al., 2006). Methods for feature extraction differ for time- and frequency-domain signals. The translation stage is similar for the two. In fact, time- and frequency-domain signals can be combined in a single classifier (Li et al., 2004).

Decoding the Spontaneous ECoG for Intent

The use of the ECoG for brain modeling can be broadly divided into two basic methodologies: evoked potentials, and spontaneous brain activity. The overwhelming majority of the research has been dedicated to evoked potential studies, where the response to stimulus is collected and analyzed under very well controlled experimental conditions. This will be the subject of a review in Chapter 7. Here we address Walter Freeman's work toward extracting information from multielectrode ECoG activity during perception. Neurophysiologic studies have revealed the generation of active state sequences in cortical activity during an act of perception evoked by the presentation of conditioned stimuli (Freeman, 2004a, b, 2005). These active states observed in the beta (12–30 Hz) and gamma (30–80 Hz) ranges, were described as "cinematographic" and can be conceived as frames of spatial patterns related to perception of stimuli. Each frame begins with sudden sharp phase resettings, followed by resynchronization, spatial pattern stabilization, and increase in intensity to a brief maximum lasting about three-to-five cycles of the carrier wave. These stable spatial amplitude modulations occur aperiodically but in synchrony among different channels. Freeman defines pragmatic information as the ratio of the instantaneous temporal pattern intensity to the rate of spatial pattern change, and demonstrated that it can detect frames of beta and gamma activity related to behavior. The instantaneous amplitude and phase are computed by the Hilbert transforms to obtain high time resolution in the frequency measurement. High values of pragmatic information indicate processing frames and they potentially can be used to asynchronously select features in the ECoG relevant to behavior. The first study applying this method to BMIs appeared in 2005 (Gunduz et al., 2005).

Time-Domain Feature Extraction

To date, feature extraction with time-domain signals has been relative simple, in many cases involving selection of specific time points from time-locked averages of signals (Donchin et al., 2000). Serby et al. (2005) compared the method used by Donchin et al. (2000) with the use of independent components analysis followed by a matched filter. The results suggest that their new procedure resulted in much better classification. Mason and Birch (2000) designed a low-frequency filter to generate features for a BCI based on motor potentials. Blankertz et al. (2004) used Fast Fourier Transform (FFT-based) band-pass filtering to classify single-trial motor potentials. Yom-Tov and Inbar (2003) have discussed the detection of motor potentials in terms of whether or not the signal is deterministic. They suggest that a matched filter would be best for deterministic signals. A more general filter may be better if the signal is not deterministic. They report that a hybrid of both methods performs best on actual motor potentials.

Frequency-Domain Feature Extraction

Feature extraction with frequency-domain signals has involved a wide variety of techniques (McFarland et al., 2006). These include methods that are time-based, space-based, and time-space methods. Time-based methods include band-pass filtering, Fourier-based spectral analysis, parametric methods such as autoregressive spectral methods, and use of wavelets. Space-based methods include Laplacian filters, principal components, independent components, and common spatial patterns. Time-space-based methods include component analysis in time and space, multivariate autoregressive models, and coherence.

Many studies have shown that various feature extraction methods do in fact produce results that can be used with a BCI. For example, Wang et al. (2004) used the envelope of band-pass filtered data to generate time-frequency patterns associated with EEGs collected during imagined hand movement. Studies comparing different methods are less common. For example, Vidaurre et al. (2005) report no difference between band power based on digital filtering and adaptive autoregressive parameters obtained by means of Kalman filtering. Boostani and Moradi (2004) compared band power, Hjorth parameters, and Fractal dimension as features for classifying motor imagery data. Band power yielded the best performance for four of five subjects, but the authors concluded that fractal dimension could be considered as an alternative to band power. Krusienski et al. (2007) compared spectral bands based on AR (autoregressive) models, the FFT, and a matched filter. In this case the matched filter outperformed other methods. This may be due to the fact that harmonics of the fundamental of the mu rhythm appear to have a preferred phase.

Although BCI signal extraction can be conceptualized as involving two distinct phases, spatial and temporal filtering, it is also possible to include both in a single process. For example, Lemm et al. (2005) used common spatial patterns with time-delay embedding. This method results in a single-step method that produces better classification than a method combining band-pass filtering and common spatial patterns.

Spatial Filtering

Various forms of spatial filtering can be used as part of the feature extraction process. McFarland et al. (1997) showed that either a Laplacian derivation or a common average reference enhanced extraction of spectral features from data collected while subjects modulated their mu rhythms to move a cursor on a video monitor. The effects of a Laplacian derivation varied with the spacing of the electrodes used in the computation. These results indicate that the spatial filtering should match the spatial characteristics of the signal of interest. As such, a uniform spatial filter would probably not be optimal for all signals.

Spatial filters can be either fixed or data driven. For example, Laplacian derivations have fixed weights, but spatial filters based on methods such as principal components or independent components have data-driven weights. Both of these methods are unsupervised. In contrast, the method of common spatial patterns is both supervised and data driven. Naeem et al. (2006) compared several independent components algorithms, Laplacian and bipolar derivations, and common spatial patterns on data derived from a four-class motor imagery task. They found that the Laplacian and independent components methods were comparable, but the method of common spatial patterns resulted in the best classification.

Feature Translation

A given feature-translation method is generally applicable to both time- and frequency-domain features. The output of the feature-translation process can be either discrete, as in the case of discriminant functions, or continuous, as in the case of regression functions (McFarland and Wolpaw, 2005). The selection of either method should be based on the nature of the signals and intended application.

Feature translation can involve use of all extracted features or a subset of these. Millan et al. (2002) provide an excellent discussion of this issue and provide data suggesting that performance is enhanced by selecting a subset of relevant features. Using spectral features collected during imagined hand movements, performance increased up to an average of 7 features and then declined. However Fabiani et al. (2004), using data collected during mu-rhythm-based cursor movement, report that performance peaked at around 10 features but did not decline afterwards.

As with feature extraction, some studies demonstrate the utility of one particular approach without comparing alternatives. For example, Peters et al. (1998) obtained EEG data from a four-class motor imagery experiment and fed resulting autoregressive parameters from 30 Laplacian filtered electrodes into an artificial neural net. The system was quite accurate in classifying the results. However, it is not clear how well other methods would perform with so many features, nor is it obvious what characteristics of the EEG were used by the classifier.

Rezaei et al. (2006) compared five classifiers on two standard archival data sets from which they extracted AR coefficients. These included a Bayesian graphical network, a neural network, a Bayesian quadratic classifier, Fisher's linear discriminant analysis, and a hidden Markov model. All of the methods produced similar results with the exception of the hidden Markov model, which produced much poorer classification than the other methods. Schlogl et al. (2005) systematically compared minimum distance analysis, linear discriminant analysis, k-nearest neighbor analysis, and a Support Vector Machine. The data were autoregressive parameters obtained from subjects performing a four-class motor imagery task. They found that the Support Vector Machine produced the best classification results.

Several BCI data competitions have been conducted (Sajda et al., 2003; Blankertz et al., 2004, 2006). These competitions involve the use of archival data sets and the efforts of many research groups. The results have been very useful in providing information on the best solutions obtained by a group of skilled data analysts. However they do not provide a means of evaluating the contributions of individual components of these methods. The winning entry involves a specific combination of feature extraction, feature selection, and classification techniques. As a result, it is not possible to evaluate the relative contribution of the individual analysis components, as well as the skill of the winning contestant. Nonetheless, conclusions can be made. From the most recent competition, (1) almost all methods were linear, (2) for oscillatory activity the winning methods tended to use common spatial patterns, and (3) several groups combined time and frequency methods (Blankertz et al., 2006).

Online Evaluation of Methods

Shenoy et al. (2006) showed that the statistics of the EEG change with the introduction of feedback. This result indicates that the potential effects of real-time, closed-loop performance should be considered. However, the vast majority of studies evaluating signal-processing methods have used offline analysis of archival data. An exception is the study of McFarland and Wolpaw (2005) where regression models that were evaluated offline were also compared online. Just as in offline analysis, the superiority of using more features was also observed in online, real-time performance. However, it should be noted that most studies have not determined whether offline performance is matched online.

SUMMARY AND CONCLUSIONS

Many studies demonstrate the feasibility of complex aggregates of methods. This makes it difficult to determine the utility of the individual components. In the future, it would be useful to compare possible alternatives for the individual components. There are many recent reports using archival data sets. It is necessary to determine whether or not results from these generalize to online systems. An additional issue concerns the magnitude of effects. Although there are small statistically significant differences between methods in some cases, often the actual size of the effect is not large. There is an important point here in that attention should be paid to which methods make only small differences and which result in larger differences. This is the issue of effect size.

Studies evaluating signal-processing methods with EEG recordings have based classification almost universally on some external stimulus, such as the presence of various targets on a computer screen or instructions to engage in certain forms of imagery. In contrast, studies dealing with multielectrode arrays have based classification on predicted limb position. Selection of the dependent variable is an interesting issue in modeling as it has ramifications for applications to users without voluntary movements.

The handful of groups in the world conducting BMI research is predominantly composed of U.S. groups, all of which were represented in the WTEC North American Baseline Workshop. During the WTEC panel's European trip we saw strong signal processing for BMI in Freiburg, Germany. The Biorobotics group in Pisa is also collaborating in BMI experiments with Freiburg using regions F4 and F5 of the premotor cortex to decode the reaching-for-food goal in monkeys.

The absence of publicly available BMI data sets is a bottleneck to attracting the interest of independent researchers in machine learning and signal processing. This is unfortunate because the complexity of the spike-train data and the differrence in scale with respect to behavior require beyond-state-of-the-art signal processing algorithms to foster, implement, and quantify new computational neuroscience theories. BMIs can indeed be experimental platforms to test brain theories in a way not possible in the past because of all the ingredients of cognitive experimentation with high resolution, synchronous measurements both at the input (spike trains, LFPs, and ECoG) and at the behavior level. If this bottleneck is not properly addressed, progress will be slow. In our opinion, the fundamental problems in BMI signal processing and modeling are the following:

1. **Nonstationarity of the neural response in time and space**. Let us look at the neurons in the motor cortex (PP, PMD, M1) as a spatial array, and let us compute the short-term cross-correlation with the desired kinematic variable of the hand (i.e., position, speed, or acceleration) in a local neighborhood of the array across time (as a movie). What we observe are spatially localized areas of correlation appearing and disappearing as the movement progresses in time, just like the bubbles in a pan of boiling water. This very rapid change in correlation over time and space defeats all the assumptions of the Wiener models or neural network training. The change is too fast to be captured appropriately by Kalman filters or other generative methods based on Bayesian statistics. In our opinion, switching local models or varying selection procedures are two candidates to help solve this problem in the binned methods, but it is still unclear how to appropriately train these switching architectures, particularly when the desired response is not available. The modeling approaches will also have to be adaptable to the patient-specific neurophysiological changes that occur during closed-loop control of a variety of dexterous movements under the influence of forces.

2. **How to train BMI models in the absence of desired response**. In a practical application of BMIs on a quadriplegic, there will be no desired response since the subject cannot move. For the communication type of applications as described in Hochberg et al. (2006), this is still tolerable because the number of degrees of freedom is few (2 degrees for a cursor on the screen). One possibility now being explored is a drastic departure from the conventional BMI architecture. The proposed BMI system works as a predictor anticipating the next move. It learns through reinforcement the relations between input patterns and actions in a closed-loop fashion. However, the complexity of the BMI system increases many-fold because it has to discover the actions that work given the user's goals. Therefore, important subquestions are how to distribute the intelligence between the subject and the BMI controller and how to cope with adaptability in both subsystems.

3. **How well can BMI generalize for other tasks beyond the ones in the training?** This is also an important question that must be addressed if BMIs are to be widely used in a variety of tasks that make up the normal interaction of a person with his environment. For example, "generalize" here means how does a BMI system trained for a reaching task perform in a cursor-tracking task? It is highly likely that there will be a need to decompose the movements in primitives and allow the BMI system to put together different pieces of the movement to achieve the end goal. As we do in speech processing, but applied to motor control, can the primitive atoms of movement be defined? This approach will benefit from the multiple-model approach discussed in item 1 and the reinforcement learning described in item 2.

4. **Are there easier ways to work directly with spikes for modeling?** The signal processing methodologies of modeling spike trains are awkward due to the stochastic and point process nature of the spike occurrence. Modeling is not trivial and many parameters are assumed. Signal processing methods work with optimal projections; therefore, it would be useful to create a metric space that will allow inner product computations directly from spike trains because all the available tools of binned methods can be applied immediately. Although spike trains are very telling of the neuron, they also are very removed from the time and macroscopic scales of behavior. Therefore, the spike-train methodology begs an answer to the question of how to optimally bridge the time scale of spike events (milliseconds) with the time scale of behavior (seconds). Currently, the relatively rudimentary method of binning is used, but so much information is lost that better, model-based, methodologies need to be devised. This is where the local field potentials (LFPs) can be useful since they are an intermediate representation of cortical activity. Therefore, how to relate spike activity with LFPs becomes a centerpiece for future research.
5. **Current BMI experimental approaches require extensive "hand tuning" of the decoding models to optimize performance**. In natural settings outside the rehabilitation clinic, neurophysiologists will not be available to spike-sort large data sets of recordings. Moreover, engineers will not be available to optimize the selection of neuronal inputs, model architecture, and learning rates to improve performance and generalization. Therefore, more sophisticated methods are needed to improve the robustness of the BCI through self-organizing adaptive principles.

REFERENCES

Arieli, A., D. Shoham, R. Hildesheim, and A. Grinvald. 1995. Coherent spatiotemporal patterns of ongoing activity revealed by realtime optical imaging coupled with single-unit recording in the cat visual cortex. *J. Neurophysiol.* 73:2072–2093.

Bashashati, A., S. Mason, R.K. Ward, and G.E. Birch. 2006. An improved asynchronous brain interface: making use of the temporal history of the LF-ASD feature vectors. *J. Neural Eng.* 3:87–94.

Berger, T.W., M. Baudry, R.D. Brinton, J.S. Liaw, V.Z. Marmarelis, A.Y. Park, B.J. Sheu, and A.R. Tanguay. 2001. Brain-implantable biomimetic electronics as the next era in neural prosthetics. *Proc. IEEE* 89:993–1012.

Blankertz, B., K.-R. Muller, D.J. Krusienski, G. Schalk, J.R. Wolpaw, A. Schlogl, G. Pfurtscheller, J. Millan, M. Schroder, and N. Birbaumer. 2006. The BCI competition III. Validating alternative approaches to actual BCI problems. *IEEE Trans. Neural Syst. Rehab. Eng.* 14:153–159.

Blankertz, B., K.-R. Muller, G. Curio, T.M. Vaughan, G. Schalk, J.R. Wolpaw, A. Schlogl, C. Neuper, G. Pfurtscheller, T. Hinterberger, M. Schroder, and N. Birbaumer. 2004. The BCI competition 2003: progress and perspectives in detection and discrimination of EEG single trials. *IEEE Trans. Biomed. Eng.* 51:1044–1051.

Boostani, R., and M.H. Moradi. 2004. A new approach in the BCI research based on fractal dimension as feature and Adaboost as classifier. *J. Neural Eng.* 1:212–217.
Brockwell, A.E., A.L. Rojas, and R.E. Kass. 2004. Recursive Bayesian decoding of motor cortical signals by particle filtering. *J. Neurophysiol.* 91:1899–1907.
Brown, E.N., D.P. Nguyen, L.M. Frank, M.A. Wilson, and V. Solo. 2001. An analysis of neural receptive field plasticity by point process adaptive filtering. *PNAS* 98:12261–12266.
Carmena, J.M., M.A. Lebedev, R.E. Crist, J.E. O'Doherty, D.M. Santucci, D.F. Dimitrov, P.G. Patil, C.S. Henriquez, and M.A. Nicolelis. 2003. Learning to control a brain–machine interface for reaching and grasping by primates. *PLoS Biol.* 1.1–16.
Chapin, J.K., K.A. Moxon, R.S. Markowitz, and M.A. Nicolelis. 1999. Real-time control of a robot arm using simultaneously recorded neurons in the motor cortex. *Nat. Neurosci.* 2(7):664–670.
Dayan, P. and L. Abbott. 2001. *Theoretical neuroscience*. Cambridge: MIT Press.
Donchin, E., K.M. Spencer, and R. Wijesinghe. 2000. The mental prosthesis: assessing the speed of a P300-based brain-computer interface. *IEEE Trans. Neural Syst. Rehab. Eng.* 8:174–179.
Donchin, O., A. Gribova, O. Steinberg, H. Bergman, S. Cardoso de Oliveira, and E. Vaadia. 2001. Local field potentials related to bimanual movements in the primary and supplementary motor cortices. *Exp. Brain Res.* 140:46–55.
Donoghue, J.P., J.N. Sanes, N.G. Hatsopoulos, and G. Gaal. 1998. Neural discharge and local field potential oscillations in primate motor cortex during voluntary movements. *J. Neurophysiol.* 79:159–173.
Eden, U.T., L.M. Frank, R. Barbieri, V. Solo, and E.B. Brown. 2004. Dynamic analysis of neural encoding by point process adaptive filtering. *Neural Comput.* 16(5):971–998.
Fabiani, G.E., D.J. McFarland, J.R. Wolpaw, and G. Pfurtscheller. 2004. Conversion of EEG activity into cursor movement by a brain computer interface. *IEEE Trans. Neural Syst. Rehab. Eng.* 12:331–338.
Fee M.S., P.P. Mitra, and D. Kleinfeld. 1996. Automatic sorting of multiple-unit neuronal signals in the presence of anisotropic and non-Gaussian variability. *J. Neurosci. Meth.* 69:175–188.
Freeman, W.J. 2004a. Origin, structure, and role of background EEG activity. Part 1. Analytic amplitude. *Clin. Neurophysiol.* 115:2077–2088.
Freeman, W.J. 2004b. Origin, structure, and role of background EEG activity. Part 2. Analytic phase. *Clin. Neurophysiol.* 115:2089–2107.
Freeman, W.J. 2005. Origin, structure, and role of background EEG activity. Part 3. Neural frame classification. *Clin. Neurophysiol.* 116(5):1118–1129.
Freeman, W.J. 2006. Origin, structure, and role of background EEG activity. Part 4. Neural frame simulation. *Clin. Neurophysiol.* 117(3):572–589.
Gao, Y., M.J. Black, E. Bienenstock, W. Wu, and J.P. Donoghue. 2003. A quantitative comparison of linear and nonlinear models of motor cortical activity for the encoding and decoding of arm motions. *The First International IEEE EMBS Conference on Neural Engineering*. Capri, Italy.
Georgopoulos, A.P., A.B. Schwartz, and R.E. Kettner. 1986. Neuronal population coding of movement direction. *Science* 233(4771):1416–1419.
Gunduz, A., J. Principe, and W.J. Freeman. 2005. Online detection of motor signatures in multichannel ECoG. In *Proceedings, 27th Annual International Conference of the IEEE Engineering in Medicine and Biology Society, IEEE-EMBS.* 3683–3686.
Haykin, S. 2002. *Adaptive filter theory*. Englewood Cliffs, NJ: Prentice-Hall.
Hochberg, L.R., M.D. Serruya, G.M. Friehs, J.A. Mukand, M. Saleh, A.H. Caplan, A. Branner, D. Chen, R.D. Penn, and J.P. Donoghue. 2006. Neuronal ensemble control of prosthetic devices by a human with tetraplegia. *Nature* 442:164–171.
Kim P., J. Sanchez, Y. Rao, D. Erdogmus, J. Carmena, M. Lebedev, M. Nicolelis, and J. Principe. 2006. A comparison of optimal MIMO linear and nonlinear models for brain-machine interfaces. *J. Neural Eng.* 3(2):145–161.

Kim S., J. Sanchez, D. Erdogmus, Y. Rao, J. Principe, and M. Nicolelis. 2003. Divide-and-conquer approach for brain machine interfaces: nonlinear mixture of competitive linear models. *Neural Networks* 16:865–871.

Krusienski, D.J., G. Schalk, D.J. McFarland, and J.R. Wolpaw. 2007. A mu-rhythm matched filter for continuous control of a brain-computer interface. *IEEE Trans. Biomed. Eng.* 54(2):273–280.

Lemm, S., B. Blankertz, G. Curio, and R.-K. Muller. 2005. Spatio-spectral filters for improving the classification of single trial EEG. *IEEE Trans. Biomed. Eng.* 52:1541–1548.

Leuthardt, E.C., G. Schalk, J.R. Wolpaw, J.G. Ojemann, and D.W. Moran. 2004. A brain–computer interface using electrocorticographic signals in humans. *J. Neural Eng.* 1:63–71.

Lewicki, M.S. 1998. A review of methods for spike sorting: the detection and classification of neural action potentials. *Network Comput. Neural Syst.* 9:R53–R78.

Li, Y., X. Gao, H. Liu, and S. Gao. 2004. Classification of single-trial encephalogram during finger movement. *IEEE Trans. Biomed. Eng.* 51:1019–1025.

Marmarelis, P.Z. and V.Z. Marmarelis. 1978. *Analysis of physiological systems: the white noise approach.* New York: Plenum.

Marmarelis, V. 1993. Identification of nonlinear biological systems using Laguerre expansions of kernels. *J. Biomed. Eng.* 21:573–589.

Mason, S.G. and G.E. Birch. 2000. A brain-controlled switch for asynchronous control applications. *IEEE Trans. Biomed. Eng.* 47:1297–1307.

McFarland, D.J., C.W. Anderson, K.-R. Muller, A. Schlogl, and D.J. Krusienski. 2006. BCI meeting 2005-workshop on BCI signal processing: feature extraction and translation. *IEEE Trans. Neural Syst. Rehab. Eng.* 14:135–138.

McFarland, D.J. and J.R. Wolpaw. 2005. Sensorimotor rhythm-based brain-computer interface (BCI): feature selection by regression improves performance. *IEEE Trans. Neural Syst. Rehab. Eng.* 13:372–379.

McFarland, D.J., L.M. McCane, S.V. David, and J.R. Wolpaw. 1997. Spatial filter selection for an EEG-based brain-computer interface. *Electroencephalogr. Clin. Neurophysiol.* 103:386–394.

Mehring, C., J. Rickert, E. Vaadia, S. Oliveira, A. Aertsen, and S. Rotter. 2003. Inference of hand movements from LFPs in monkey motor cortex. *Nat. NeuroSci.* 6(12):1253–1254.

Millan, J.R., F. Renkens, J. Mourino, and W. Gerstner. 2002. Noninvasive brain-actuated control of a mobile robot by human EEG. *IEEE Trans. Biomed. Eng.* 51:1026–1033.

Moran, D.W. and A.B. Schwartz. 1999. Motor cortical representation of speed and direction during reaching. *J. Neurophysiol.* 82(5):2676–2692.

Naeem, M., C. Brunner, R. Leeb, B. Graimann, and G. Pfurtscheller. 2006. Seperability of four-class motor imagery data using independent components analysis. *J. Neural Eng.* 3:208–216.

Nicolelis, M. 2001. Action from thoughts. *Nature* 409:403–407.

Paninski, L., S. Shoham, M.R. Fellows, N.G. Hatsopoulos, and J.P. Donoghue. 2004. Superlinear population encoding of dynamic hand trajectory in primary motor cortex. *J. Neurosci.* 24(39):8551–8561.

Peters, B.O., G. Pfurtscheller, and H. Flybjerg. 1998. Mining multi-channel EEG for its information content: an ANN-based method for a brain-computer interface. *Neural Networks* 11:1429–1433.

Pfurtscheller, G., G.R. Muller-Putz, A. Schlogl, B. Graimann, R. Scherer, R. Leeb, C. Brunner, C. Keinrath, F. Lee, G. Townsend, C. Vidaurre, and C. Neuper. 2006. Fifteen years of BCI research at Graz University of Technology: current projects. *IEEE Trans. Neural Syst. Rehab. Eng.* 14:205–210.

Rezaei, S., K. Tavakolian, A.M. Nasrabadi, and S.K. Setarehdan. 2006. Different classification techniques considering brain-computer interface applications. *J. Neural Eng.* 3:139–144.

Rickert, J., S. Cardoso de Oliveira, E. Vaadia, A. Aertsen, S. Rotter, and C. Mehring. 2005. Encoding of movement direction in different frequency ranges of motor cortical local field potentials. *J. Neurosci.* 25:8815–8824.

Sajda, P., A. Gerson, K.-R. Muller, B. Blankertz, and L. Parra. 2003. A data analysis competition to evaluate machine learning algorithms for use in brain-computer interfaces. *IEEE Trans. Neural Syst. Rehab. Eng.* 11:184–185.

Sanchez J., J. Carmena, M. Lebedev, M. Nicolelis, J. Harris, and J. Principe. 2003. Ascertaining the importance of neurons to develop better brain machine interfaces. *IEEE Trans. Biomed. Eng.* 61:943–953.

Sanchez J., D. Erdogmus, J. Principe, J. Wessberg, and M. Nicolelis. 2002a. A comparison between nonlinear mappings and linear state estimation to model the relation from motor cortical neuronal firing to hand movements. Workshop on Motor Control, Edinburg, Scotland, 59–65.

Sanchez, J.C., P.R. Carney, and J.C. Principe. 2006. Analysis of amplitude modulated control features for ECoG neuroprosthetics. *Twentyeighth Annual International Conference of the IEEE Engineering in Medicine and Biology Society, IEEE-EMBS.* 5468–5471.

Sanchez, J.C., S.P. Kim, D. Erdogmus, Y.N. Rao, J.C. Principe, J. Wessberg, and M. Nicolelis. 2002b. Input-output mapping performance of linear and nonlinear models for estimating hand trajectories from cortical neuronal firing patterns. International Workshop on Neural Networks for Signal Processing. Martigny, Switzerland: IEEE.

Schlogl, A., F. Lee, H. Bischof, and G. Pfurtscheller. 2005. Characterization of four-class motor imagery data for the BCI-competition 2005. *J. Neural Eng.* 2:14–22.

Schwartz, A., D. Taylor, and S. Tillery. 2001. Extraction algorithms for cortical control of arm prosthetics. *Curr. Opin. Neurobiol.* 11:701–708.

Serby, H., E. Yom-Tov, and G.F. Inbar. 2005. An improved P300-based brain-computer interface. *IEEE Trans. Neural Syst. Rehab. Eng.* 13:89–98.

Serruya, M.D., N.G. Hatsopoulos, L. Paninski, M.R. Fellows, and J.P. Donoghue. 2002. Brain-machine interface: instant neural control of a movement signal. *Nature* 416:141–142.

Shenoy, K.V., D. Meeker, S. Cao, S.A. Kureshi, B. Pesaran, C.A. Buneo, A.P. Batista, P.P. Mitra, J.W. Burdick, and R.A. Andersen. 2003. Neural prosthetic control signals from plan activity. *NeuroReport* 14:591–597.

Shenoy, P., M. Krauledat, B. Blankertz, R.P.N. Rao, and R.-K. Muller. 2006. Towards adaptive classification for BCI. *J. Neural Eng.* 3:13–23.

Simoncelli, E.P., L. Paninski, J. Pillow, and O. Schwartz. 2004. Characterization of neural responses with stochastic stimuli. In: *The New Cognitive Neuroscience*, 3rd ed. Ed. Gazzaniga, M. Cambridge: MIT Press.

Soderstrom, T. and P. Stoica. 1989. *System identification.* Englewood Cliffs, NJ: Prentice Hall.

Song, D., V.Z. Marmarelis, and T.W. Berger. 2002. Parametric and nonparametric models of short-term plasticity. *Proceedings of the Second Joint EMBS/BMES Conference* 3:1964–1965.

Taylor, D.M., S.I.H. Tillery, and A.B. Schwartz. 2002. Direct cortical control of 3D neuroprosthetic devices. *Science* 296(5574):1829–1832.

Tillery, S.I.H., D.M. Taylor, and A.B. Schwartz. 2003. Training in cortical control of neuroprosthetic devices improves signal extraction from small neuronal ensembles. *Rev. NeuroSci.* 14(1–2):107–119.

Vidaurre, C., A. Schlogl, R. Cabeza, R. Scherer, and G. Pfurtscheller. 2005. Adaptive on-line classification for EEG-based brain computer interfaces with AAR parameters and band power estimates. *Biomed. Technik* 50:350–354.

Wang, T., J. Deng, and B. He. 2004. Classifying EEG-based motor imagery tasks by means of time-frequency synthesized spatial patterns. *Clin. Neurophysiol.* 115:2744–2753.

Wang, Y., A.R.C. Paiva, and J.C. Principe. 2006. A Monte Carlo sequential estimation for point process optimum filtering. *Proc. IJCNN 2006*, Portland.

Wessberg, J., C.R. Stambaugh, J.D. Kralik, P.D. Beck, M. Laubach, J.K. Chapin, J. Kim, S.J. Biggs, M.A. Srinivasan, and A. Nicolelis. 2000. Real-time prediction of hand trajectory by ensembles of cortical neurons in primates. *Nature* 408(6810):361–365.

Wolpaw, J.R., N. Birbaumer, D.J. McFarland, G. Pfurtscheller, and T.M. Vaughan. 2002. Brain-computer interfaces for communication and control. *Clin. Neurophysiol.* 113:767–791.

Wood, F., M.J. Black, C. Vargas-Irwin, M. Fellows, J.P. Donoghue. 2004. *IEEE Trans. Biomed. Eng.* 51(6):912–918.

Wu, W., M.J. Black, Y. Gao, E. Bienenstock, M. Serruya, and J.P. Donoghue. 2002. Inferring hand motion from multicell recordings in motor cortex using a Kalman filter. *SAB Workshop on Motor Control in Humans and Robots: on the Interplay of Real Brains and Artificial Devices*. University of Edinburgh, Scotland.

Wu, W., Y. Gao, E. Bienenstock, J.P. Donoghue, and M.J. Black. 2005. *Neural Comput.* 18:80–118.

Yom-Tov, E. and G.F. Inbar. 2003. Detection of movement-related potentials from the electroencephalogram for possible use in a brain-computer interface. *Med. Biol. Eng. Comput.* 41:85–93.

CHAPTER 5

Hardware Implementation

John K. Chapin

INTRODUCTION: RESTORING MOVEMENT IN PARALYSIS PATIENTS

Paralysis, which adversely affects millions of people worldwide, has many causes, including trauma, stroke, infection, and autoimmune diseases. The primary damage can be manifested in the brain, spinal cord, spinal nerves, or the muscles themselves. In general, "paralysis" refers to severe or complete loss of motor function, whereas "paresis" refers to relatively minor loss. Severe paralysis is a major problem, not just because patients lose their ability to have a normal life, but also because of the tremendous cost of patient maintenance. Patients with tetraplegic spinal cord injuries, for example, lose virtually all voluntary motor function below the neck, but also lose somatosensation, i.e., their senses of touch, pain, temperature, and limb position. Severe amyotrophic lateral sclerosis (ALS) is even worse, in that patients can lose all motor function throughout their bodies.

Fortunately, recent publicity about spinal cord injury (SCI) has spurred increased efforts to find a cure for SCI and other paralyzing diseases. Over the past 20 years, most research has involved cell biology approaches, which, it is hoped, will develop neurotrophic factors, transformed somatic cells, or embryonic stem cells to stimulate damaged nerve fibers to regrow beyond the site of spinal injury, and hopefully, to make synaptic contact on the correct motor neurons. Although such approaches may one day provide the ultimate cure for paralysis, this area of research is only in its infancy. Based on current progress and the number of break-throughs that will be required, it will be many years before an actual cure is available. Thus, these methods may not be helpful to SCI patients in the near term.

EEG-Based Brain-Computer Interfaces

On the other hand, because of recent breakthroughs in brain-computer interfaces (BCI), it has now been shown that modern electronic and computer technologies

T.W. Berger et al., Brain-Computer Interfaces, 65–79.

may provide a viable near-term solution to help restore some motor function. In general, BCIs use direct recordings from the brain to allow paralyzed patients to control external devices, such as a cursor on a computer screen (Wolpaw et al., 1998; Wolpaw and McFarland, 2004). This chapter reports on a variety of different methods for BCI-based control, as well as different devices that can be controlled from the brain.

There are two different types of BCIs in current use, one of which is the "noninvasive" BCI that is a computer-actuated electronic recording system using electrode contacts on the scalp to record electroencephalographic (EEG) signals from the subject's cerebral cortex. Typically, the subject must use biofeedback methods to learn how to control his own "brain waves" with sufficient accuracy to move a cursor on a computer screen. After much practice, the cursor control becomes sufficiently accurate to spell words by using the BCI to control the computer cursor to pick out letters of the alphabet. This BCI approach has been highly successful and seems poised to provide an excellent near-term solution for providing severe SCI patients with some means of controlling external devices. It also demonstrates the potential usefulness of using electronic approaches to restore motor function in SCI. There are drawbacks, however, in that subjects require substantial training and intense concentration to perform this control. Moreover, the noninvasive BCI approach is limited in terms of the number of degrees of freedom that can be controlled (currently, one to two) by using EEG recordings.

Direct Brain-Computer Interfaces

An alternative approach to EEG-based BCI control is to use implanted multi-electrode arrays to directly record from the motor control circuitry within the brain. The feasibility of this direct-BCI approach has been demonstrated over the past eight years, beginning in animals (Chapin et al., 1999; Wessberg et al., 2000) and more recently progressing to humans (Hochberg et al., 2006). Since this method uses multielectrode arrays to record from populations of neurons in the subject's motor cortex, it can be used to extract the brain's own motor commands as they are being processed. The fact that the arrays directly tap into the brain's intrinsic processing functions allows the recorded signals to be used to directly drive a BCI whose output manifests the internal motor command.

These "invasive," direct BCIs are thus different from noninvasive BCIs for two reasons. First, the subjects do not require extensive practice to control the outputs, although many do learn how to improve their accuracy. Second, the number of tasks (i.e., degrees of freedom) that can be performed is limited only by the number of electrodes that can be implanted in different brain areas. Thus, a subject's brain should be able to simultaneously direct several different movements, as could the brain of any nonparalyzed subject. For example, the paralyzed subject should be able to control the complex interplay between arm and finger movements required for real manipulation and grasping.

Direct BCIs Record Neural Information from Electrodes in the Motor Cortices

Direct BCIs obtain their information by recording from arrays of fine electrodes (a tenth the thickness of a human hair) chronically implanted within the motor control areas of the cerebral cortex, broadly including most of the frontal and parietal cortical regions. Aside from their small size, these implants are similar to other brain-implanted devices, such as deep brain stimulators for Parkinson's disease. Though modern surgical techniques have substantially reduced the risks associated with chronic brain implants, a current debate is whether it is best to record from electrode arrays implanted within the actual brain circuitry, or from larger electrodes implanted just above the cortical surface where electrocorticograms (ECoGs) can be recorded (Leuthardt et al., 2004). Most researchers in this field tend to prefer the intracortical approach because single neurons in the motor cortex are known to encode the direction of intended arm movement, whereas local field potentials and ECoG recordings generally do not. However, recent studies at the University of Freiburg suggest that careful mathematical analysis of ECoG recordings may reveal some intended movement direction.

DIFFERENT APPROACHES TO BCI RESEARCH WORLDWIDE

Direct BCI research was initiated in the United States, where it continues to grow rapidly as a research area. The U.S. lead in this research area is attributable to many factors, including timely funding from NIH and DARPA and also to the relatively liberal climate for use of nonhuman primates in research. As a result, the United States now has a number of excellent scientists performing different kinds of research in the general area of direct BCIs. Although the original ideas for these BCIs were proposed in the 1970s (Humphrey et al., 1970; Schmidt, 1980; Dormont et al., 1982), the first working demonstrations of brain-controlled robots emerged in the period since about 1999 (Chapin et al., 1999; Wessberg et al., 2000; Schwartz et al., 2001; Taylor et al., 2002; Donoghue, 2002; Shenoy et al., 2003). More recently, this area has seen an explosive surge of interest as investigators probe a variety of new approaches to the problem. Despite the differences in approach, all are focused on the idea of using this technology to restore movement in paralysis victims, especially victims of spinal cord injury and ALS.

Recording, Extracting, and Decoding Neural Motor Commands

All direct BCIs depend on the ability of a multielectrode recording system to detect the subject's intended movements by sampling from large populations of neurons in the motor cortices. Each electrode can record action potentials from several nearby neurons, and each of these provides some specific information about the intended movement's direction. The same electrodes are also used to record the intensity of the local field potentials, which can be used to provide information on the velocity or intensity of the intended movement (Hermer-Vazquez et al., 2003). As these neural signals are recorded in the brain, they are immediately amplified, filtered, discriminated, and transmitted to a computer, which decodes all of this information. The computer then translates this information into an output format that controls a computer cursor or a robot.

Why is it Necessary to Record from Multineuron Ensembles?

Though these recordings can precisely measure the action potentials from a single neuron, the average spiking rates are relatively slow (about 10 Hz), and thus the rate code does not specify enough information to be used to control an output device. In our development of this method, we discovered that one could statistically average these multineuron signals and increase the accuracy of the brain information that we wished to measure (Chapin and Patel, 1987; Shin and Chapin, 1990; Nicolelis et al., 1993, 1998). This allowed us to utilize mathematical methods such as principal components analysis or multivariate linear regression to maximize our ability to statistically predict experimental parameters (Chapin and Nicolelis, 1999). For example, our early studies showed that use of factor analysis to evaluate multineuron recordings in the somatosensory cortex allowed one to predict with good accuracy the locations of sensory stimuli applied to different points on the body or face of experimental animals. Later we used multineuron recordings to predict limb movements using a multivariate regression approach that treated each neuron as a variable and weighted these variables to define a linear filter whose accuracy improved with increased numbers of neurons.

Use of Multivariate Regression Analysis (MRA) to Predict Limb Movement Kinematics

We have found that MRA is a robust method for translating multineuron recordings into a continuous prediction of ongoing limb movement in a subject and/or for control of a robot arm. Though many approaches to this problem have been developed, most multineuron-controlled devices continue to utilize variants of MRA, which generate linear models capable of predicting a range of dependent (output) variables from a set of independent (input) variables. MRA is the statistical technique of choice for defining the linear mathematical relationship between a set of independent predictor variables (e.g., a population of motor system neurons) and a dependent variable (e.g., a motor output function).

In its simplest form, MRA will yield a linear equation of the form

$$Y = a + b_1X_1 + b_2X_2 + ... + b_pX_p$$

in which Y is the dependent variable (e.g., lever angle), X_n are the independent variables (e.g., neuronal population activity), a is an offset, and b_n are coefficients for weighting the independent variables. This relation is easily calculated from a sample data set by linear least squares estimation, which computes a line through the observed data points such that the sum of the squared deviations of the observed points and that line is minimized. This calculation also yields the coefficient of determination (R^2) which quantifies the predictive reliability of the model (in a range from 0 to 1). Beyond this, it is very important to test the model through application of a number of residual analysis procedures, which can identify outliers and gross violations of the assumptions of MRA. Of course, there has been a marked recent increase in interest in different methods for improving the mathematical approaches to neural encoding for direct BCIs (see Chapter 4).

ORIGINAL FEASIBILITY DEMONSTRATIONS FOR BRAIN-CONTROLLED ROBOTICS

In our original feasibility demonstration, multineurons in the motor cortex (MI) and VL thalamus were recorded in rats trained to press a lever in order to move a robot arm that retrieved water from a dropper (Chapin et al., 1999). Population encodings of the brain's "motor signals" were electronically implemented in real time, allowing the robot arm to be moved in direct proportion to the population function amplitude. The rat thereafter obtained its daily water by using this neural signal alone to control the robot's movement to the water dropper. Over time, the rat was able to obtain its water without actually pressing the lever, suggesting that the motor cortical neurons had learned a direct representation of the robot arm, independent of the real arm. Though the robot arm was moved independently of the brain's normal control of the arm movement, the animals were subsequently able to move their real arms with no apparent problems. This general finding has since been corroborated in many laboratories. It has now been replicated on animals in other laboratories (Carmena et al., 2003; Taylor et al., 2002; Donoghue, 2002) and humans (Hochberg et al., 2006).

Extraction of Motor Commands from the Monkey Brain

The next milestone was the demonstration that multineuron recordings in the motor cortex of monkeys could be used to extract neural information capable of encoding movements in multiple directions. Monkeys were used not only because they are more relevant for understanding the human motor cortex, but also because their larger brain affords more space for implantation of electrodes. Figure 5.1 shows the setup for this experiment, in which the monkey grasped a handle and spontaneously moved it left and right in order to receive juice rewards. At the same time, multineuron recordings were captured from 32-channel electrode arrays previously implanted in several brain regions, including the motor, premotor, somatosensory, and parietal cortices on both sides of the brain. An offline computer was used to calculate the neuronal weightings for multivariate linear regression predictive filters. An online computer then used these filters to convert real-time neuronal activity into neural population codes that simultaneously specified three dimensions of hand position, velocity and/or force. These population-coded outputs were then used to control a robot arm and/or a cursor on a computer screen with good accuracy ($R^2 = \sim 0.8$) compared with the monkey's arm itself.

Although the quality and quantity of useful coding information obtained in these recordings varied across the sensorimotor cortical areas, all areas contributed at least some useful information to the control. This is consistent with previous findings that cortical neurons tend to be widely tuned and thus can be active in a

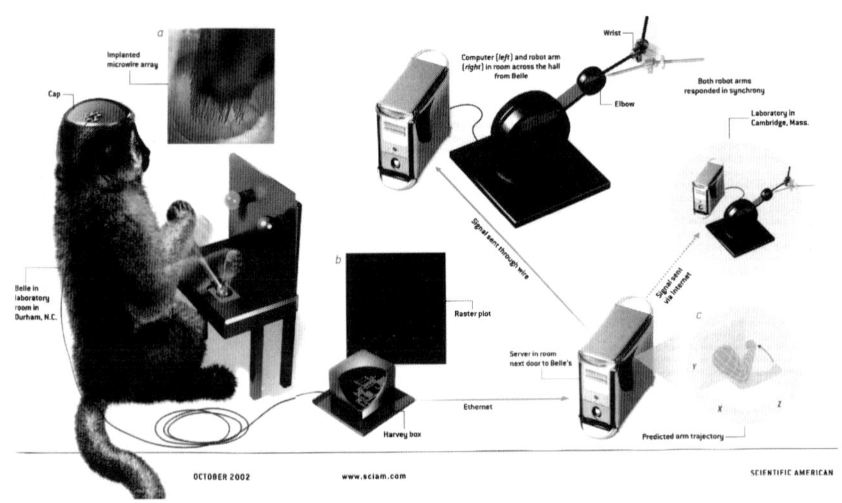

Figure 5.1. Extraction of motor commands from the monkey brain (Nicolelis and Chapin, 2002; © Scientific American, Inc.).

wide variety of experimental conditions. Much of the success of multielectrode recording is attributable to this wide tuning, because the investigator can depend on obtaining useful signals from the majority of recorded neurons. This was expected from previous reports in the literature that cortical motor neurons are broadly tuned to the direction of intended movement. This was true for both naïve and overtrained owl monkeys. Consequently, the same sample of neurons exhibited increased firing prior to movements in two different directions.

The first important observation regarding the potential use of this approach for the design of brain-controlled prosthetic devices was that our cortical recordings remained stable and viable for over 18 months. This finding further confirmed our previous data obtained in the somatosensory cortex of owl monkeys and suggested that chronically implanted microwires are still the best electrodes available for long-term single- and multiunit recordings in freely behaving animals. Indeed, to this date, despite the introduction of very elaborate designs, no other electrode has been able to match the long-term stability and yield of microwires. This allows one to utilize the weighted discharges of multiple neurons to predict movement. A model that predicts movement of the real arm can then be used to drive the motion of a robot arm.

Biofeedback Changes Coding of Robot Arm Movement

After switching from arm control to brain control of a robot, most investigators have corroborated the finding that the robot arm movement dissociates from the real arm. Since the subject is rewarded only for accurate control of the robot arm, the real arm is moved less reliably over time.

Meanwhile, monkeys' accuracy usually continues to improve. In fact, Carmena et al. (2003) showed that the actual neuronal direction coding changes over continued training. This dissociation appears to be completely dependent on the experimental context, however, because the subjects use their real arms normally immediately after they are returned to their cages. Since it is clear that animal subjects can improve their accuracy of control, it compels us to think that human subjects may also be able to learn to modify the properties of their own neurons.

This is important because the neural activity that might be measured in paralysis victims cannot be decoded by asking the subjects to move their arms. Though Hochberg et al. (2006) reported that their human patient was able to learn how to control a computer cursor, it is not clear that humans have an unlimited ability to mentally modify their own neural coding. At some point the subject's ability to maintain a high level of attention will be lost, leading to the same problems of "cognitive load" that obtain in noninvasive BCIs. It has been shown useful, however, to employ fMRI to locate the brain's focus of arm movement control in patients who are mentally imagining movements of their own arms (Kennedy and Bakay, 1998). Electrode implants in those areas should enable neurons to be recorded during mentally rehearsed movements, thus allowing encoding of their directional codings.

BRAIN CONTROL OF MULTIPLE-OUTPUT FUNCTIONS

Now that brain control of movement has been definitively demonstrated, the next step will be to investigate the use of this approach to control a multiplicity of output functions. Though most applications so far have only controlled kinematics (e.g., hand position), most real movements also involve control of movements against various forces. Thus, Carmena et al. (2003) and Francis and Chapin (2006) have demonstrated that large ensembles of motor cortex neurons can be used to generate linear models that, depending on their weightings, encode different outputs such as kinematics, forces, or work. Moreover, when such linear models are trained using data recorded from animals moving against two or more forces, the model is able to predict all of the forces with correlations approaching 0.9. When such linear models are trained using data from animals moving against two or more different types of force fields (i.e., constant, spring), the model is able to predict all of the forces with good accuracy (Figure 5.2).

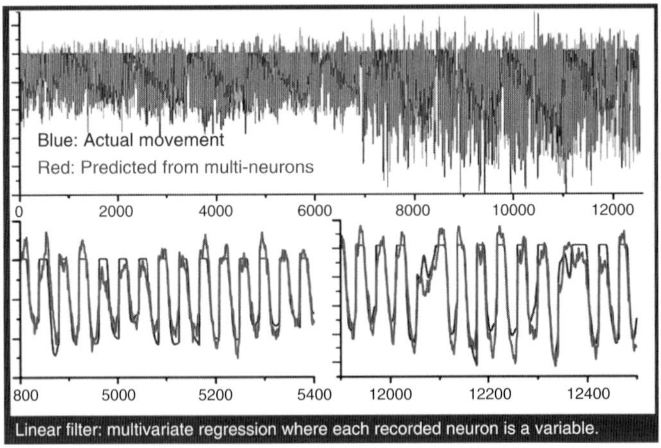

Figure 5.2. Results from an experiment in which an animal was trained to make reaching movements while grasping a robotic manipulandum, which produced two distinct constant forces, either a 3 g or a 10 g force. The blue traces are the actual work produced by the animal, and the red traces are produced via the animal's neural activity being filtered by a weight matrix that was constructed via a regression model. The weights were determined by fitting the neural activity in previous trials to the desired variable of interest, in this case, work. Note the high correlation values between the predicted trajectories (red) and the real trajectories (blue).

Linear models designed to simultaneously predict multiple parameters, such as force, position, and work, are being analyzed to understand subtle differences in the neural encoding of these parameters. Simultaneous neural recordings obtained in the motor cortex and proprioceptive system are being analyzed to determine the importance of proprioceptive feedback in determining the forces exerted.

Need for Sensor Feedback to Facilitate Control of Robotic Movements

Heretofore, most brain-controlled robots have been guided by visual feedback, either from watching a cursor on a computer screen or the movement of an actual robot arm. Reliance on such visual feedback is problematic in that it is slow and does not replace somatosensation, which normally provides the tactile and force feedback necessary for efficient manipulation of objects. One solution to this problem could be to develop a somatosensory neuroprosthesis consisting of stimulating electrode arrays implanted in the somatosensory system. If tactile and proprioceptive sensors were placed on a robot arm/hand, these sensors could be used to drive the somatosensory neuroprosthesis in the brain, which would theoretically produce appropriate feedback sensation in the subject. Of course, little is known about how effectively such stimuli would reproduce natural tactile or proprioceptive sensation. As a noninvasive alternative, one could develop a tactile array consisting of many small, tactile, computer-controlled stimulators. This computer could receive inputs from sensors on the robot arm and reconfigure these signals to send to the tactile array which would be placed on an available area of skin. As with other noninvasive BMI devices, the patient would need to learn how to associate the different vibratory patterns on the tactile array with real skin locations on the hand and arm.

Need to Control More Degrees of Freedom in Brain-Controlled Robots

The major rationale for developing brain-controlled robots for paralysis victims is the hope that these prostheses will allow them to perform neurorobotic motor functions with dexterity sufficient to reach and grasp objects and manipulate or transport them. Though such movements are commonplace for most people, further development will be required to emulate that capability using brain-controlled neural prostheses. First, it will be necessary to demonstrate the ability to control more degrees of freedom. Based on previous studies that utilized multiple linear models, it may be possible to use implanted electrodes in the motor cortical hand and finger areas to record information capable of controlling independent movements of these body segments. Each of these hand/finger movements could be controlled by a different appropriately weighted linear model. This will require recording from each of a large number of neurons in each of several motor cortical zones. Based on the distributed processing known to exist in the motor cortex, it is likely that many of these neurons will be able to provide some information to control the different movements.

Need for High Degree-of-Freedom (DOF) Biomimetic Arm/Hand and Leg Prosthetic Robots

Though robotics research is a huge area in technology research worldwide, only recently has there been much impetus for development of prosthetic robots for paralysis victims. That may be due to the fact that robots are normally developed to perform particular tasks, whereas the human body has evolved to carry out a wide variety of tasks. Indeed, the "overcompleteness" of the human motor system has always been a disincentive to biomimetic design. When designing a robotic arm/hand prosthesis for a human paralysis victim, however, there are obvious reasons for designing in as many human attributes as possible. At the very least, such a prosthetic device should allow independent movements of the thumb and index fingers (3 DOF), rotation and flexion-extension of the hand (2 DOF), and arm/shoulder movements (3 DOF). This minimal 8-DOF configuration could allow some rudimentary manipulation, but increased degrees of freedom would improve it further.

There is also a need to develop brain-controlled legged robots to restore locomotory functions for paraplegic and quadriplegic patients. Professors Yongji Wang and Xu Qi at Huazhong University in Wuhan, China, are collaborating with Dr. Jeping He at Arizona State University. Though their general approach is similar to that of other invasive BCI researchers, they are actively investigating new algorithms (Fang et al., 2006) for motor control. They are also developing ind-welling flexible electrodes for epidural, spinal cord stimulation to assist paralyzed patients.

BIOMIMETIC ROBOT RESEARCH AT THE SCUOLA SUPERIORE SANT'ANNA

Professor Paolo Dario and his colleagues at the Scuola Superiore Sant'Anna in Pisa, Italy, have long been a major force in the development of biomechatronic and biomorphic control systems. Their Advanced Robotics and Technology Systems (ARTS) lab has developed a number of biomimetic and anthropomorphic robotic devices for rehabilitation and assistive robotics. This group has developed colla-borations with other universities in Europe, as well as Waseda in Japan. Together they have formed "Robocasa," a model for educational, research, and industrial cooperation between Italy and Japan in the field of robotics.

They have developed a number of important robotic systems that not only include actuators, but also proprioceptive, tactile, and visual sensors. For example, their ARTS humanoid robot features a head, arm, and hand and includes a total of 25 DOFs, 2 visual sensors, 39 proprioceptive sensors, and 135 tactile sensors. Its control involves a biologically inspired multinetwork architecture that uses progressive and adaptive learning for object manipulation. Their new biomechatronic Cyberhand (Figure 5.3.) is a five-digit biomimetic hand prosthesis that includes at least one actuator for each digit plus embedded biomimetic force and proprioceptive sensors. Though the digits are underactuated, the hand's compliance and embedded closed-loop control confers excellent gripping characteristics for an artificial hand. This device is intended for use by amputees, and therefore its control will come from an interface with nerves in the patients' remaining arm.

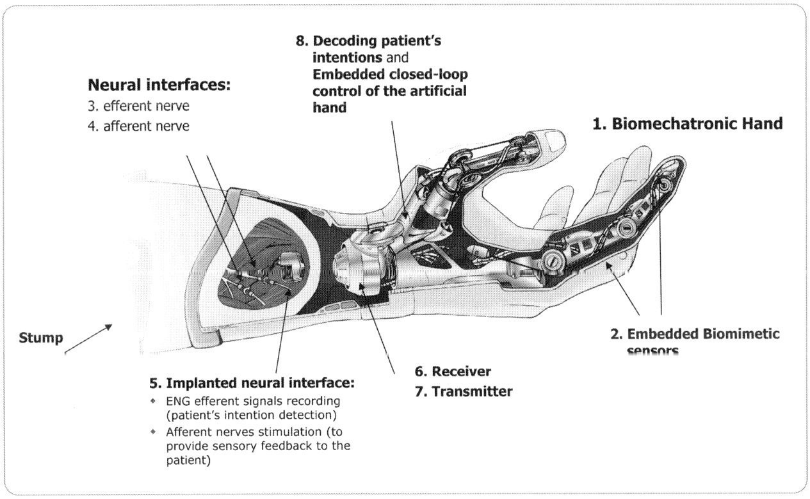

Figure 5.3. Cyberhand system.

Rationale for Biomimetic Hand Prostheses

Until recently, typical hand prostheses have consisted of metal hooks controlled through a myoelectric interface. Thus it is important for both functional and aesthetic reasons to improve these prosthetic hands. Thus the Cyberhand (Citi et al., 2006) is inspired by understanding of the kinesiology of hand and finger movements (Figure 5.4). Since it will be impossible to build a robot that includes all of the features that exist in the real human, the challenge is to design modifications that combine similar functionality with less complexity.

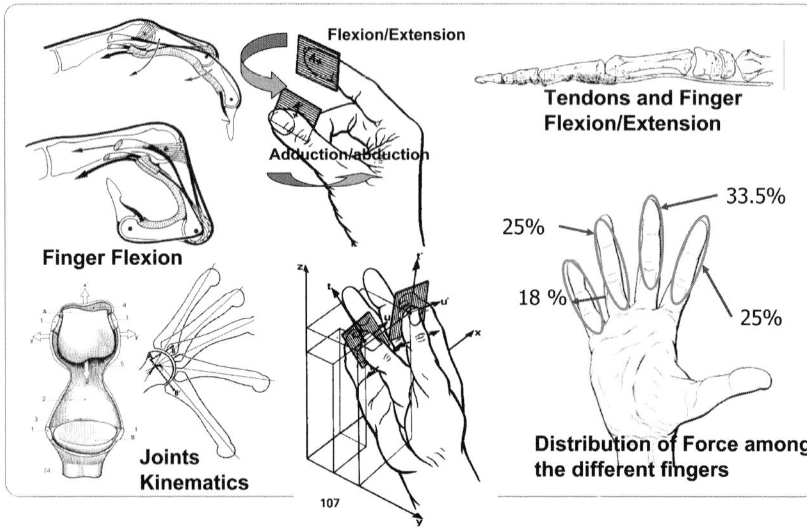

Figure 5.4. Kinesiology of hand and digit actions during manipulation.

Research Approach to Biomechatronics at SSSA

Professor Dario stresses the importance of utilizing anatomy, physiology, and neuroscience to inspire robotic development (Dario et al., 2005). As outlined in Figure 5.5, this laboratory favors a biomimetic approach that goes beyond simple mechanics, but instead considers the whole system that is normally involved in controlling the hand. That includes implementation of neuroscience models not just for controlling the hand, but higher order motor control models involving grasp primitives, posture, and feedforward control. Moreover, the importance of proprioceptive, tactile, and other forms of sensor feedback is recognized as critical for closed-loop control of the robotic device.

Using Direct BCIs to Control Biomimetic Robotic Prostheses

Though this work at SSSA is focused on controlling hand prostheses by using signals recorded from peripheral nerves (Navarro et al., 2005), one can also envision the use of direct BCIs that would allow a prosthesis to be controlled directly from the brain. Our current BCI experiments in monkeys involve use of a bidirectional brain interface that includes a motor prosthesis (actuated by neural recordings in the motor cortices) and also a tactile/proprioceptive neuroprosthesis actuated by stimulating electrode arrays in the somatosensory pathways. The next step would be to show feasibility for controlling such a "closed-loop BCI" in many

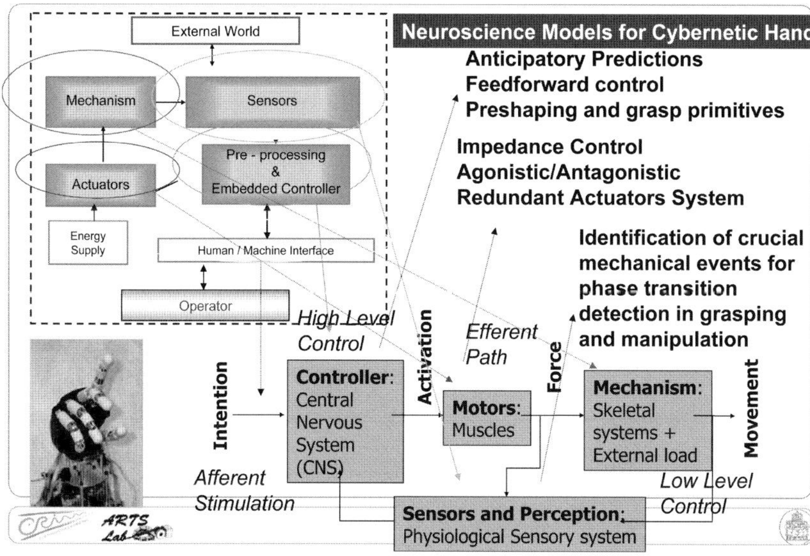

Figure 5.5. Biomechatronic approach to duplicating the natural hand.

degrees of freedom. This possibility is enabled only by the parallel development of high-DOF prosthetic robots such as the Cyberhand. It is important to observe that that work in many different disciplines ultimately causes synergistic advances across the board. This has occurred at the SSSA because it emphasizes research that exploits the give and take between neuroscientific hypothesis-driven research and the technological model of continual development and testing of new technological devices.

Worldwide, many other laboratories are working to develop more naturalistic, humanoid, robotic devices. For example, Dr. Yoshinobu Tonomura at the NTT Communications Laboratories in Kanagawa, Japan, directs a group focusing on "parasitic humanoid," wearable robots for modeling human movements. These robots can sense eye movements, finger touching, posture, and locomotion. They use an internal computer to process these sensor data, allowing continuous learning of the user's sensorimotor patterns. The purpose is to predict and "optimize" human sensorimotor behavior. A BCI system is expected to be included to enhance the subject's ability to control these functions. At the Advanced Technology Research Institute in Kyoto, Japan, Professor Mitsuo Kawato is working on an "enlightened" future vision for BCIs. This involves both noninvasive and invasive approaches, including neural prostheses to restore neurological function in patients with sensory, motor, and cognitive deficits. But it also goes beyond purely "medical" issues by contemplating the use of BCI-enabled, humanoid robotics for a wider variety of concerns such as academics, entertainment, telecommunications, control of aircraft and other machinery, smart homes, and disaster management.

REFERENCES

Carmena, J.M., M.A. Lebedev, R.E. Crist, J.E. O'Doherty, D.M. Santucci, D.F. Dimitrov, P.G. Patil, C.S. Henriquez, and M.A. Nicolelis. 2003. Learning to control a brain-machine interface for reaching and grasping by primates. *PLoS Biology* 1:E42.

Chapin, J.K. and I.M. Patel. 1987. Effects of ethanol on arrays of single cortical neurons recorded for several days. *Social Neuroscience Abstract.*

Chapin, J.K., R.A. Markowitz, K.A. Moxon, and M.A.L. Nicolelis. 1999. Direct real-time control of a robot arm using signals derived from neuronal population recordings in motor cortex. *Nature Neuroscience* 2:664–670.

Chapin, J.K and M.A.L. Nicolelis. 1999. Population coding in simultaneously recorded neuronal ensembles in ventral posteromedial (VPM) thalamus: Multidimensional sensory representations and population vectors. *J. Neurosci. Methods* 94:121–140.

Citi, L., J. Carpaneto, K. Yoshida, K.P. Hoffmann, K.P. Koch, P. Dario, and S. Micera. 2006. Charterization of tfLIFE neural response for the control of a cybernetic hand. *BioRob.* 477–482.

Dario, P., M.C. Carrozza, E. Guglielmelli, C. Laschi, A. Menciassi, S. Micera, and F. Vecchi. 2005. Robotics as a future and emerging technology: Biomimetics, cybernetics and neurorobotics in European projects. *IEEE Robot. Autom. Mag.* 12:29–45.

Donoghue, J.P. 2002. Connecting cortex to machines: recent advances in brain interfaces. *Nature Neurosci. Suppl.* 5:1085–1088. Review.

Dormont, J.F., A. Schmied, and H. Condé. 1982. Motor command in the ventrolateral thalamic nucleus: neuronal variability can be overcome by ensemble average. *Experimental Brain Research* 48:315–322.

Fang, H., Y. Wang, J. Huang, and J. He. 2006. Multiclass support vector machines for brain neural signals recognition. Intelligent Control and Automation (WCICA 2006). *The Sixth World Congress* 2:9940–9944.

Francis, J.T. and J.K. Chapin. 2004. Force field apparatus for investigating movement control in small animals. *IEEE Trans. Biomed. Eng.* 51:963–965.

Francis, J.T. and J.K. Chapin. 2006. Neural ensemble activity from multiple brain regions predicts kinematic and dynamic variables in a multiple force field reaching task. *IEEE Trans. Neural Rehab. Eng.* 14:172–174.

Georgopoulos, A.P., R.E. Kettner, and A.B. Schwartz. 1986. Neuronal population coding of movement direction. *Science* 233:1416–1419.

Hermer-Vazquez, L., R. Hermer-Vazquez, and J.K. Chapin. 2003. Methods for studying the neural basis of voluntary movement. Part 1: Correlations among behavior, multiunit activity and local field potential fluctuations. *Proceedings of the Twentyfifth Annual International Conference of the IEEE Engineering in Medicine and Biology Society* 3:2152–2155.

Hochberg, L.R., M.D. Serruya, G.M. Friehs, J.A. Mukand, M. Saleh, A.H. Caplan, A. Branner, D. Chen, R.D. Penn, and J.P. Donoghue. 2006. Neuronal ensemble control of prosthetic devices by a human with tetraplegia. *Nature* 442:164–171.

Humphrey, D.R., E.M. Schmidt, and W.D. Thompson. 1970. Predicting measures of motor performance from multiple cortical spike trains. *Science* 170:758–762.

Kakei, S., D.S. Hoffman, and P.L. Strick. 1999. Muscle and movement representations in the primary motor cortex. *Science* 285:2136–2139.

Kennedy, P.R. and R.A. Bakay. 1998. Restoration of neural output from a paralyzed patient by a direct brain connection. *Neuroreport.* 9:1707–1711.

Leuthardt, E.C., G. Schalk, J.R. Wolpaw, J.G. Ojemann, and D.W. Moran. 2004. A brain-computer interface using electrocorticographic signals in humans. *J. Neural Eng.* 1:63–71.

Navarro, X., T.B. Krueger, N. Lago, S. Micera, P. Dario, and T. Stieglitz. 2005. A critical review of interfaces with the peripheral nervous system for the control of neuroprostheses and hybrid bionic systems. *J. Periph. Nerv. Syst.* 10:229–258.

Nicolelis, M.A.L., C.-S. Lin, D.J. Woodward, and J.K. Chapin. 1993. Peripheral block of ascending cutaneous information induces immediate spatio-temporal changes in thalamic networks. *Nature* 361:533–536.

Nicolelis, A.L., A.A. Ghazanfar, C.R. Stambaugh, L.M. Oliveira, M. Laubach, J.K. Chapin, R.J. Nelson, and J.H. Kaas. 1998. Simultaneous representation of tactile information by distinct primate cortical areas rely on different encoding strategies. *Nat. Neurosci.* 1:621–630.

Nicolelis, M.A.L. and J.K. Chapin. 2002. Building brain-machine interfaces to restore and augment human motor performance. *Sci. Am.* 4:45–52.

Scott, S.H. and J.F. Kalaska. 1995. Changes in motor cortex activity during reaching movements with similar hand paths but different arm postures. *J. Neurophysiol.* 73:2563–2567.

Schmidt, E.M. 1980. Single neuron recording from motor cortex as a possible source of signals for control of external devices. *Annals Biomed. Eng.* 8:339–349.

Schwartz, A.B., D.M. Taylor, and S.I. Tillery. 2001. Extraction algorithms for cortical control of arm prosthetics. *Curr. Opin. Neurobiol.* 11:701–707 Review.

Shin, H.-C. and J.K. Chapin. 1990. Movement-induced modulation of afferent transmission to single neurons in the ventroposterior thalamus and somatosensory cortex in rat. *Exp. Brain Res.* 81:515–522.

Shenoy, K.V., D. Meeker, S. Cao, S.A. Kureshi, B. Pesaran, C.A. Buneo, A.P. Batista, P.P. Mitra, J.W. Burdick, and R.A. Andersen. 2003. Neural prosthetic control signals from plan activity. *Neuroreport* 14:591–596.

Taylor, D.M., S.I. Tillery, and A.B. Schwartz. 2002. Direct cortical control of 3D neuroprosthetic devices. *Science* 296:1829–1832.

Wessberg, J., C.R. Stambaugh, J.D. Kralik, J.K. Chapin, J. Kim, S.J. Biggs, M.A. Srinivasan, and M.A.L. Nicolelis. 2000. Real-time prediction of hand trajectory by ensembles of cortical neurons in primates. *Nature* 408:361–365.

Wolpaw, J.R., H. Ramoser, D.J. McFarland, and G. Pfurtscheller. 1998. EEG-based communication: improved accuracy by response verification. *IEEE Trans. Rehab. Eng.* 6:326–333.

Wolpaw, J.R. and D.J. McFarland. 2004. Control of a two-dimensional movement signal by a noninvasive brain-computer interface in humans. *Proc. Natl. Acad. Sci.* USA 101:17849–17854.

CHAPTER 6

Functional Electrical Stimulation and Rehabilitation Applications of BCIs

Dawn M. Taylor

OVERVIEW OF FUNCTIONAL ELECTRICAL STIMULATION

Functional Electrical Stimulation (FES) is the controlled application of electrical current to the peripheral nerves for the purpose of generating useful muscle contractions in people with nervous system dysfunction. Over the last several decades, many different applications of FES technology have been developed (Figure 6.1), and these can be divided into two main categories. The first category includes those systems that save lives by restoring essential autonomic functions. Probably the most well-known and widespread example of commercial FES technology is the cardiac pacemakers used to reliably activate heart muscles in people with damage to the neural circuitry of the heart. Other commercial technologies, such as the Vocare® system, are used to restore bladder function after spinal cord injury. FES diaphragm-pacing systems have the potential to eliminate need for a ventilator in severely paralyzed individuals. Also, methods to stimulate nerves that coordinate breathing and swallowing reflex pathways are being developed to treat sleep apnea or to facilitate swallowing after stroke.

The other large class of FES applications includes those systems designed to control movement and maintain posture by generating contractions in skeletal muscles. These types of systems have traditionally been developed for individuals with paralysis after spinal cord injury or stroke. However, FES technologies are also being investigated to improve the lives of people with cerebral palsy, multiple sclerosis, and other neuromuscular disorders.

People paralyzed by stroke can have a wide range of isolated impairments that can benefit from FES technologies. A number of commercial FES systems are available for "foot drop," the inability of a patient to raise the foot at the ankles. Foot drop is a common problem among stroke survivors. Hand systems to restore

T.W. Berger et al., Brain-Computer Interfaces, 81–94.

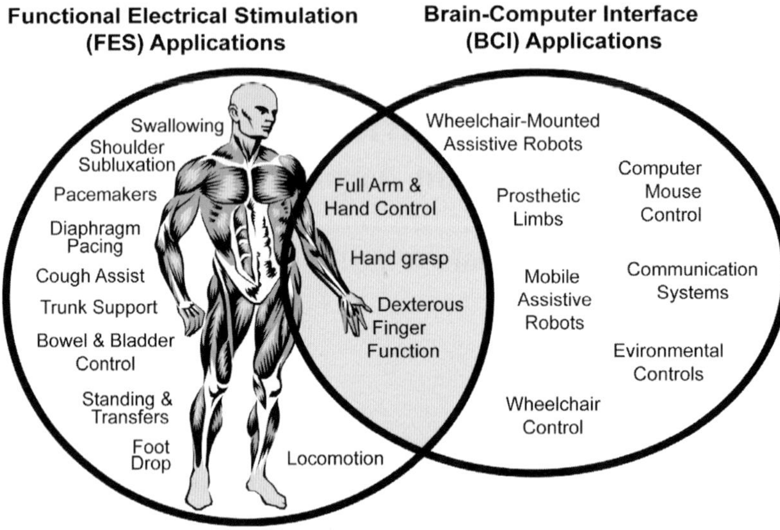

Figure 6.1. Overview of FES applications and BCI applications. Intersection (grey) shows research areas most often targeted for combined FES-BCI system development.

grasp after stroke are also being developed. Shoulder pain can also be a problem following stroke if key muscles are weakened and can no longer support the weight of the arm within the shoulder socket. Stimulation of these weakened paralyzed muscles is being evaluated for the ability to maintain normal shoulder-joint configuration and reduce pain due to this subluxation.

For people with spinal cord injuries, the level of spinal cord injury determines the extent of the paralysis and the FES system most needed. For low-level spinal cord injuries that have left people paralyzed below the waist, FES systems have been developed that facilitate standing from a seated position and can keep the body upright in a standing position if the person holds on to a walker for balance. These "standing-and-transfer" systems enable individuals to transfer from their wheelchair to another seating location or to reach objects that could not be reached while seated. Patterned stimulation of the lower limbs can also enable locomotion with the aid of a walker. For people whose injury has left them with trunk instability, stimulation of paralyzed trunk muscles while seated can improve posture to enhance breathing and expand the workspace for reaching.

For people with injuries of the cervical spine, arm and hand function can also be impaired. For people with C5–C6 level injuries, hand function is compromised, but most arm function remains. These individuals are able to place their hands where they want in space but are unable to generate a functional grasp once they get there. By restoring hand grasp via an FES system, such as the commercial Neuro-Control Freehand System, these individuals can perform the useful reach and grasp functions necessary for independence. However, for injuries located higher in the cervical spine, more of the arm becomes paralyzed as well. Stimulation of muscles for restoring hand grasp needs to be accompanied by stimulation of paralyzed arm muscles in order to restore any useful function (for a review of clinical applications of FES, see Creasey et al., 2004).

Electrode technologies for activating the peripheral nervous system cut across these various applications. The challenges of reliably stimulating and/or recording from axon bundles in the nerves differ from the challenges one faces when interacting with layers of neurons in the cortex. During muscle contractions, the peripheral nerves and muscles often move significantly relative to the surrounding tissues. This motion exerts much greater mechanical stress on the electrode, which can be damaging to the delicate neural tissue. Selectively activating specific axon bundles within a nerve is also a challenge, because most nerves often contain multiple fascicles that innervate different muscles. Many different types of electrodes have been developed for interacting with the peripheral nerves. Stimulating electrodes are, of course, essential for activating paralyzed muscles. However, recording electrodes are also being incorporated into some FES systems as a way to access proprioceptive and cutaneous information. These decoded sensory signals can then be fed back to the FES control system directly for improved closed-loop control. Sensory signals can also be relayed to the conscious perception of the user by stimulating somatosensory areas of the brain directly or through sensory substitution where one type of sensory information is conveyed through another modality (e.g., the rate of stimulation of the skin on the neck is proportional to the amount of grip force a FES-activated hand is generating).

The type of electrodes used for peripheral nerve stimulation will have an impact on the ability to control an FES system with a BCI. This is because different types of stimulating electrodes require different levels of current; high levels of stimulation current can generate artifacts in the recorded brain signals used by certain BCI systems. Compounding this problem, optimal stimulation frequencies for activating muscles are 12–16 Hz. This stimulation range can overlap with the field potential frequencies that are used in many BCI applications. Both current levels and electrode location will affect the amount of stimulation artifact recorded at the brain. Closely spaced bipolar configurations minimize current spread and reduce artifacts over configurations that use a distant ground. Surface electrodes, where the current must travel through the skin to activate the peripheral nerves, require the most current and generate the largest stimulus artifacts. Epimesial and intramuscular electrodes are stitched onto, or injected into, the muscle itself and activate the nerve as it enters the muscle. These muscle-based electrodes require current levels that are lower than those used for surface stimulation but are still an order of magnitude higher than electrodes placed in or on the peripheral nerves directly. Figure 6.2 shows a wide range of electrode technologies for stimulating/recording in the peripheral nervous system including many designed to interact with the nerve directly using low levels of current (Figure 6.2b–d).

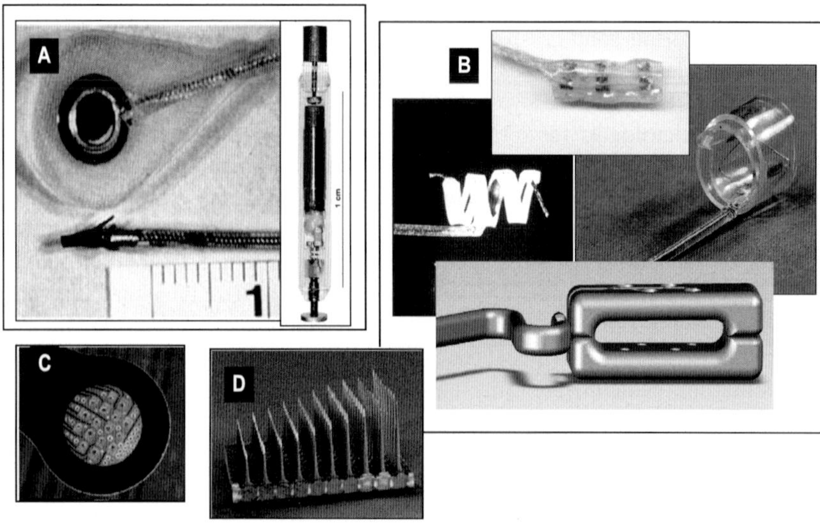

Figure 6.2. Examples of electrodes for activating muscles. (a) Electrodes placed directly on or in the muscle: epimesial (a-top) & intramuscular (a-bottom) electrodes developed at Case Western Reserve University, BION® from Advanced Bionics (a-right) developed in conjunction with the Alfred Mann Foundation; (b) electrodes designed to be placed around the nerve: cuff electrode used in the ActiGait® foot-drop system, Aalborg (b-top), spiral nerve cuff (b-right) and the flat interface nerve electrode (FINE) (b-bottom) developed at Case Western Reserve University, Huntington spiral nerve cuff electrode (b-left); (c) sieve electrodes trigger cut ends of nerves to grow through the electrode itself (example shown from the University of Michigan); (d) slant silicon microelectrode array (University of Utah) with multiple needle-like electrodes that penetrate the nerve itself.

FES APPLICATIONS OF BCI TECHNOLOGY AROUND THE WORLD

Combining BCI technology with FES technology could enable paralyzed individuals, once again, to move their own body just by thinking about doing so. This is a laudable goal, but the various technological advances needed to make this happen are just now beginning to come together. The most likely first candidate application for direct brain control of FES is restoration of arm and hand function by "thought" in people with spinal cord injury. Although BCI technologies could be applied to a number of other FES applications (e.g., volitional control of standing, locomotion, posture, and bladder), virtually all combined FES-BCI research has focused on using brain signals to drive various upper-limb FES systems. This upper-limb focus is due to several factors: (1) the need for BCIs that generate reach and grasp commands spans prosthetics and robotics applications as well as FES applications; (2) lower limb FES systems for standing, transferring, and locomotion can be easily controlled with a number of other convenient non-brain-based command signals; and (3) relatively large brain areas are associated with arm and hand function and are easily accessible with both noninvasive and invasive recording technologies.

A strong interest to link BCI and FES technology exists in both Europe and North America. The introduction sections of many BCI papers allude to how the authors' findings could be applied to the control of paralyzed limbs in "the future." Although many labs are working to extract intended arm and hand movement information from recorded brain signals, only a few BCI groups have formed the necessary clinical collaborations to actually implement a brain-controlled FES system in the lab. Currently, no combined BCI-FES system has been adopted for use on a regular basis by any paralyzed individual as part of his or her everyday routine. One reason for the limited progress in BCI-driven FES stems from the limited number of clinical facilities around the world that are actively deploying FES systems. In spite of the well-documented functional and quality of life improvements that accompany these upper limb FES systems (Peckham et al., 2001; Taylor et al., 2002b), only a limited number of facilities have the teams of physicians, surgeons, and therapists with the knowledge and experience needed to effectively deploy FES systems to the people who can benefit from them.

The other reason why combined BCI-FES systems have not yet been adopted for use outside of the lab is that practical, portable, and cosmetically acceptable BCI systems are still under development. Unlike severely locked-in individuals who could benefit from a large, stationary, in-home BCI system for communication or computer control, FES users are generally wheelchair-mobile and use their FES systems at different locations throughout their homes as well as in their work environments and other public places. Portability, power, and cosmesis (the user's appearance) are issues that must be addressed before combined BCI-FES systems can be used outside of a laboratory setting.

BCI systems that utilize action potentials recorded on multiple intracortical microelectrodes have been almost exclusively developed and tested in North America. Real-time, intracortical, microelectrode-based BCI systems have now been demonstrated in both humans and nonhuman primates in the United States. However, EEG-based BCI systems have been the primary focus in a majority of the European research labs. This difference in BCI technologies between North America and Europe stems from a long research history in EEG-based BCIs in Europe. Also, the extensive pressure from animal rights groups in Europe has discouraged many researchers from undertaking studies with nonhuman primates.

The limited work in BCI-driven FES systems reflects these historical differences between the United States and Europe. In 2003, Pfurtscheller (Graz, Austria) demonstrated that EEG signals could be used to control a sequence of hand movements to generate a functional grasp in a spinal-cord-injured person (Pfurtscheller et al., 2003). This demonstration utilized surface electrodes to activate the paralyzed muscles of the hand and forearm. In 2005, this same team published a similar demonstration (Muller-Putz et al., 2005) using a commercial, implanted,

FES-hand-grasp system—the Freehand® System (Neurocontrol, Cleveland, OH) (Figure 6.3, middle and bottom images). In this case, the experimenters overrode the system's normal command input, a shoulder-mounted joystick, and used decoded EEG commands to control the implanted FES system through the system's wireless interface.

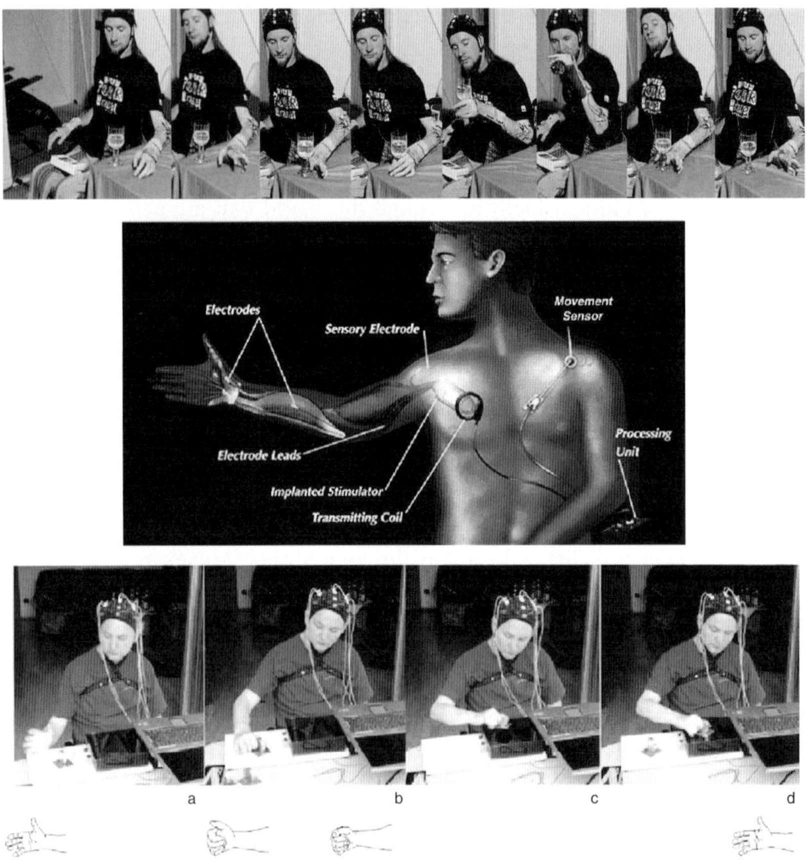

Figure 6.3. EEG-triggered hand grasp via FES. Top: hand grasp sequence views of an individual using EEG to trigger surface stimulation of the hand muscles; middle: diagram showing the commercial Freehand® hand-grasp FES system that has an external shoulder-mounted joystick to control degree of hand opening and closing; Bottom: individual using implanted Freehand® system but with EEG signals overriding normal joystick controls. In both top & bottom cases, the EEG signal was used as a discrete switch to activate three preprogrammed functions that enabled the grasp to be generated (shown by the hand diagrams in (a)–(d)). These users had retained arm function and used their own muscle activity to position their hand around the object and move their arm from one location to another once the FES system had enabled the grasp of the object.

Multiple examples of real-time BCIs based on intracortical microelectrodes in nonhuman primates have been extensively implemented in the United States (Wessberg et al., 2000; Serruya et al., 2002; Taylor et al., 2002a, 2003; Carmena et al., 2003; Musallam et al., 2004; Santhanam et al., 2006), and work is under way to incorporate intracortical BCIs with FES systems in nonhuman primate models (Santhanam et al., 2005; Morrow and Miller, 2003). Also in the United States, both telemetered and percutaneous, intracortical, microelectrode systems have been developed by startup companies and approved by the FDA for chronic use in humans. Chronic, glass-cone, intracortical microelectrodes and their associated telemetry systems (NeuralSignals, Inc.) were approved for use in severely locked-in individuals to run typing software for communication (Kennedy et al., 2000).

In the spring of 2004, a second U.S. company, Cyberkinetics, Inc., received FDA approval to implant chronic, intracortical microelectrode arrays in people with spinal cord injuries at the C4 level or above. Cyberkinetics implanted its first participant in the summer of 2004 and has released its initial findings. Preliminary results from the Cyberkinetics study showed that units could still be recorded in the motor cortex of a spinal-cord-injured person several years after the injury, and the person could still modulate the firing patterns of those neurons with attempted movements. The first subject was able to use his neural activity to move a cursor on a computer screen. Using the brain-controlled, computer cursor and customized software, the subject was able to change the channel on his TV or open a file by moving the cursor to the appropriate icon. Most notably, the subject was able to carry on conversations with the researchers while moving the brain-controlled cursor. This example demonstrated that willful modulation of the recorded neurons did not require such focused attention as to be impractical for applications such as FES (Hochberg et al., 2006).

The National Institutes of Health recently awarded two contracts in an effort to further the development of cortically-controlled arm and hand FES systems for people with high-level spinal cord injuries. The National Institute of Child Health and Human Development awarded a contract to Case Western Reserve University to jointly develop with Cyberkinetics the first intracortically controlled FES system for restoration of full arm and hand function in people with high-level spinal cord injuries at the C4 level or above (R. Kirsch, PI). As part of this contract, a virtual model of an FES-activated arm is being developed that paralyzed individuals participating in the Cyberkinetics intracortical implant trial can practice controlling with their decoded brain signals. The use of the brain-controlled virtual arm model will enable these researchers to refine decoding and control algorithms appropriate for control of an FES-activated arm and hand prior to implementing a combined intracortical BCI/FES system in future study participants. To further develop effective brain-based controllers for upper-limb FES systems, the National Institute of Neurological Disorders and Stroke funded a second contract to expand the virtual FES system model and make it available to the research community. Once the virtual FES arm model is available, any research group can test and refine its BCI system specifically for the application of controlling an FES-activated arm and hand.[1]

HOW DIFFERENT TYPES OF BCI COMMAND SIGNALS CAN BE APPLIED TO FES

Although FES applications currently make up only a small portion of the BCI research around the world, virtually all the brain-signal decoding algorithms being developed for other applications can be applied to the control of FES systems. The invasive work on decoding movement trajectories and fine details of motor parameters is advancing our knowledge of how the nervous system encodes reach and grasp movements. As our ability to extract intended movement details improves, we move closer to being able to restore movement by "natural thought" via a BCI-controlled FES system. However, more generic proportional and discrete signals used in BCIs for spelling or cursor control can be applied to FES control as well. This is because FES systems can be configured to utilize virtually any type of command signal that is convenient for the user to generate. For example, the Freehand hand grasp system depicted in the middle of Figure 6.3 normally uses the motion of the contralateral shoulder, transduced via a joystick, to generate a proportional signal to control the degree of hand opening and closing. However, these same hand-grasp systems can be controlled using an analogous EEG-based command.

Many EEG-based BCIs are designed to activate a binary switch by using classifier functions to choose between two or more different brain states. These simple discrete signals can be used to trigger different preprogrammed movements. For example, the bottom of Figure 6.3 shows how the user's EEG-based brain "switch" was used to cycle the hand grasp pattern through the three different phases needed to generate a useful grasp. The motor imagery used to drive the FES system often does not match the motor action produced by the FES system. For example, the person in the top of Figure 6.3 used imagined foot movement to trigger the FES system to open/close the hand because the user was already skilled at using imagined foot movement to trigger a BCI-based switch. The person in the bottom of Figure 6.3 triggered the sequence to cycle through different phases of a grasp pattern using imagined left-hand movements to trigger grasp in the right hand. In addition, only coarse "hand movement" versus "no hand movement" brain states were detected with the EEG. However, these coarse binary switch signals were able to trigger the controller to cycle through each phase of the grasp in the other hand.

---

[1] Further information on accessing this research tool will be posted at http://taylorbmilab.case.edu/when it becomes available.

Reach and grasp tasks requiring control of many degrees of freedom can be accomplished even with a single discrete switch. This can be achieved by sequencing through control of each degree of freedom one at a time and using the switch to turn on and then off a constant velocity motion until that degree of freedom has been moved the desired amount. These "gated ramp" type systems work well if the user cannot generate fine proportional command signals, but can control the fine *timing* of switch execution. Switch-based, gated ramp-control systems are commonly used in many FES applications and are also used in some high-degree-of-freedom assistive robots intended to provide reach and grasp function to the severely paralyzed (e.g., Assistive Robot Manipulator (ARM) from ExactDynamics).

Discrete, EEG-based classifier functions can be used to generate pseudo-proportional movement commands by rapidly applying the classifier at each time step to generate movement trajectories that appear smooth over time but are really made up of a sequence of small fixed movements. For example, a classifier that can move a cursor either 45 degrees left or right can generate a continuous trajectory that can go in *any* angle by rapidly combining many of these small left or right movements one after the other in the appropriate proportions. The right-versus left-movement classifier schemes used in the EEG-based BCIs by Klaus Müller's group in Berlin is an example of how a two-state classifier can be used to generate pseudoproportional movements that can continuously vary over a complete range of different directions.

True proportional commands from EEG-based BCIs developed for one- or two-dimensional computer mouse control can also be mapped to various FES movements such as proportional hand opening/closing and elbow extension. Again, these mappings between imagined movement and computer cursor can be quite abstract (e.g., imagine hand movement vs. rest to move cursor left or right and imagine foot movement or rest to move the cursor up or down). Similarly, the mapping between imagined movement and FES function are unlikely to exactly match the imagined or attempted movement in EEG-based systems due to the limited ability to precisely decode intended movement details from the low-resolution EEG.

APPLICATION AREAS OF BCI-CONTROLLED FES SYSTEMS

There are three primary focus areas where using BCI systems to drive FES technology could have an advantage over using other types of command signals. These include: (1) situations where other command signal options are limited,

(2) when the BCI system can generate better, more natural control signals than other options, and (3) for potential therapeutic benefits.

When Additional Command Options are Needed

Using abstract or unrelated imagined movements to control a real arm and hand via FES may seem to defeat the purpose of tapping into the brain to achieve "natural movement by thought." However, in certain situations even unnaturally generated command signals are an improvement over the alternatives. Efforts are under way to develop FES systems for restoring both arm and hand function in individuals with high-level, spinal cord injuries that leave individuals essentially paralyzed below the neck (C4 and above). In order to restore useful function, the user must be able to generate enough command signals to adequately control both arm and hand function (i.e., the user has to be able to position the hand in the workspace, orient the hand appropriately, and then generate a functional grasp). However, people with spinal cord injuries at the C4 level or above are limited to generating device commands from the neck up (e.g., voice commands, tongue-touch keypads, chin-operated joysticks, or facial muscle commands). Accessing movement commands directly from the brain will increase the command options available to this population and could enable these people to control reach and grasp functions while retaining normal use of their face and mouth.

Recently, the first person with a high-level spinal cord injury (C1-level motor complete) received a FES system with 24 channels of stimulation to restore arm and hand movements. This implanted system included four bipolar channels for recording EMGs to detect neck, shoulder, and scalp muscle contractions, which the person uses to control the FES system (Figure 6.4). However, this individual has difficulty activating these four implanted muscles independently. Other muscle options have been considered, but the available facial muscles are more risky to implant and would require unnatural facial movements to control the limb, which can negatively affect social interactions. Currently, this individual can use neck-muscle activity to position the hand within a limited workspace, and then can activate a "mode switch" via patterned, scalp-muscle activity to switch the system into hand-configuration mode. The individual then uses the neck muscles to grasp and acquire an object and then again uses the mode switch to put the system back in hand-position-control mode to move the object to another location.

This example points out two things. First, significant function can be restored, even with limited and unnatural command signals. Second, many more command signals are needed to utilize these expanded FES systems to their full potential. By augmenting this system with additional brain-based commands, users should be able to simultaneously control more upper limb degrees of freedom and be able to more effectively accomplish functional tasks.

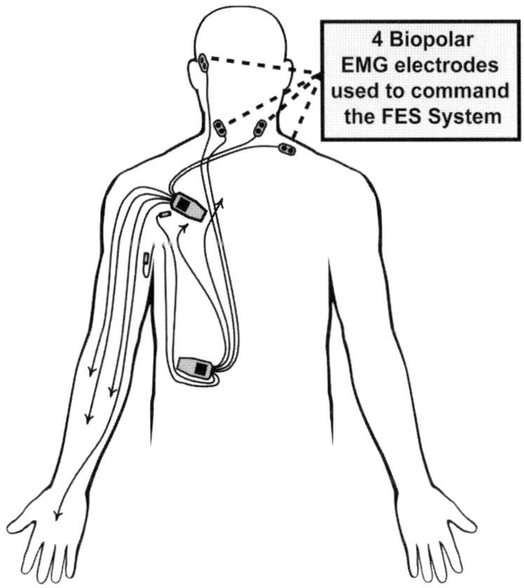

Figure 6.4. First FES system implanted to restore full arm and hand function in an individual with high tetraplegia. Two 12-channel stimulators were implanted (only some of the stimulating electrode leads are included in the diagram). A total of four bipolar EMG electrodes were included to record activity in the right and left neck muscles (platysma), the left trapesius, and the auricularis muscle behind the right ear. Contractions of these muscles are used to command the FES system (By permission, Cleveland Functional Electrical Stimulation Center).

Generation of Better, More Natural Control

In the example above, the person effectively used unnatural actions (i.e., activation of neck and head muscles) to generate useful arm and hand movements. Although these unnatural mappings can be robust and effective, generating movement just by thinking of making that movement is preferable as long as the quality of the decoded-movement intent is accurate. Even if we can decode intended movement precisely, actually generating the intended movement by activating selected muscles is a challenging problem. One way around this is to use invasive BCIs to decode intended muscle activations directly instead of decoding kinematic parameters that then have to be converted into the muscle activations needed to achieve those movements. Directly decoding muscle activations simplifies the FES system control task in that it eliminates the need for a "middle man" to reverse-engineer the muscle activations from the prescribed limb configuration. By putting the user's brain in direct control of individual muscle activations, the user should have much more flexibility over what movements are generated and may learn over time to generate a wider repertoire of useful arm and hand movements. Initial testing in direct brain control of muscle activations has begun in monkeys at Northwestern University, where the animal's limb is temporarily paralyzed via a local pharmacological block.

Potential Therapeutic Applications

Repetitive movement therapy is regularly used to promote natural recovery after a stroke or spinal cord injury. Movements can be generated via FES or by robotically driven, orthotic devices. Many researchers are now speculating that driving movements of the paralyzed limbs with the natural brain commands will enhance recovery of function by reinforcing the neural pathways connecting the brain with the appropriate lower motor circuits. A number of labs around the world are either starting or making plans to start BCI-triggered movement therapy studies. In Tübingen, MEG is being used to detect intended hand opening/closing at a higher spatial resolution than can be achieved with EEG. These detected hand open/close commands are then used to drive a pneumatic hand orthosis to generate the intended movement in the paralyzed hand. A similar study at the Cleveland Veteran's Administration is investigating the use of EEG to trigger robotically controlled arm movements in stroke subjects as well. EEG-triggered hand function activated via FES is also being looked into by a number of labs including Aalborg, the Essex group in collaboration with people from Southampton, and two groups in Cleveland. The Graz group is taking a more direct approach to stroke rehabilitation. They are simply using a virtual reality system to display opening and closing of an EEG-controlled virtual hand. The goal here is to provide feedback of brain activity to help the user learn to generate stronger hand opening/closing signals from the parts of the brain damaged by stroke.

Although more and more of these studies are getting under way, nothing conclusive has surfaced on enhanced recovery with brain-triggered therapy. Non-brain-triggered movement therapy has already been shown to promote recovery alone. This makes it difficult to demonstrate increased efficacy with brain-triggered therapy because brain- and non-brain-triggered therapy cannot be compared within the same subject without one treatment confounding another. Therefore, larger groups of subjects receiving either brain-triggered or non-brain-triggered therapy will need to be compared. This may be a daunting task because many therapy sessions are likely to be needed across a large number of subjects before any significant results are seen. Setting up and conducting each EEG-based BCI therapy session in the lab can be quite time consuming, making this type of study extremely challenging. Simplified, user-friendly, at-home BCI-FES systems could enable convenient daily therapy sessions and potentially make EEG-triggered therapy a realistic option for promoting functional recovery after stroke or spinal cord injury.

PRACTICAL CONSIDERATIONS

Although direct brain control of FES systems is an exciting and achievable goal, BCI technologies have to be evaluated against other available command options. Users will choose whichever technologies are the most practical and effective for use in their everyday lives. The shoulder joystick used to control hand opening and closing in the Freehand® system (see the middle of Figure 6.3) is robust and accurate. It feels "natural" to the users after they have been using it for a while. None of the EEG-based systems that have been tested with the Freehand® system in the lab has persuaded any users to switch from their shoulder joystick to EEG-based BCI control. However, for other applications, such as with restoration of full arm and hand function in high-tetraplegia, viable command alternatives are more limited and the need to explore these BCI options is high. Still, for BCI technology to be adopted for use by wheelchair-mobile individuals, the technology must be portable with low power requirements; it must be easy to don and maintain or else be fully implanted; and it must be comfortable and cosmetically acceptable.

For wheelchair-mobile FES users, the BCI software and hardware systems must be designed to work reliably in a wide variety of noisy environments. Most laboratory studies to date have not addressed this issue. Electromagnetic noise in the environment can potentially lead to artifacts in the recorded signals, which is why many research labs conduct their BCI studies inside large Faraday cages to avoid these very real problems in the lab. Also, decoding systems should accommodate brain signals from uncontrolled sensory inputs from the environment—you do not want your arm to fly up every time the telephone rings. Finally, the effects of cognitive load on different BCI systems need to be addressed. The initial report from Cyberkinetics is promising in that its study participant could carry on a conversation while controlling the cursor with motor activity recorded via intracortical microelectrodes. However, EEG-based BCIs may require a higher level of concentration, which could make them less attractive for use in distracting, real-world environments.

REFERENCES

Carmena, J.M., M.A. Lebedev et al. 2003. Learning to control a brain-machine interface for reaching and grasping by primates. *Public Library Sci.* 1(2):1–16.

Creasey, G.H., C.H. Ho, et al. 2004. Clinical applications of electrical stimulation after spinal cord injury. *J. Spinal Cord Med.* 27(4):365–375.

Hochberg, L.R., M.D. Serruya, G.M. Friehs et al. 2006. Neuronal ensemble control of prosthetic devices by a human with tetraplegia. *Nature* 442(7099):164–171.

Jackson, A., C.T. Moritz, J. Mavoori, T.H. Lucas, and E.E. Fetz. 2005. The Neurochip BCI: towards a neural prosthesis for upper limb function. *IEEE Trans. Neural Syst. Rehab. Eng.* 14(2):187–190.

Kennedy, P.R., R.A.E. Bakay et al. 2000. Direct control of a computer from the human central nervous system. *IEEE Trans. Rehab. Eng.* 8(2):198–202.
Morrow, M.M. and L.E. Miller. 2003. Prediction of muscle activity by populations of sequentially recorded primary motor cortex neurons. *J. Neurophysiol.* 89(4):2279–2288.
Muller-Putz, G., R. Scherer et al. 2005. EEG-based neuroprosthesis control: A step towards clinical practice. *Neurosci. Lett.* 382:169–174.
Musallam, S., B.D. Corneil et al. 2004. Cognitive control signals for neural prosthetics. *Science* 305:258–262.
Peckham, P.H., M.W. Keith et al. 2001. Efficacy of an implanted neuroprosthesis for restoring hand grasp in tetraplegia: a multicenter study. *Arch. Phys. Med. Rehab.* 82(10):1380–1388.
Pfurtscheller, G., G.R. Muller et al. 2003. "Thought"—control of functional electrical stimulation to restore hand grasp in a patient with tetraplegia. *Neurosci. Lett.* 351:33–36.
Serruya, M.D., N.G. Hatsopoulos et al. 2002. Instant neural control of a movement signal. *Nature* 416(6877):141–142.
Santhanam, G., S.I. Ryu, B.M. Yu, A. Afshar, and K.V. Shenoy. 2006. A high-performance brain-computer interface. *Nature* 442:195–198.
Taylor, D.M., S.I.H. Tillery, et al. 2002a. Direct cortical control of 3D neuroprosthetic devices. *Science* 296:1829–1832.
Taylor, D.M., S.I.H. Tillery et al. 2003. Information conveyed through brain-control: cursor versus robot. *IEEE Trans. Neural Syst. Rehab. Eng.* 11(2):195–199.
Taylor, P., J. Esnouf, et al. 2002b. The functional impact of the Freehand System on tetraplegic hand function, clinical results. *Spinal Cord* 40(11):560–566.
Wessberg, J., C.R. Stambaugh et al. 2000. Real-time prediction of hand trajectory by ensembles of cortical neurons in primates. *Nature* 408(6810):361–365.

CHAPTER 7

Noninvasive Communication Systems

Dennis J. McFarland

INTRODUCTION

Conditions such as amyotrophic lateral sclerosis (ALS), brainstem stroke, and severe brain or spinal cord injury can impair the neural pathways that control muscles or impair the muscles themselves. Individuals most severely affected may lose all voluntary muscle control, including eye movements and respiration, and may be completely locked in to their bodies, unable to communicate in any way. A variety of studies over the past 15 years have shown that the scalp-recorded electroencephalogram (EEG) can be used as the basis for a brain-computer interface (Wolpaw et al., 2002). BCI can provide an alternative method of communication and control for those severely affected individuals.

A BCI system consists of sensors that record neural activity, signal processing that extracts features, and a translation algorithm that creates device commands to operate an external device (Wolpaw et al., 2002). The loop is completed with feedback from the external device to the BCI system user. These basic elements of a BCI communication system are illustrated in Figure 7.1. As can be seen in the figure, there is a flow of information through each of these elements which ultimately feeds back to the user. A functioning BCI system is by necessity a closed-loop, real-time system. In the case of BCI communication systems, the external device serves as a means for the user to communicate.

There have been a number of BCI communication systems that have been designed to demonstrate proof of principle. These are based on a variety of neural features such as slow cortical potentials (Birbaumer et al., 1999), motor potentials (Mason et al., 2004), event-related synchronizations and desynchronizations (Pfurtscheller et al., 1993; Wolpaw et al., 1991), steady-state evoked potentials (Jones et al., 2003), and P300 potentials (Farwell and Donchin, 1988). These systems have generally used surface-recorded EEG. Noninvasive EEG recordings provide a safe alternative to invasive methods that may provide useful BCI communication devices for individuals with disabilities.

T.W. Berger et al., Brain-Computer Interfaces, 95–108.

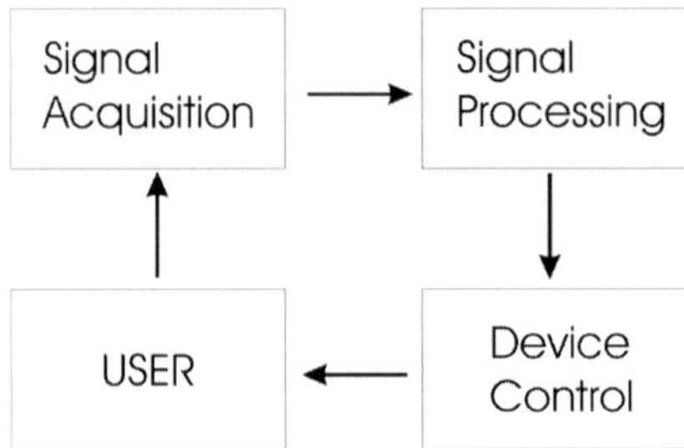

Figure 7.1. Basic parts of a BCI communication system. Signals flow from the user to signal acquisition, signal processing, device control, and then back to the user. Efficient operation requires that this process be completed in real time.

SLOW CORTICAL POTENTIALS

Slow cortical potentials (SCPs) are low frequency potentials (e.g., less than 2 Hz, at times referred to as DC potentials) recorded from the surface that are associated with various cognitive or sensory-motor events. A classic example is the contingent negative variation (Birbaumer et al., 1990), a negative shift that occurs during the interval when an individual anticipates some event. Increased cortical activation is associated with scalp negativity and decreased activation is associated with positivity (Birbaumer et al., 1990). Birbaumer and colleagues (1999) trained individuals to modify SCPs based on feedback and used this paradigm for BCI-based communication.

The SCP-based communication device (i.e., the thought translation device, or TTD) is conceptualized as depending upon the principles of operant conditioning. Training proceeds through several stages. Users first learn by trial and error to move a cursor vertically on a video monitor that moves at a constant horizontal rate under computer control. The goal is to select targets at either the top or bottom edge of the screen. Next, the user works with a binary, five-level matrix of letters presented on the screen. The alphabet and punctuation are split into two parts, each with 16 symbols. The user selects among these, and the resulting selection is next split into two. Users first learn this task with error-free copy spelling, followed by free spelling. Birbaumer et al. (1999) found that a severely disabled user was able to compose text using this system. Although slow, these results demonstrate the feasibility of SCP-based communication.

Motor Potentials

Movements are accompanied by transient potentials on the scalp surface. A negativity called the readiness potential, or *Bereitschaftspotential*, precedes self-paced movements. In addition, the actual execution of a movement is also associated with transient potentials. Imagined movement also elicits similar potentials (Nielson et al., 2006; Yom-Tov and Inbar, 2003). The scalp location of these movement-related transient potentials varies with the nature of the movement in question. For example, hand movements are generally accompanied by peak potentials over the contralateral hemisphere.

Mason and Birch (2000) have used motor-related potentials as the basis of a BCI-communication system. They emphasized use of a BCI within subject-paced paradigms rather than as a response to an external cue. They refer to this user-initiated paradigm as an asynchronous BCI. Training a classifier with a self-paced response is a challenge, since it is either necessary to allow actual movements or to provide some means of verifying the user's intentions. Use of overt movements is not feasible with severely motor-impaired users. Recently, Mason et al. (2004) have used a sip-and-puff switch to verify intent in quadriplegic users.

Blankertz et al. (2003) have used movement-related potentials resulting from actual left and right key-strokes. They have shown that useful information is available from the scalp prior to useful information being available from the EMG. Blankertz et al. (2006a) have also demonstrated the existence of detectable motor potentials from phantom limb commands recorded from patients with amputations.

Motor potentials represent a potential rich source of information to drive a BCI communication system. However, to date they have not been used extensively in real-time, closed-loop communication systems.

Event-Related Synchronizations and Desynchronizations

Both movements and motor imagery are accompanied by changes in oscillatory activity that can be recorded from the scalp. An increase in synchronous activity in response to some event is referred to as an event-related synchronization (ERS); a decrease in synchronous activity is referred to as an event-related desynchronization (ERD). Pfurtscheller and his colleagues have pioneered the study of the basic science of ERD/ERS phenomena. Pfurtscheller and Aranibar (1977) described the basic phenomenon of ERD as it is related to motor movements. Subsequently, Pfurtscheller described the topographic specificity of ERD and ERS phenomena (see Pfurtscheller and Lopus da Silva [1999] for an excellent review). For example, right- or left-hand movements are associated with desynchronization of mu activity predominantly over the contralateral scalp. Foot movements are associated with desynchronization predominantly over the central midline. These observations illustrate how ERDs are topographically related to areas of sensory motor cortex associated with the particular movement in question. In addition, Pfurtscheller and Lopus da Silva (1999) described center-surround phenomena where areas lateral or medial to sensor motor areas controlling the movement in question actually show an increase in mu-rhythm synchronization (ERS).

Wolpaw et al. (1991) showed that ERD-related phenomena could be used for a BCI based on a two-target, cursor-movement task. The mu rhythm has also been used for tasks involving multiple targets in one dimension (McFarland et al., 2003), answering questions (Miner et al., 1998), two-dimensional cursor movement (Wolpaw and McFarland, 1994, 2004), spelling devices (Blankertz et al., 2006a; Scherer et al., 2004), and control of an orthosis (Muller-Putz et al., 2005).

A number of studies have examined alternative feature extraction and classification algorithms that might improve mu-based BCI performance. For example, McFarland et al. (1997) and Babiloni et al. (2000) showed that spatial filters, such as the surface Laplacian, greatly improve the signal-to-noise ratio of the mu-rhythm signal. Ramoser et al. (2000) showed that an empirically derived spatial filter, referred to as common spatial patterns, also improved the signal-to-noise ratio of mu-rhythm control signals.

Milan et al. (2002) and Fabiani et al. (2004) showed that increasing the number of features used to define a mu-rhythm-based control signal improved target prediction by a classifier. These two, and other studies (e.g., data competitions, such as that described by Blankertz et al. [2004, 2006b]), have been based on offline analysis of data, so that it is not certain that the results obtained would generalize to actual online performance. Exceptions are studies by McFarland and Wolpaw (2005) and Krusienski et al. (2007). In the McFarland and Wolpaw (2005) study, regression models evaluated in offline data sets were subsequently applied online in real time. The online results replicated effects observed in the offline simulations. In the study of Krusienski and colleagues (2007), a matched filter that was found to outperform AR modeling in offline analysis also produced similar effects online and in real time.

STEADY-STATE EVOKED POTENTIALS

Attended stimuli presented at a constant rate entrain certain brain rhythms. For example, when individuals attend to short-duration visual stimuli presented at a steady rate of 13 Hz, rhythms appear over posterior visual areas with a fundamental frequency of 13 Hz and harmonics of this fundamental. This steady-state visual-evoked potential (SSVEP) has served as the basis of several BCI designs. Middendorf et al. (2000) used a SSVEP-based system to allow users to select one of two virtual buttons flashing at different rates on a computer screen. The user selected the desired button simply by looking at it. Muller-Putz et al. (2005) used an SSVEP-based system to allow users to select one of four flashing lights on a video screen. Cheng et al. (2002) used a SSVEP-based system to allow users to select one of 12 buttons flashing at different rates on a computer screen. Eight of thirteen users could dial a mobile phone with this system.

Jones et al. (2003) compared an SSVEP-based system to selection with a mouse. They note that although the SSVEP-based system is slower, it is less restricted on target distance and may be useful when the operator's hands are not free. These considerations apply to use by individuals without motor impairments. Trejo et al. (2006) designed a SSVEP-based two-dimensional cursor-movement system based on flickering checkerboard stimuli at each border of the screen.

SSVEP-based systems usually involve monitoring the spectral peak corresponding to the frequency that the steady-state visual stimulus is presented. Muller-Putz et al. (2005) found that use of harmonics increased accuracy of classification.

The SSVEP appears to depend upon users controlling their attention. This might mean that SSVEP-based BCI systems depend upon the user having good voluntary control of eye movements (Wolpaw et al., 2002). Tactile stimuli have also been presented at a steady rate to provide the basis of a steady-state somatosensory-evoked potential (SSSEP) (Muller-Putz et al., 2006) that does not depend upon the ability to control eye movements.

Wang et al. (2006) have noted that the SSVEP depends upon intact eye movements. However, they also suggest that for most people, the SSVEP-based BCI is more feasible than other systems. This is due to advantages of high information transfer rate and the fact that little user training is required. They are developing a practical SSVEP-based system that uses only a single bipolar electrode and is simple to use. This requires careful selection of the channel location, stimulus frequency, and speed of selection.

P300 Evoked Potentials

The P300 is a large, positive potential over midline areas that has been studied extensively within the context of the oddball paradigm. This potential occurs with latency around 300 ms in response to target stimuli that occur infrequently and that subjects are instructed to respond to in some manner. Donchin and colleagues (Donchin et al., 2000; Farwell and Donchin, 1988) first reported the use of the P300 for BCI communication. Their paradigm involved a 6×6 matrix of grey symbols on a dark background. Rows and columns of the matrix were randomly intensified. A P300 was produced when the attended row or column flashed. The attended symbol was selected by averaging responses for rows and columns. Accurate performance was obtained in users with and without disabilities. However, users attended only to the letter "P" in these studies; although demonstrating proof of principle, these initial studies did not actually involve communication.

Sellers and Donchin (2006) showed that both users without motor impairments and users with ALS were able to use the P300-based, single-stimulus system using either auditory or visual presentations. Sellers et al. (2007) showed that matrix size and ISI are both important for P300-based matrix performance. In addition, they report the results of a simulation showing that accuracy less than 60% may be associated with information transmission as measured by bits, but not in terms of useful communication as measured by the time required to select the letters in a word.

Kaper et al. (2004) analyzed data offline from a single subject using the 6 × 6 P300-based matrix. This report describes the winner of the BCI 2003 competition for data set IIb and showed excellent classification in a single subject with a support vector machine.

Serby et al. (2005) showed improved offline performance with the 6 × 6 matrix by using independent components analysis and a matched filter. Subsequent online work resulted in performance better than that reported by Donchin et al. (2000), but not as good as their offline results. Krusienski et al. (2006) showed that by supplementing the classical midline electrodes with posterior locations online performance can be significantly improved.

Vaughan et al. (2006) describe the daily use of an in-home P300 system by an individual with ALS. This system consists of a reduced set of electrodes, a portable amplifier, and a laptop computer. The software is a specific instance of BCI2000. This home system is shown in Figure 7.2.

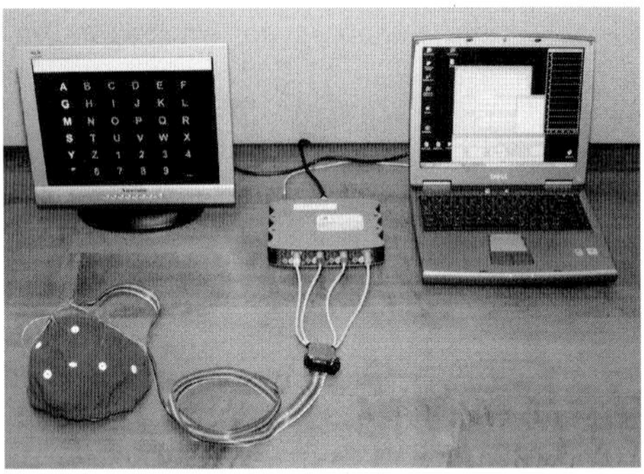

Figure 7.2. The hardware for a P300-based BCI home system. The basic P300 matrix is shown on the left monitor and an instance of the BCI2000 software system is shown on the right monitor. The portable 16-channel amp appears in the middle with an attached electrocap.

Adaptation

There are at least three distinct orientations toward BCI development. The first of these views sees BCI as an operant conditioning paradigm (Birbaumer et al., 2003). This view regards the process as one in which the experimenter, or trainer, shapes the desired output by means of reinforcement. The training process then consists of guiding or leading the user. The second of these views, expressed best by the statement "let the machines learn" (Blankertz et al., 2003), considers BCI to be mainly a problem of machine learning. This view implicitly sees the user as producing a predictable signal that needs to be discovered. For example, Blankertz et al. (2006a) state that they use "well-established motor competences" to operate an EEG-based communication system that does not require subject training. A third possibility views the user and system as the interaction of two dynamic processes (Taylor et al., 2002; Wolpaw et al., 2002). By this view, the goal of the BCI system is to select those signal features that the user can best control and optimize the translation of these signals into device control. This optimization facilitates further learning by the user, which in turn leads to further changes in the BCI system. These three views are illustrated in Figure 7.3.

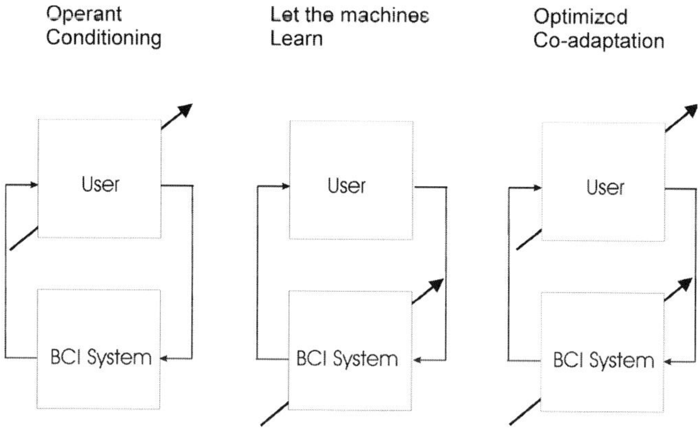

Figure 7.3. Three views of BCI systems. In the first, operant conditioning is viewed as a way of shaping the user's EEG. In the second, machine learning algorithms are used to optimize existing EEG signals. The third view sees the process as involving the interaction between user and the BCI system.

Although the need for adaptation has been noted for some time now (e.g., Neat et al., 1990), empirical support for this concept is generally lacking. In addition, it is not clear which aspects of a BCI system should be adapted and which should not be adapted. There have been a number of successful demonstrations of adaptive BCI systems. Ince et al. (2006) report that an adaptive feature extraction procedure resulted in improved offline classification of motor imagery data. Sykacek et al. (2004) report that an adaptive Bayesian classifier outperformed a static classifier on the offline identification of cognitive tasks.

The nonstationary nature of the BCI signal provides a rationale for the design of an adaptive BCI system. Vidaurre et al. (2006) describe a BCI based on an adaptive classifier. They used a quadratic discriminant analysis with adaptive estimation of the covariance matrix. Vidaurre et al. (2006) present some interesting illustrations of how two-dimensional projections of the feature distributions change from one session to the next. Similar results were reported by Shenoy et al. (2006). These studies illustrate the nonstationary nature of ERD/ERS statistical data. Demonstrating this provides an objective rationale for use of adaptive methods. Unfortunately, these studies have been rather short-term to date.

The Graz group (Krausz et al., 2003) found that fast adaptation of parameters during training was not necessary. They suggest that a classifier could be updated at the beginning of each session. The issues of how to adapt and at what rate are complex and will require more investigation.

McFarland et al. (2006) have discussed aspects of adaptation within the context of sensorimotor rhythm training. They note that there are actually several parameters that could be adaptively adjusted in a BCI paradigm. Not all parameters should necessarily be adjusted by the same outcome measure or according to the same time constant. For example, a slow process that is based on classification accuracy, a form of feedback control, might adjust EEG feature weights. In contrast, normalization of the classifier output could use a faster process based on signal statistics, a form of feedforward control. Thus, different aspects of signal processing and translation might be adjusted according to different criteria.

Sellers et al. (2007) have discussed the possibility that BCI paradigms might differ with respect to the need for adaptation. For example, mu-rhythm control appears to benefit from user feedback, which, in turn, may change signal statistics. In contrast, the P300 paradigm may rely much less on learning. Accordingly, these two paradigms may differ in the extent to which adaptation is advisable. It should be noted, however, that hard, empirical data on any of these issues are currently lacking.

ONLINE EVALUATIONS

Many studies that involve investigations of neurophysiologic or psychophysiological phenomena, like the basic cellular mechanisms of motor control (e.g., Sergio et al., 2005) or scalp potentials associated with target detection (e.g., Allison and Pineda, 2003), could be construed as being related to BCI development. However, BCI research is concerned with the development of complete systems that can provide alternative means of communication and control by directly accessing information from the brain and using it to perform functions directed by the user (Wolpaw et al., 2002). Human communication and movement control occur in real time and involve feedback to the user. This requires closing the loop in real time among brain sensors, signal processing, and the user's perceptual apparatus.

The real-time requirements of a BCI system require certain design considerations. There have been several data sets used in BCI competitions (e.g., Blankertz et al., 2004; Schlogl et al., 2005) that provide a convenient means of evaluating alternative prediction algorithms. However, real-time prediction algorithms need to estimate parameters in a causal manner (i.e., only the data collected up to the present time are available rather than the entire session, as is the case with offline analysis). Offline prediction algorithms may estimate the statistics of the data from observations across an entire session and can perform these computations over a protracted period of time. In addition, the analyst may review the results and make modifications to the process. This is not possible for a system operating in real time. Moreover, users of BCI systems change over time as a result of, for example, learning, fatigue, or changes in motivation. Consequently, an adaptive BCI system co-evolves with an adaptive user (Wolpaw et al., 2002; Taylor et al., 2002). To further complicate the issue, it is extremely difficult to evaluate or fine-tune new signal processing algorithms offline using data collected from an adaptive or closed-loop system. This is because the user is no longer in the control loop and it is impossible to model exactly how the user would react to the feedback produced by a new algorithm. Thus, both online experiments, as well as intelligently designed offline simulations, are necessary for effective algorithm development in an adaptive or closed-loop system.

BCI research has undergone an explosive growth in recent years (Vaughan and Wolpaw, 2006). Although many of the earlier BCI investigations were online, real-time studies, a greater proportion of the more numerous recent studies use archival data. The use of archival data is convenient, but it does not provide a means of addressing real-time issues. Certainly, collecting data with a closed-loop system is a technically difficult task. However, this problem is partially alleviated by shared software systems such as BCI2000 (Schalk et al., 2004).

PROSPECTS FOR PRACTICAL BCI COMMUNICATION SYSTEMS

In contrast to invasive systems, noninvasive BCI systems are currently at a point in their development where they could provide the most severely motor-impaired individuals with an alternative means of communication. Vaughan et al. (2006) have already installed a P300-based system in the home of a 47-year-old man with ALS. They report that he is using this device 4–6 hours/day for tasks such as email correspondence. They also note that he found the BCI system to be superior to an eye-gaze system that he had been using.

Moving from the laboratory to the home requires training caregivers in the application of sensors. Although Donoghue (2002) has stated that “multielectrode EEG systems can take an hour to attach…,” this has not been the experience of the Wadsworth group. We find that the setup time for our 64-channel system is usually around ten minutes. The reduced eight-channel montage we are developing for home use takes even less time. The actual problem is not the time required for application but in training the BCI users’ caregiver to apply the cap and to recognize problems with the recording. Nonetheless, noninvasive BCI systems are currently ready for practical use, in contrast to invasive systems, which still face problems such as the long-term stability of recording electrodes (McFarland, 2007).

The population of potential users of BCI communication devices will ultimately depend upon the communication rate that can be realized as well as the ease of using these systems. Current BCI communication devices, such as the P300-based matrix speller, can support acceptable communication rates for individuals with few other options. Expansion of the potential user base depends upon advances that result in increased rates and accuracy.

Donoghue (2002) has stated that EEG systems "typically allow only a few choices per minute…"; however, this has not been the experience of the Wadsworth group.

Hotchberg et al. (2006) state that EEG-driven BCIs requires "concentration to the exclusion of other actions…" and "two-dimensional tasks appear to engage all controllable signals…" However, they do not cite any references that support these statements. In fact, it has yet to be determined how many independent channels of information can be extracted from surface EEG recordings. As of this time, noninvasive methods have produced roughly the same information transfer rates as invasive systems (Wolpaw and McFarland, 2004). Developing noninvasive BCI devices with more than two independent channels of control is a challenge that could expand the range of potential BCI users. Use of current BCI systems on a routine basis by individuals who can actually benefit from these devices currently requires caregiver training in system use. It is necessary to train caregivers in electrode application since current noninvasive methods require continual application of the sensors. This process includes insuring that caregivers can recognize problems with the quality of the EEG recordings and make appropriate adjustments. This instruction takes some time and is acquired more quickly by some caregivers than others. One solution to this problem would be to have more foolproof sensors, such as dry, capacitance-based electrodes (Alizadeh-Taheri et al., 1996; Harland et al., 2002). Currently, the useful density of surface sensors is limited by the use of jells (Greischar et al., 2004). Improved sensor technology could also provide superior signal-to-noise ratios that might ultimately lead to faster and more accurate communication devices.

SUMMARY AND CONCLUSIONS

There have been many demonstrations of proof of principle for BCI communication devices using a variety of EEG features. Some of these have involved online and real-time systems, but there is a recent trend to perform simulations with archival data. Current technology is at a point that practical systems could be available to individuals who could actually benefit from their use. At the same time, future developments in sensor technology, signal processing, and identification of useful features could expand the potential population that could benefit from BCI communication systems.

Development of practical BCI communication systems must deal with two major issues. One is the limited bandwidth of current systems. The other is the technical difficulty inherent in the use of current-generation systems. Currently, there are individuals and their caregivers who are willing and able to deal with these problems. Increasing the bandwidth of BCI communication systems depends upon the continuing innovative exploration of new methods and signals, as well as a deeper understanding of the phenomena to which these methods are applied. Reducing the technical complexity of current BCI communication systems is a matter of applying existing technology.

In the short term, severely disabled individuals may begin to benefit from existing BCI technology. With continuing development in this area, the population of individuals using practical BCI systems may gradually expand toward a higher number where even individuals without disabilities may benefit from this technology. The ultimate limit to this technology can only be ascertained by research that explores these possibilities.

REFERENCES

Allison, B.Z. and J.A. Pineda. 2003. ERPs evoked by different matrix sizes: Implications for a brain computer interface (BCI) system. *IEEE Trans. Neural Syst. Rehab. Eng.* 11:110–113.

Alizadeh-Taheri, B., R.L. Smith, and R.T. Knight. 1966. An active, microfabricated, scalp electrode array for EEG recording. *Sens. Actuators* 54:606–611.

Babiloni, F., F. Cincotti, L. Lazzarini, J. Millan, J. Mourino, M. Varsta, J. Heikkonen, L. Bianchi, and M.G. Marciani. 2000. Linear classification of low-resolution EEG patterns produced by imagined hand movements. *IEEE Trans. Rehab. Eng.* 8:186–188.

Birbaumer, N., A.G. Elbert, A.G. Canavan, and B. Rockstroh. 1990. Slow potentials of the cerebral cortex and behavior. *Physiol. Rev.* 70:1–40.

Birbaumer, N., N. Ghanayim, T. Hinterberger, I. Iversen, B. Kotchoubey, A. Kubler, J. Perlmouter, E. Taub, and H. Flor. 1999. A spelling device for the paralyzed. *Nature* 398:297–298.

Birbaumer, N., T. Hinterberger, A. Kubler, and N. Newman. 2003. The thought-translation device: neurobehavioral mechanisms and clinical outcome. *IEEE Trans. Neural Syst. Rehab. Eng.* 11:120–123.

Blankertz, B., G. Dornhege, C. Schafer, R. Krepki, J. Kohlmorgen, R.K. Muller, V. Kunzmann, F. Losch, and G. Curio. 2003. Boosting bit rates and error detection for the classification of fast-paced motor commands based on single-trial EEG analysis. *IEEE Trans. Neural Syst. Rehab. Eng.* 11:100–104.

Blankertz, B., R.K. Muller, G. Curio, T.M. Vaughan, G. Schalk, J.R. Wolpaw, A. Schlogl, C. Neuper, G. Pfurtscheller, T. Hinterberger, M. Schroder, and N. Birbaumer. 2004. The BCI competition 2003: progress and perspectives in detection and discrimination of EEG single trials. *IEEE Trans. Biomed. Eng.* 51:1044–1051.

Blankertz, B., G. Dornhege, M. Krauledat, K-R. Muller, V. Kunzmann, F. Losch, and G. Curio. 2006a. The Berlin brain-computer interface: EEG-based communication without subject training. *IEEE Trans. Neural Syst. Rehab. Eng.* 14:147–152.

Blankertz, B., R.K. Muller, D.J. Krusienski, G. Schalk, J.R. Wolpaw, A. Schlogl, G. Pfurtscheller, J.R. Millan, M. Schroder, and N. Birbaumer. 2006b. The BCI competition III. validating alternative approaches to actual BCI problems. *IEEE Trans. Neural Syst. Rehab. Eng.* 14: 153–159.

Cheng, M., X. Gao, S. Gao, and D. Xu. 2002. Design and implementation of a brain-computer interface with high transfer rates. *IEEE Trans. Biomed. Eng.* 49:1181–1186.

Donchin, E., K.M. Spencer, and R. Wijesinghe. 2000. The mental prosthesis: assessing the speed of a P300-based brain-computer interface. *IEEE Trans. Rehab. Eng.* 8:174–179.

Donoghue, J.P. 2002. Connecting cortex to machines: recent advances in brain interfaces. *Nature Neuroscience* (Supplement) 5:1085–1088.

Fabiani, G.E., D.J. McFarland, J.R. Wolpaw, and G. Pfurtscheller. 2004. Conversion of EEG activity into cursor movement by a brain computer interface (BCI). *IEEE Trans. Rehab. Eng.* 12:331–338.

Farwell, L.A. and E. Donchin. 1988. Talking off the top of your head: a mental prosthesis utilizing event-related potentials. *Electroencephalogr. Clin. Neurophysiol.* 70:510–523.

Greischar, L.L., C.A. Burghy, C.M. van Reekum, D.C. Jackson, D.A. Pizzagalli, C. Mueller, and R.J. Davidson. 2004. Effects of electrode density and electrolyte spreading in dense array electroencephalography recording. *Clin. Neurophysiol.* 115:710–720.

Harland, C.J., T.D. Clark, and R.J. Prance. 2002. Remote detection of human electroencephalograms using ultrahigh input impedance electric potential sensors. *Appl. Phys. Lett.* 81:3284–3286.

Hotchberg, L.R., M.D. Serruya, G.M. Friehs, J.A. Mukand, M. Saleh, A.H. Caplan, A. Branner, D. Chen, R.D. Penn, and J.P. Donoghue. 2006. Neuronal ensemble control of prosthetic devices by a human with tetraplegia. *Nature* 442:164–171.

Ince, N.F., S. Arica, and A. Tewfik. 2006. Classification of single trial motor imagery EEG recordings with subject adapted non-dyadic arbitrary time-frequency tilings. *J. Neural Eng.* 3:235–244.

Jones, K.S., M. Middendorf, G.R. McMillan, G. Calhoun, and J. Warm. 2003. Comparing mouse and steady-state visual evoked response-based control. *Interact. Computers* 15:603–621.

Kaper, M., P. Meinicke, U. Grossekathoefer, T. Ligner, and H. Ritter. 2004. BCI competition 2003- Data set IIb: Support vector machines for the P300 speller paradigm. *IEEE Trans. Biomed. Eng.* 51:1073–1076.

Krausz, G., R. Scherer, G. Korisek, and G. Pfurtscheller. 2003. Critical decision-speed and information transfer in the Graz brain-computer interface. *Appl. Psychophysiol. Biofeedback* 28:233–240.

Krusienski, D.J., G. Schalk, D.J. McFarland, and J.R. Wolpaw. 2007. A mu-rhythm matched filter for continuous control of a brain-computer interface. *IEEE Trans. Biomed. Eng.* 54(2): 273–80.

Krusienski, D.J., E.W. Sellers, D.J. McFarland, T.M. Vaughan, and J.R. Wolpaw. 2006. Toward enhanced P300 speller performance. *J. Neurosci. Methods* 3:299–305.

Mason, S.G. and G.E. Birch. 2000. A brain-controlled switch for asynchronous control applications. *IEEE Trans. Biomed. Eng.* 47:1297–1307.

Mason, S.G., R. Bohringer, J.F. Borisoff, and G.E. Birch. 2004. Real-time control of a video game with a direct brain-computer interface. *J. Clin. Neurophysiol.* 21:404–408.

McFarland, D.J. 2007. Invasive BCI approaches: preface. In *Towards brain-computer interfacing.* Eds. Dornhege, G., J.R. Millan, T. Hinterberger, D. McFarland, and K-R Muller. Cambridge: MIT Press.

McFarland, D.J., D.J. Krusienski, and J.R. Wolpaw. 2006. Brain-computer interface signal processing at the Wadsworth Center: Mu and sensorimotor beta rhythms. *Prog. Brain Res.* 159:421–429.

McFarland, D.J., L.M. McCane, S.V. David, and J.R. Wolpaw. 1997. Spatial filter selection for EEG-based communication. *Electroencephalogr. Clin. Neurophysiol.* 103:386–394.

McFarland, D.J., W.A. Sarnacki, and J.R. Wolpaw. 2003. Brain-computer interface (BCI) operation: optimizing information transfer rates. *Biol. Psychol.* 63:237–251.

McFarland, D.J. and J.R. Wolpaw. 2005. Sensorimotor rhythm-based brain-computer interface (BCI): feature selection by regression improves performance. *IEEE Trans. Neural Syst. Rehab. Eng.* 13:372–379.

Middendorf, M., G. McMillian, G. Calhoun, and K. Jones. 2000. Brain-computer interfaces based on the steady-state visual-evoked response. *IEEE Trans. Rehab. Eng.* 8:211–214.

Millan, J.R., M. Franze, J. Mourino, F. Cincotti, and F. Babiloni. 2002. Relevant features for the classification of spontaneous motor-related tasks. *Biol. Cybernet.* 86:89–95.

Miner, L.A., D.J. McFarland, and J.R. Wolpaw. 1998. Answering questions with an electroencephalogram-based brain-computer interface. *Arch. Phys. Med. Rehab.* 79:1029–1033.

Muller-Putz, G.R., R. Scherer, C. Brauneils, and G. Pfurtscheller. 2005. Steady-state visual evoked potential (SSVEP)-based communication: impact of harmonic frequency components. *J. Neural Eng.* 2:123–130.

Muller-Putz, G.R., R. Scherer, C. Neuper, and G. Pfurtscheller. 2006. Steady-state somatosensory evoked potentials: suitable brain signals for brain-computer interfaces? *IEEE Trans. Neural Syst. Rehab. Eng.* 14:30–37.

Muller-Putz, G.R., R. Scherer, G. Pfurtscheller, and R. Rupp. 2005. EEG-based neuroprothesis control: a step towards clinical practice. *Neurosci. Lett.* 382:169–174.

Neat, G.W., D.J. McFarland, C.A. Forneris, and J.R. Wolpaw. 1990. EEG-based brain-to-computer communication: system description. *IEEE Conference in Medicine and Biology*.

Nielson, K.D., A.F. Cabrera, and O.F. Nascimento. 2006. EEG-based BCIz: towards a better control. Brain-computer interface research at Aalborg University. *IEEE Trans. Neural Syst. Rehab. Eng.* 14:202–204.

Pfurtscheller, G. and A. Aranibar. 1977. Event-related cortical desynchronization detected by power measurements of scalp EEG. *Electroencephalogr. Clin. Neurophysiol.* 42:817–826.

Pfurtscheller, G., D. Flotzinger, and J. Kalcher. 1993. Brain-computer interface: a new communication device for handicapped persons. *J. Microcomput. Appl.* 16:293–299.

Pfurtscheller, G. and F.H. Lopes da Silva. 1999 Event-related EEG/MEG synchronization and desynchronization: basic principles. *Clin. Neurophysiol.* 110:1842–1857.

Ramoser, H., J. Muller-Gerking, and G. Pfurtscheller. 2000. Optimal spatial filtering of single trial EEG during imagined hand movement. *IEEE Trans. Rehab. Eng.* 8:441–446.

Schalk, G., D.J. McFarland, T. Hinterberger, N. Birbaumer, and J.R. Wolpaw. 2004. BCI2000: a general purpose brain-computer interface (BCI) system. *IEEE Trans. Biomed. Eng.* 51:1034–1043.

Scherer, R., G.R. Muller, C. Neuper, B. Graimann, and G. Pfurtscheller. 2004. An asynchronously controlled EEG-based virtual keyboard: Improvement of the spelling rate. *IEEE Trans. Biomed. Eng.* 51:979–984.

Schlogl, A., F. Lee, H. Bischof, and G. Pfurtscheller. 2005. Characterization of four-class motor imagery EEG data for the BCI-competition 2005. *J. Neural Eng.* 2(4):L14–L22.

Sellers, E.W. and E. Donchin. 2006. A P300-based brain-computer interface. Initial tests by ALS patients. *Clin. Neurophysiol.* 117:538–548.

Sellers, E.W., D.J. Krusienski, D.J. McFarland, T.M. Vaughan, and J.R. Wolpaw. 2006. A P300 event-related potential brain-computer interface (BCI): the effects of matrix size and inter stimulus interval on performance. *Biol. Psychol.* 73(3):242–252.

Sellers, E.W., D.J. Krusienski, D.J. McFarland, and J.R. Wolpaw. 2007. Noninvasive brain-computer interface research at the Wadsworth Center. In *Towards brain-computer interfacing*. Eds. Dornhege, G., J.R. Millan, T. Hinterberger, D. McFarland, and K-R Muller. Cambridge: MIT Press.

Serby, H., E. Yom-Tov, and G.F. Inbar. 2005. An improved P300-based brain-computer interface. *IEEE Trans. Neural Syst. Rehab. Eng.* 13:89–98.

Sergio, L.E., C. Hamel-Paquet, and J.F. Kalaska. 2005. Motor cortex neural correlates of output kinematics and kinetics during isometric-force and arm-reaching tasks. *J. Neurophysiol.* 94: 2353–2378.

Shenoy, P., M. Krauledat, B. Blankertz, R.P.N. Rao, and K-R. Muller. 2006. Towards adaptive classification for BCI. *J. Neural Eng.* 3:13–23.

Sykacek, P., S. Roberts, and M. Stokes. 2004. Adaptive BCI based on variational Bayesian Kalman filtering: an empirical evaluation. *IEEE Trans. Biomed. Eng.* 51:719–727.

Taylor, D.M., S.I.H. Tillery, and A.B. Schwartz. 2002. Direct cortical control of 3D neuroprosthetic devices. *Science* 296:1829–1832.

Trejo, L.J., R. Rosipal, and B. Mathews. 2006. Brain-computer interfaces for 1D and 2D cursor control: designs using volitional control of the EEG spectrum or steady-state visual evoked potentials. *IEEE Trans. Neural Syst. Rehab. Eng.* 14:225–229.

Vaughan, T.M., D.J. McFarland, G. Schalk, W.A. Sarnacki, D.J. Krusienski, E.W. Sellers, and J.R. Wolpaw. 2006. The Wadsworth BCI research and development program: at home with BCI. *IEEE Trans. Neural Syst. Rehab. Eng.* 14:229–233.

Vaughan, T.M. and J.R. Wolpaw. 2006. The third international meeting on brain-computer interface technology: making a difference. *IEEE Trans. Neural Syst. Rehab. Eng.* 14:126-127.

Vidaurre, C., A. Schlogl, R. Cabeza, R. Scherer, and G. Pfurtscheller. 2006. A fully online adaptive BCI. *IEEE Trans. Biomed. Eng.* 53(6):1214–1219.

Wang, Y., R. Wang, X. Goa, B. Hong, and S. Gao. 2006. A practical VEP-based brain-computer interface. *IEEE Trans. Neural Syst. Rehab. Eng.* 14:234–239.

Wolpaw, J.R., N. Birbaumer, D.J. McFarland, G. Pfurtscheller, and T.M. Vaughan. 2002. Brain-computer interfaces for communication and control. *Clin. Neurophysiol.* 113:767–791.

Wolpaw, J.R., D.J. McFarland, G.W. Neat, and C.A. Forneris. 1991. An EEG-based brain-computer interface for cursor control. *Electroencephalogr. Clin. Neurophysiol.* 78:252–259.

Wolpaw, J.R. and D.J. McFarland. 1994. Multichannel EEG-based brain-computer communication. *Electroencephalogr. Clin. Neurophysiol.* 90:444–449.

Wolpaw, J.R. and D.J. McFarland. 2004. Control of a two-dimensional movement signal by a noninvasive brain-computer interface in humans. *PNAS* 101:17849–17854.

Yom-Tov, E. and G.F. Inbar. 2003. Detection of movement-related potentials from the electroencephalogram for possible use in a brain-computer interface. *Med. Biol. Eng. Comput.* 41:85–89.

CHAPTER 8

Cognitive and Emotional Neuroprostheses

Walid V. Soussou and Theodore W. Berger

INTRODUCTION

The preceding chapters describe BCI applications that restore motor or communication functions of the brain to patients paralyzed by spinal cord injuries or muscular dystrophies. These BCIs extract electrophysiological signals from healthy motor cortices and process them into control commands for computers, robotic machines, or communication devices. The brain can suffer damage directly, however, from genetic disorders or injuries from stroke or disease. Damage to the brain can lead to numerous cognitive impairments, such as memory loss, mood or personality alterations, and even behavioral changes that include motor or communication dysfunction. This chapter presents some of the developments of neuroprostheses that aim to address such cognitive or emotional dysfunction.

A major challenge for cognitive prostheses is that the neural code for their intended tasks is not yet elucidated. Unlike motor cortex signals where neuronal activity is tuned to desired motion direction, speed, and even grip force that can be decoded (Chapter 5), or large EEG components that can be readily classified (Chapter 7), the coding for cognitive processes is still being deciphered. The following cognitive prostheses thus have devised ingenious strategies to overcome or bypass this obstacle in using higher cognitive functions for BCI applications. The main feature of these prostheses is that they extract cognitive state information from neural signals to produce appropriate feedback to the user.

VOLITIONAL PROSTHESES

Among higher level cognitive processes, volition is the will to execute an action and includes several conscious and unconscious functions such as attention, intention, motivation, expectation, and state of being. Volitional prostheses can decode these

T.W. Berger et al., Brain-Computer Interfaces, 109–123.

functions from their appropriate brain regions and combine them with somatomotor-prostheses commands to refine control of robots or machines.

Goal and Intent

Frontal and parietal cortices are involved in planning actions as well as movement execution. Goal information, which can encode intent to reach for or look at an object, has been extracted from neurons in these areas in conjunction with trajectory information, which encodes direction of arm movement. Intended targets were successfully decoded from neural recordings in the parietal reach region (PRR) of monkeys (Musallam et al., 2004). Figure 8.1 shows the spike trains of neurons during a reach trial compared to a brain control trial in which the monkeys only intended to reach for the target, and therefore the spikes do not encode movement information. Fewer PRR neurons were necessary to predict the intended target than would be needed from motor cortex recordings.

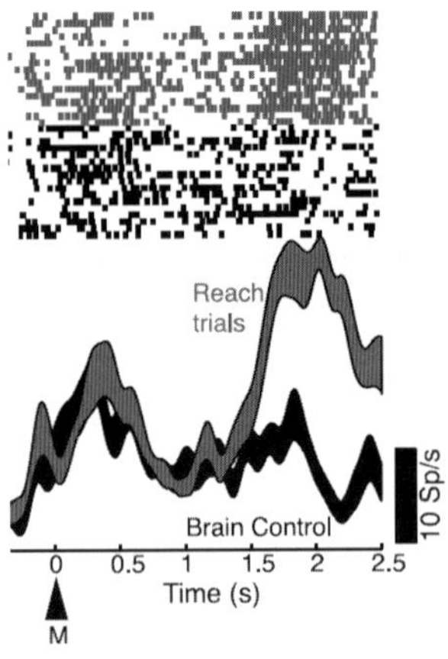

Figure 8.1. Raster plots and poststimulus time histogram of neuronal spike activity in monkey parietal reach region during reaching (red) and brain control (black) trials (Musallam et al., 2004; © American Association for the Advancement of Science).

The limited information content of electroencephalogram (EEG) recordings for robotic control can be compensated for with such compact decoding of intent rather than movement (Millán, 2007). Very high-frequency oscillations in scalp EEG have been demonstrated to enable accurate discrimination of movement intent in a two-class classification task (Gonzalez et al., 2006).

Expected Value

The expected value of an action can also be decoded in the PRRs of monkeys during their brain control tasks (Musallam et al., 2004). Neuron firing was found to increase spatial tuning during tasks when the outcome was preferred, whether in type, probability, or magnitude of reward (Figure 8.2). The expected value could thus be decoded simultaneously with the intended goal from the same neurons. Decoding and classifying these decision variables of expected value could therefore be used to communicate the preferences and motivations of patients using BCI.

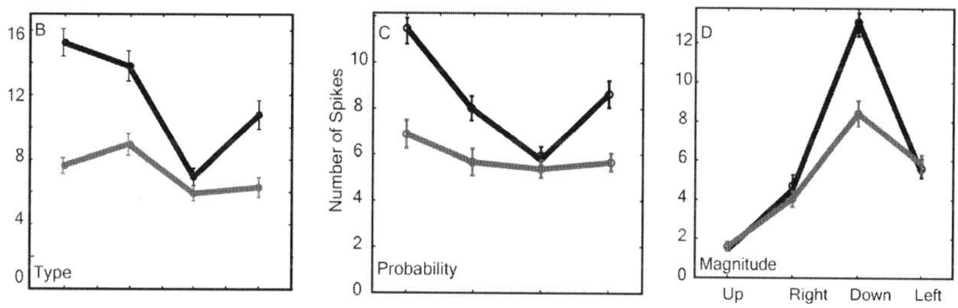

Figure 8.2. Tuning curves of a neuron for preferred (black) and nonpreferred rewards (red). Type: orange juice vs. water; probability: 0.80 vs. 0.40; magnitude: 0.12 vs. 0.05 ml. The down direction is easier to distinguish in the preferred reward conditions (Musallam et al., 2004; © American Association for the Advancement of Science).

Cognitive States

BCI users may have different needs depending on their cognitive states such as awake, alert, attentive, restive, frustrated, planning a movement, or moving. Decoding these states could provide useful information to a BCI-controlled device, such as whether an action was intended or not, or if an error was made, or feedback on the timing of the action.

Cognitive state can be simultaneously decoded with movement or goal information from the same recording electrodes. In monkey eye saccade experiments, local field potentials (LFPs) were used to simultaneously decode the goal (intended direction of movement) and state of the action (planning vs. movement) (Pesaran et al., 2006). LFPs were reported to be better than spike trains for decoding cognitive states of the animal, with the information about state being carried at a different frequency band (0–20 Hz) than the information about goal (25–90 Hz). LFPs from PRR enabled discrimination of five different cognitive states: a baseline state, planning a saccade, executing a saccade, planning an arm reach, and executing an arm reach (Scherberger et al., 2005).

EEG signals also carry information about intent and cognitive states such as errors, alarms, attention, frustration, or confusion. The EU project at Switzerland's IDIAP Research Institute (initially referred to as "Institut Dalle Molle d'Intelligence Artificielle Perceptive") titled Mental Augmentation through determination of Intended Action (MAIA) (Millán, 2007), is harnessing such higher-level cognitive states to fine-tune BCI control and reduce decisionmaking errors. The investigators have been able to discriminate single-trial EEG error potentials generated in response to decoding errors made by the interface (Buttfield et al., 2006). These signals can then be fed back to the BCI to correct mistakes and improve performance. In the future, BCIs should be able to decipher more cognitive parameters, including emotions.

EMOTIONAL COMPUTERS AND ROBOTS

Emotions are high-level cognitive states that encode subjective feelings to a situation or environment. As such, emotions can carry large amounts of information in a compact form. For example, anger or frustration at continued BCI errors could lead to user rejection of a system. An emotionally aware BCI would, however, realize the irritation and adapt by attempting to adjust its output. The ability to decode emotional states could therefore empower BCIs to interact with their users in a state-dependent manner or to express their users' emotions on their behalf.

The Human-Machine Interaction Network on Emotion (HUMAINE) (Schroeder and Cowie, 2007), is a consortium of EU researchers developing systems that can register, model, and influence human emotional states and processes. Their emotion-oriented computing is based on psychobiological investigations of emotion and is designed to interface with human users on an emotional level. Robocasa (Takanishi et al., 2007), a collaboration between the Scuola Superiore Sant'Anna in Italy and Waseda University in Japan, is creating emotional humanoid robots with expressive gestures, capable of expressing several human emotions with face and arms (Figure 8.3) (Miwa et al., 2002). Combining the ability to decode emotions from EEG or other brain recordings with such emotion-oriented computing or emotional robots would enable BCI users to express their emotions or enable the BCI to respond appropriately to the user's emotional state.

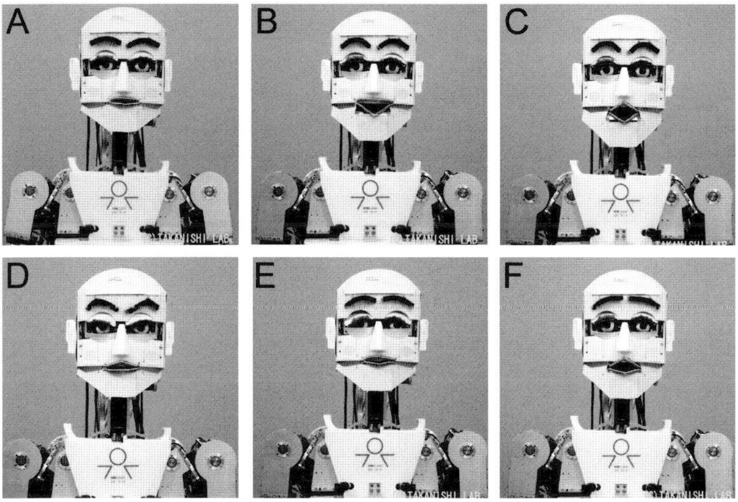

Figure 8.3. Emotional robots (Waseda and SSSA). Six emotions are expressed by the WE-4RII. humainoid robot: A. Neutral expression, B. Happinesss, C. Surprise, D. Anger, E. Sadness, F. Fear (Source: Waseda Head Robot Team, Takanishi Lab: http://www.takanishi.mech.waseda.ac.jp/research/we/we-4rII/index.htm).

MEMORY PROSTHESES

There is only one major attempt that can be identified to develop a neural prosthesis for replacement of memory function lost due to central brain region damage or disease. That project first started at the University of Southern California (USC) and now involves collaborative efforts with Wake Forest University (WFU) and the University of Kentucky (UK). The project focuses on the hippocampus, the part of the brain responsible for long-term memories. Compromised structural and functional properties of the hippocampus are consistently associated with stroke, epilepsy, and Alzheimer's disease. Patients with severely damaged hippocampi are incapable of forming new long-term memories, leaving them highly dependent on family or health staff to manage daily life.

The goal is to replace damaged regions of the hippocampus with microchip-based systems that mimic the functional properties of the lost tissue (Berger et al., 2001). The replacement silicon systems would have functional properties specific to those of the damaged hippocampal cells, and would both receive as inputs and send as outputs electrical activity to regions of the brain with which the hippocampus previously communicated (Figure 8.4).

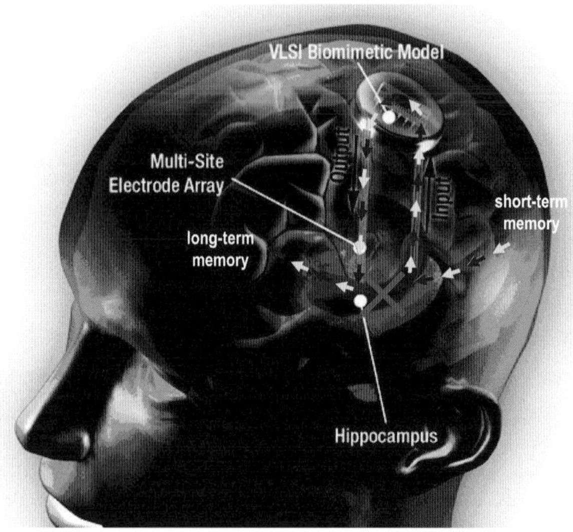

Figure 8.4. Concept for a cortical prosthesis that utilizes a biomimetic model of hippocampal function and bypasses damaged regions of that structure to restore long-term memory formation.

Specifically, multisite electrode arrays would record activity of neuronal populations that normally provide input to the damaged region and transmit that information to the "biomimetic" prosthetic device. A second set of multisite electrode arrays would transmit the output from the biomimetic device to brain regions that normally receive efferents from the damaged region, and as electrical stimulation, would drive those target regions to the required output state. Thus, the prosthesis would replace the computational function of the damaged region of hippocampus and restore the transmission of that computational result to appropriate regions of the brain.

Proof of Concept in the Hippocampal Slice

Given the complexity of this task, the first step taken was to attempt a "proof of concept" in a reduced preparation of the rat hippocampus—the hippocampal slice. The basic objective is illustrated in Figure 8.5. The major intrinsic circuitry of the hippocampus consists of an excitatory cascade of the dentate, CA3, and CA1 subregions (dentate→CA3→CA1) (Figure 8.5A) and is maintained in a transverse slice preparation. Our proof-of-concept hippocampal prosthesis consists of (i) surgically eliminating the CA3 subregion; (ii) replacing the biological CA3 with a VLSI-based model of the nonlinear dynamics of CA3 (Figure 8.5B and C); and

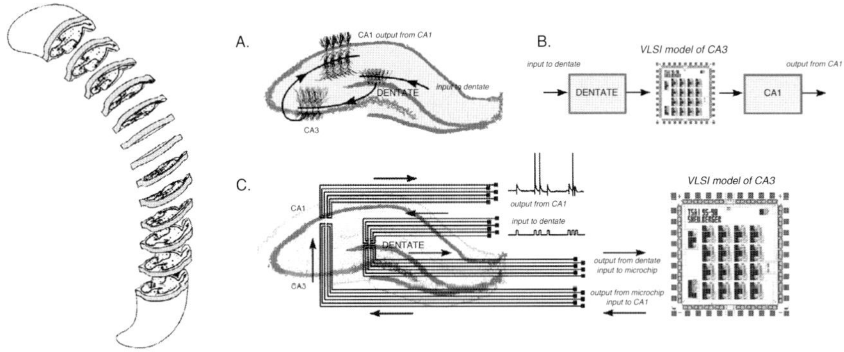

Figure 8.5. Left: illustration of the rat hippocampus and the orientation of slices prepared from the hippocampus. Right: strategy for replacing subfield CA3 of the hippocampus with a VLSI-based model of CA3 nonlinear dynamics.

(iii) through a specially designed multisite electrode array, transmitting dentate output to the VLSI model and transferring VLSI-model output to the inputs of CA1 (Figure 8.5C). The definition of a successful implementation of the prosthesis is the propagation of temporal patterns of activity from dentate→VLSI model→CA1, which reproduces what is observed experimentally in the biological dentate→CA3→CA1 circuit.

The USC-WFU-UK group was able to accomplish all of the steps outlined above. One important point is that the core of the prosthesis is a nonlinear dynamic model of CA3. This model utilizes a combined experimental-theoretical approach to capture the input-output properties of the neural system studied. An important assumption is that information is carried in the time between spikes, i.e., in a temporal pattern, so that the response of a given neuron depends not just on the most current input, but also on the time since prior inputs. For characterization of the hippocampus, the USC-WFU-UK investigators electrically stimulated the inputs to the dentate with a random interval impulse train and simultaneously recorded outputs from the dentate, CA3, and CA1. Both the inputs and the outputs of CA3 were recorded, and their relationship was modeled using a Volterra functional power series approach (Berger et al., 2005). The result is a model that allows the output of CA3 to be accurately predicted for any arbitrary CA3 input (sequence of impulse intervals, or temporal pattern). The USC-WFU-UK group went on to show that, in response to random interval impulse stimulation of dentate input, the output of CA1 was nearly identical for normal, intact slices and "hybrid" slices in which the CA3 region was replaced with a hardware model of CA3 dynamics (VLSI field programmable gate array [FPGA]) (see Figure 8.6).

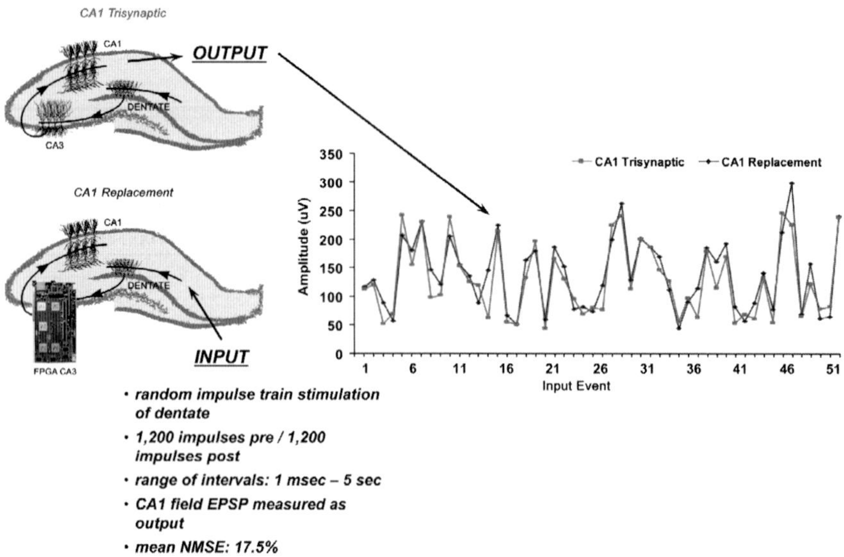

Figure 8.6. Data showing the amplitudes of population EPSPs (excitatory postsynaptic potentials) recorded from the molecular layer of CA1 in response to electrical stimulation of inputs to the dentate gyrus. Characteristics of the stimulation are shown. After formulating a nonlinear input-output model for CA3, CA3 was removed surgically and replaced with an FPGA-based input-output. The FPGA hardware and the slice communicated bidirectionally via the multisite electrode arrays illustrated in Figure 8.5. Population EPSP amplitudes are shown here for 50 of the 1,200 responses evoked in one random train of stimulation. Inter-impulse intervals varied from 1 milliseconds to 5 seconds, but these intervals are not shown; each stimulation and its associated amplitudes are represented simply as "input events."

A Hippocampal Neural Prosthesis for the Behaving Animal

With this proof of concept completed, the group became focused on developing a hippocampal prosthesis for the behaving rat. This essentially requires extending the input-output model to multiple slices, or circuits, along the longitudinal axis of hippocampus (Figure 8.5, Left). Achieving this goal also requires developing the input-output model from recordings of population single-cell activity (extracellular "spikes") in the behaving rat as the animal performs a memory task that demands normal hippocampal function.

To this end, we extended our approach to *in vivo* multielectrode recording during a "delayed non-match-to-sample" memory task in the rat. During this task, a rat is presented with one of two "sample" stimuli; the rat must remember that stimulus and provide evidence of that memory by responding after a variable delay period (0–60 seconds) to the *opposite* stimulus of the sample. Multiple single-cell recordings were obtained from an array of electrodes in CA3 and a second array of electrodes in CA1. The modeling task was to determine the nonlinear input-output properties for the CA3 (input)—CA1 (output) population data, where both the input and the output are multiple-point processes. In other words, the goal is for the model to predict how the activity of each output neuron depends on (i) the temporal pattern of activity of each of the input neurons, and on (ii) the *interactions* between the temporal patterns of the input spike streams.

Again, the USC-WFU-UK group successfully developed a multiple-input, multiple-output model for transformation of population CA3 to population CA1 spatiotemporal patterns (Song et al., 2007). Figure 8.7 shows one such result, in this case for a 16-input, 7-output neuron recording. Because there were 7-output neurons, 7 multiple-input, single-output models were constructed.

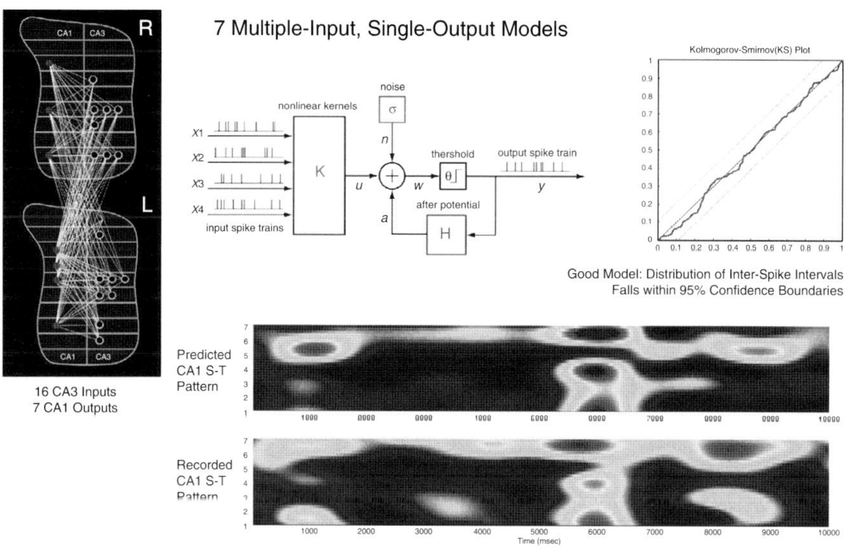

Figure 8.7. A 16-input, 7-output neuron recording.

Each model included a multiple-input, third-order kernel component representing the effects of mechanisms of synaptic transmission and dendritic integration (K), the somatic membrane potential (u), a noise term to represent spontaneous activity (σ), the spike threshold (Θ), and a spike-triggered after-potential (H). Results showed that the model faithfully predicts the spatiotemporal pattern of action potentials in CA1 of the behaving animal based on the spatiotemporal pattern of action potentials in CA3: see color plots in Figure 8.7 and matching distributions of interspike intervals for observed and predicted data sets (upper-right plot).

With this established, the USC-WFU-UK group is now developing a preparation in which the CA3-CA1 connections are transected. A prosthesis based on the above model then will be used in an attempt to reinstate memory function in this task.

NEUROFEEDBACK

Neurofeedback is operant conditioning of brain activity with sensory feedback instead of reward-based training to guide control. In most neurofeedback paradigms, subjects are trained with visual or auditory feedback of their EEG, or more recently, with fMRI BOLD (blood-oxygen-level-dependent) signals, to control their brain activity to a desired state. Early studies in the late-1960s demonstrated the ability of animals (and later humans) to regulate their EEG signals in specific frequency bands (see Sterman and Egner, 2006 for a review). A controllable rhythm with a spectral peak between 12 and 14 Hz was first distinguished and termed "sensorimotor rhythm" (SMR) due to its mapping to sensorimotor cortex and correlation with a drop in muscle tone and immobility. Non-oscillatory EEG activity has also been shown to be amenable to volitional control; slow cortical potentials (SCPs <2 Hz) can be modulated through neurofeedback training (Elbert et al., 1980). Both SMR and SCP reflect states of neuronal excitability, and their control can be used to treat conditions marked by neuronal hyperexcitability such as epilepsy and attention deficit hyperactivity disorder (ADHD). Neurofeedback devices can therefore act as noninvasive neural prostheses that help patients control their brain activity to maintain stable neural states and socially functional behavior.

Neurofeedback for Epilepsy

Epilepsy is a neural hyperactivity disease, where the excitation threshold of neurons is decreased to the point where they fire in synchronous and often oscillatory bursts, leading to seizures. Over the last 30 years, many reports described significant reduction in seizures with neurofeedback therapy that can last over a year after treatment (Monderer et al., 2002; Walker and Kozlowski, 2005; Egner and Sterman, 2006; Sterman and Egner, 2006). Epilepsy reduction by SMR training is attributed to an increased threshold of excitation in thalamocortical somatosensory and somatomotor pathways, and its endurance beyond training is suggested to be consequent to a form of long-term potentiation that is consolidated by post-synchronization reinforcement oscillations (Sterman and Egner, 2006).

Dr. Niels Birbaumer's group at the University of Tübingen demonstrated that SCP control could also be used to reduce seizures in epileptic patients (Rockstroh et al., 1993; Kotchoubey et al., 1996). In a combined EEG and fMRI study, a reduction in BOLD signal was spatially and temporally correlated to a positive SCP, reflecting a state of decreased activation (Strehl et al., 2006a). In their paradigm, patients are trained to produce positive SCPs during neurofeedback training, and then transfer their acquired skills to non-guided sessions. The neurofeedback device is thus used as a training tool before the skill is transferred to real-life situations. Epileptic patients who received SCP self-regulation training showed a decrease in epileptic frequency comparable to a matched group who received anti-convulsive drugs (Kotchoubey et al., 2001). Dr. Birbaumer's group is combining behavioral therapy treatment to teach temporal lobe epilepsy patients to increase their aura sensitivity, with neurofeedback to control SCP and prevent seizure-reinforcing contingencies. There is considerable variability in the responses of patients to SCP neurofeedback; however, in the group's most recent experiments, one half of the patients showed significant decrease of seizure frequency with neurofeedback.

Neurofeedback for ADHD

ADHD is a psychiatric disorder that affects 5% of children before the age of 19 and causes them to have a short attention span and be hyperactive, thereby affecting their scholastic performance and social life. Neurofeedback's reduction in brain hyperexcitability in epileptic patients appeared to simultaneously reduce ADHD symptoms. These early observations led Lubar and Shouse (1976) to investigate EEG biofeedback training on a hyperkinetic child. They reported an increase in motor inhibition when the child succeeded in producing SMR of 12–14 Hz without 4–7 Hz activity.

Since then, several studies have reported on the efficacy of neurofeedback for treating children with ADHD. The spectral power of EEG activity in ADHD children is characterized by elevated Theta rhythms (4–7 Hz) and reduced Alpha (8–12 Hz) and Beta (12–22 Hz) rhythms. Theta-band suppression and SMR and Beta rhythms enhancement through neurofeedback training were found to have ameliorative effects comparable to stimulant treatments with methylphenidate on several attentional and behavioral tests for children with ADHD (Fuchs et al., 2003). SCPs are also affected in ADHD patients, exhibiting reduced negativity during task anticipation. Children with ADHD were successfully trained to control their negative SCP with neurofeedback (Figure 8.8), and were able to transfer that control to non-training conditions (Strehl et al., 2006b).

Neurofeedback training led to increased intelligence scores and academic achievement, as well as to reduced hyperactivity, impulsivity, and frequency of conflicts at home. These behavioral effects are reported to last as long as six months after training.

Neurofeedback with SCP control is reported to produce the same behavioral amelioration as control of Theta/Beta ratio (Leins et al., 2007). Figure 8.9 shows the handheld device's user interface with which patients practiced to control their SCPs or Theta/Beta ratios in daily environments before applying this control to other tasks such as doing homework.

On the commercial side, among other for-profit companies, CyberLearning Technology, a NASA technology spinoff based in San Marcos, California, is marketing neurofeedback games through its S.M.A.R.T. BrainGames subsidiary to help children with ADHD.

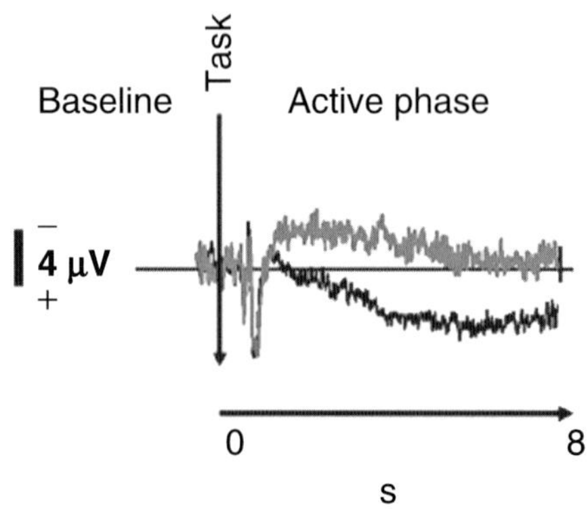

Figure 8.8. Mean EEG traces of SCP during neurofeedback sessions. The upper (gray) trace shows a negative (activation) potential, and the lower trace (black) shows a positive (deactivation) response (Strehl et al., 2006b; © American Academy of Pediatrics).

Figure 8.9. Neurofeedback task screen display. The upper panels show a task where the objective is the highlighted upper rectangle, and the subject must create negative SCP potentials to move the ball upward while it traverses the screen from left to the right. A successful trial is rewarded with a smiley and a reinforcing tone. The lower panel shows a transfer session, where the user must create a similar negative potential, without the visual feedback (Strehl et al., 2006b; © American Academy of Pediatrics).

Neurofeedback for Control of Emotions and Antisocial Personality Disorders

EEG handheld devices work well to train certain cortical signals, but they are not well suited to localize specific brain areas. Real-time fMRI neurofeedback is, however, being used to self-regulate local brain activity and therefore control associated functions. For example, regulation of insula affects subjective rating of emotional response images, while regulation of amygdala—which often shows hippocampal components—affects emotional recall. Thus, self-regulation of BOLD signal can be used to affect emotional perception.

Control of BOLD signals to self-induce cognitive changes is now being tested for treatment of personality disorders. Psychopaths and social phobics respond differently to aversive stimuli than healthy people, and their brain activity patterns differ. A hypoactive frontolimbic circuit correlates with psychopathic behavior, whereas an overactive frontolimbic system underlies social fear (McCloskey et al., 2005). Dr. Birbaumer's group is currently training ex-prisoners to self-regulate their prefrontal cortex BOLD signal to enable them to modify their criminal behavior. The success of these neurofeedback experiments may present applications to the treatment of other personality disorders such as obsessive compulsion and schizophrenia.

SUMMARY AND CONCLUSIONS

Cognitive prostheses present significant enhancements to current somatomotor BCIs as well as several new treatments for brain injury. High-level cognitive processes encode information very densely. For example, fewer neurons are required to encode intended-reach targets than arm-movement trajectories. Decoding such goal or intent signals therefore has the potential of reducing the computational load necessary to control a robot arm, while movement details could still be decoded from neurons or coded into the robot software and hardware. In addition, decoding cognitive states can provide feedback on errors, user motivation, or emotional state. This information could be useful for error correction, adjusting to needs and moods of users, or enabling emotional expression.

Furthermore, many cognitive functions are not linear processes whose neural coding is elucidated. The demonstration of a memory prosthesis that can replace the functionality of a damaged hippocampus therefore provides a powerful implantable solution that might generalize to other regions and functions. Similarly, albeit noninvasively, neurofeedback presents a generic computer interface that is enabling patients to self-regulate their brain states to control epileptic seizures, ADHD, and even personality or emotional states.

It is worth mentioning that also under development are deep-brain stimulation (DBS) treatments for some cognitive disorders that affect mood and behavior, such as depression (Mayberg et al., 2005) and obsessive-compulsive disorders (Abelson et al., 2005); however, these and other DBS applications are outside the scope of this BCI study. The main difference between BCI and DBS is that the latter deliver fixed stimulation paradigms to activate or inactivate certain pathways in specific brain regions, whereas in the former, the neural prostheses record specific neural activity and respond with appropriate feedback.

In conclusion, cognitive prostheses are currently demonstrating their usefulness as BCIs. They can complement existing somatomotor prostheses by providing higher-level command signals, or information on users' cognitive or emotional states. Moreover, neurofeedback applications enable patients to regain control over their brains' activity, and memory prostheses can replace lost hippocampal functionality.

REFERENCES

Abelson, J.L., G C. Curtis, O. Sagher, R C. Albucher, M. Harrigan, S.F. Taylor, B. Martis, and B. Giordani. 2005. Deep brain stimulation for refractory obsessive-compulsive disorder. *Biol. Psychiatr.* 57(5):510–516.

Berger, T.W., A. Ahuja, S.H. Courellis, S.A. Deadwyler, G. Erinjippurath, G.A. Gerhardt, G. Gholmieh, J.J. Granacki, R.E. Hampson, M-C. Hsiao, J. LaCoss, V.Z. Marmarelis, P. Nasiatka, V. Srinivasan, D. Song, A.R. Jr. Tanguay, and J.D. Wills. 2005. Hippocampal-cortical neural prostheses to restore lost cognitive function. *IEEE EMBS Special Issue: Toward Biomimetic Microelectronics as Neural Prostheses* 24:30–44.

Berger, T.W., M. Baudry, R.D. Brinton, J.S. Liaw, V.Z. Marmarelis, A.Y. Park, B.J. Sheu, and A.R. Tanguay, Jr. 2001. Brain-implantable biomimetic electronics as the next era in neural prosthetics. *Proc. IEEE* 89(7):993–1012.

Buttfield, A., P.W. Ferrez, and J. del R. Millán. 2006. Towards a robust BCI: error potentials and online learning. *IEEE Trans. Neural Syst. Rehab. Eng.* 14(2):164–168.

Egner, T., and M.B. Sterman. 2006. Neurofeedback treatment of epilepsy: from basic rationale to practical application. *Expert Rev. Neurother.* 6(2):247–57.

Elbert, T., B. Rockstroh, W. Lutzenberger, and N. Birbaumer. 1980. Biofeedback of slow cortical potentials. *J. Electroencephalogr. Clin. Neurophysiol.* 48(3):293–301.

Fuchs, T., N. Birbaumer, W. Lutzenberger, J.H. Gruzelier, and J. Kaiser. 2003. Neurofeedback treatment for attention-deficit/hyperactivity disorder in children: a comparison with methylphenidate. *Appl. Psychophysiol. Biofeedback* 28(1):1–12.

Gonzalez, S.L., R. Grave de Peralta, G. Thut, J. del R. Millán, P. Morier, and T. Landis. 2006. Very high frequency oscillations (VHFO) as a predictor of movement intentions. *Neuroimage* 32(1):170–179.

Kotchoubey, B., D. Schneider, H. Schleichert, U. Strehl, C. Uhlmann, V. Blankenhorn, W. Froscher, and N. Birbaumer. 1996. Self-regulation of slow cortical potentials in epilepsy: a retrial with analysis of influencing factors. *Epilepsy Res.* 25(3):269–276.

Kotchoubey, B., U. Strehl, C. Uhlmann, S. Holzapfel, M. Konig, W. Froscher, V. Blankenhorn, and N. Birbaumer. 2001. Modification of slow cortical potentials in patients with refractory epilepsy: a controlled outcome study. *Epilepsia* 42(3):406–416.

Leins, U., G. Goth, T. Hinterberger, C. Klinger, N. Rumpf, and U. Strehl. 2007. Neurofeedback for children with ADHD: a comparison of SCP and Theta/Beta protocols. *Appl. Psychophysiol. Biofeedback* 32(2):73–88.

Lubar, J.F., and M.N. Shouse. 1976. EEG and behavioral changes in a hyperkinetic child concurrent with training of the sensorimotor rhythm (SMR): a preliminary report. *Biofeedback Self-Regul.* 1(3):293–306.

Mayberg, H.S., A.M. Lozano, V. Voon, H.E. McNeely, D. Seminowicz, C. Hamani, J.M. Schwalb, and S.H. Kennedy. 2005. Deep brain stimulation for treatment-resistant depression. *Neuron* 45(5):651–660.

McCloskey, M.S., K.L. Phan, and E.F. Coccaro. 2005. Neuroimaging and personality disorders. *Curr. Psychiatr. Rep.* 7(1):65–72.

Millán, J.D.R. 2007. Mental augmentation through determination of intended action (MAIA) Website at http://www.maia-project.org/.

Miwa, H., T. Okuchi, H. Takanobu, and A. Takanishi. 2002. Development of a new human-like head robot WE-4. Proceedings, IEEE/RSJ International Conference on Intelligent Robots and Systems 3:2443- 2448.

Monderer, R.S., D.M. Harrison, and S.R. Haut. 2002. Neurofeedback and epilepsy. *Epilepsy Behav.* 3(3):214–218.

Musallam, S., B.D. Corneil, B. Greger, H. Scherberger, and R.A. Andersen. 2004. Cognitive control signals for neural prosthetics. *Science* 305(5681):258–262.

Pesaran, B., S. Musallam, and R.A. Andersen. 2006. Cognitive neural prosthetics. *Curr. Biol.* 16(3):R77–80.

Rockstroh, B., T. Elbert, N. Birbaumer, P. Wolf, A. Duchting-Roth, M. Reker, I. Daum, W. Lutzenberger, and J. Dichgans. 1993. Cortical self-regulation in patients with epilepsies. *Epilepsy Res.* 14(1):63–72.

Scherberger, H., M.R. Jarvis, and R.A. Andersen. 2005. Cortical local field potential encodes movement intentions in the posterior parietal cortex. *Neuron* 46(2):347–354.

Schroeder, M., and R. Cowie. 2007. HUMAINE Emotion Research 2007. Website at http://emotion-research.net.

Song, D., R.H. Chan, V.Z. Marmarelis, R.E. Hampson, S.A. Deadwyler, and T.W. Berger. 2007. Nonlinear dynamic modeling of spike-train transformations for hippocampal-cortical prostheses. *IEEE Trans. Biomed. Eng.* 54(6, Pt. 1):1053–1066.

Sterman, M.B., and T. Egner. 2006. Foundation and practice of neurofeedback for the treatment of epilepsy. *Appl. Psychophysiol. Biofeedback* 31(1):21–35.

Strehl, U., U. Leins, G. Goth, C. Klinger, T. Hinterberger, and N. Birbaumer. 2006a. Self-regulation of slow cortical potentials: A new treatment for children with attention-deficit/hyperactivity disorder. *Pediatrics* 118(5):e1530–1540.

Strehl, U., T. Trevorrow, R. Veit, T. Hinterberger, B. Kotchoubey, M. Erb, and N. Birbaumer. 2006b. Deactivation of brain areas during self-regulation of slow cortical potentials in seizure patients. *Appl. Psychophysiol. Biofeedback* 31(1):85–94.

Takanishi, A., S. Hashimoto, P. Dario, and M.C. Carrozza. 2007. Robocasa Project. Website at http://www.robocasa.net

Walker, J.E. and G.P. Kozlowski. 2005. Neurofeedback treatment of epilepsy. *Child Adolesc. Psychiatr. Clin. N. Am.* 14(1):163–176, viii.

CHAPTER 9

Research Organization-Funding, Translation-Commercialization, and Education-Training Issues

Theodore W. Berger

BCI RESEARCH ORGANIZATION AND FUNDING

Europe

The WTEC panel was highly impressed with Europe's large-scope and long-term commitment to BCI development through multicountry, multiuniversity, interdisciplinary teams. This level of commitment clearly has its roots in the European Community (EC) 6th Framework Program philosophy. The Information Technology Society (IST) in the 6th Framework (FP6) is characterized as follows:

> The focus of IST in FP6 is on the future generation of technologies in which computers and networks will be integrated into the everyday environment, rendering accessible a multitude of services and applications through easy-to-use human interfaces. This vision of "ambient intelligence" places the user at the centre of future developments for an inclusive knowledge-based society for all.
>
> This research effort will therefore reinforce and complement the [European] objectives and look beyond them to the 2010 goals of the Union [the i2010 initiative] of bringing IST applications and services to everyone, every home, every school, and to all businesses (European Commission 2005).

To help reach this vision, the EC is supporting multiple "networks of excellence" that bring together researchers of different countries and diverse backgrounds. The WTEC panel could visit only a subset of the "nodes" in these networks, but it became evident to panelists how instrumental these networks are in promoting large-scale BCI research and in generating momentum toward BCI goals. The panel found substantial funding in Europe intended to capitalize on an infrastructure of expertise for high-risk, paradigm-shift, long-term, interdisciplinary research on BCIs and fundamental research related to BCIs. In general, the EU-sponsored

T.W. Berger et al., Brain-Computer Interfaces, 125–140.

interdisciplinary research programs that the WTEC panel observed were characterized by (1) high-level vision with collective credibility (top-to-bottom buy-in), (2) levels of support appropriate to the vision, and (3) vision-generated mechanisms for implementation.

European BCI Research Programs

Examples follow of the EU-sponsored interdisciplinary research programs that the WTEC panel saw.

HUMAINE. The Human-Machine Interaction Network on Emotion (HUMAINE), funded by the EU Information Society Technology program, aims toward the development of systems that can register, model, and/or influence human emotional and emotion-related states and processes. (See http://emotion-research.net/).

BrainNet. BrainNet Europe is a network of excellence funded by the European Commission in the 6th Framework Program Life Science. It consists of 19 established brain banks across Europe and is coordinated by the Centre for Neuropathology and Prion Research at Ludwig-Maximilians-University, Munich, Germany. Its main goal is the collection and distribution of well-characterized, high-quality, post mortem brain tissue for basic research in neuroscience. (See http://www.brainnet-europe.org/).

Bernstein Centers for Computational Neuroscience. Germany's ministry of research and education, *Bundesministerium für Bildung und Forschung* (BMBF), has established four centers to integrate neurobiology, cognitive science, systems biology, and information sciences to advance brain research. (See http://www.bccn-berlin.de/).

The *Fraunhofer-Gesellschaft*. Fifty-eight Fraunhofer Institutes at over forty different locations throughout Germany undertake applied research of direct utility to private and public enterprise and of wide benefit to society. Ninety percent of its annual research budget of over one billion euros is generated through contract research. (See http://www.fraunhofer.de/fhg/EN/).

EURON. The EUropean RObotics research Network (EURON) consists of European robotics groups and resources in research, industry, and education joined by a common interest in working to make better robots. According to the member map on the EURON website (see Figure 9.1), the organization has some 210 members widely spread out in 28 countries of Europe. (See http://www.euron.org/.)

MAIA. Mental Augmentation through determination of Intended Action (MAIA) is a project for brain-wave control of robots involving multiple European universities and institutes: IDIAP, Katholieke Universiteit Leuven, University Hospital of Geneva, Fondazione Santa Lucia (Rome), and Helsinki University of Technology. (See http://www.maia-project.org/).

BACS. Bayesian Approach to Cognitive Systems (BACS) is an integrated project under the EC Sixth Framework Program that has been allocated

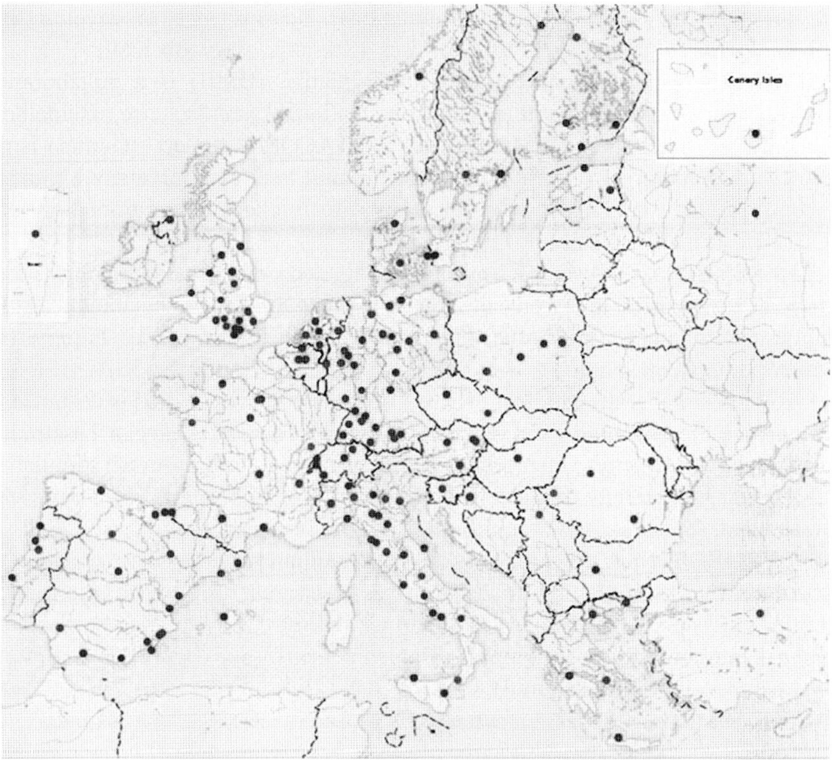

Figure 9.1. Locations of EURON members (EURON n.d.).

€7.5 million in funding. It brings together researchers and commercial companies working on artificial perception systems to model neuronal functions and cognitive processes, to optimize existing learning algorithms, and to realize intelligent artificial systems. (See http://www.bacs.ethz.ch/).

Cyberhand. Cyberhand is a project funded by the EU Future Emerging Technology Program to develop a hierarchical, distributed-control, multiple-degrees-of-freedom robotic hand for replacement of lost limbs. The hand is designed to respond to signals from the human nervous system. It is included in the DARPA Revolutionizing Prosthetics program. (See http://www.cyberhand.org/).

Blue Brain Project. The Blue Brain Project is a massive cooperative project of EPFL (*Écoles Polytechniques Fédérale de Lausanne*, Switzerland) and IBM. It involves state-of-the-art experimental, theoretical, modeling, database, computational, and visual display technologies to realize a biologically based representation of neocortical neurons, microcircuitry, and systems-level structure and function using IBM's Blue Gene supercomputer. (See http://bluebrain.epfl.ch/).

BBCI. The Berlin Brain Computer Interface (BBCI) project is a collaboration between the *Fraunhofer-Institut für Rechnerarchitektur und Softwaretechnik* (FIRST) Institute, Charité University of Medicine, Technical Institute of Berlin, and the Bernstein Institute for Computational Neuroscience to develop BCI technology for commercial and medical uses. (See http://www.bbci.de/).

BMII. The Brain Machine Interfacing Initiative (BMII) is a collaboration between the Bernstein Institute for Computational Neuroscience, Heidelberg Academy of Sciences and Humanities, METACOMP project (German-Israel Project Cooperation), University Hospital Freiburg, and the University Klinikum Freiburg to study neural dynamics in relation to fundamental neurobiology and BCIs. (See http://www.bmi.uni-freiburg.de/).

PRESENCCIA. This is a €7-million EC-funded collaboration among fifteen different laboratories in seven countries for the purpose of developing virtual reality environments with substantial BCI applications. (See http://www.presenccia.org/).

GRIP. A collaborative project of five European countries, GRIP is designed to demonstrate the feasibility of a regeneration-type of neural connector based on a micromachined structure incorporating through-holes for improved stimulation and recording selectivity and mechanical stability for FES control of a paralyzed human hand. (See http://www-ti.informatik.uni-tuebingen.de/~grip/).

NEUROBOTICS. This is a 2004–2008 €6.7 million project under the European Sixth Framework Programme focused on basic research fusing neuroscience and robotics to design, develop, and test tele-operated robotic systems to help restore personal autonomy to sensory-motor-disabled persons. (See Figure 9.2 and http://www.neurobotics.org/index.html).

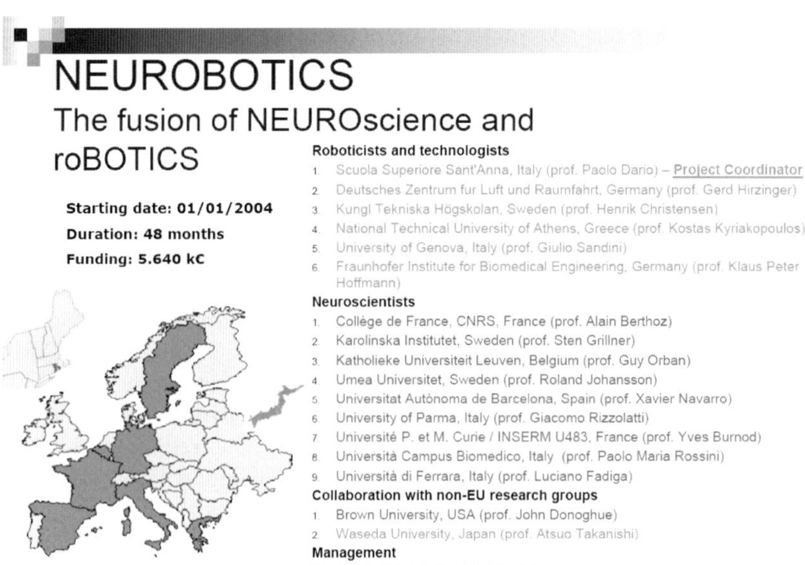

Figure 9.2. The NEUROBOTICS program.

Components of the European Model

The WTEC BCI panel concluded that the success of the European research model is attributable to the components listed in Figure 9.3. The panelists felt there were several components of the European model for research that could benefit the U.S. system, including commitments to (1) long-term goals, (2) large-scale funding, (3) high-risk projects, and (4) fundamental research. In the United States, it is primarily DARPA and the NSF Engineering Research Centers (ERCs) that support multidisciplinary, high-risk, visionary projects. The time scale of DARPA projects (18-month deliverables) is shorter than optimal; however, the administrative load for NSF ERCs can be overwhelming, and the level of NSF support per investigator is minimal given the challenges of BCIs and the high level of expected outcomes. NIH centers and program projects fund at too low a level to compete with the scope of EU Programs. The United States needs to reexamine its organizational and support mechanisms if U.S. BCI research is to advance significantly.

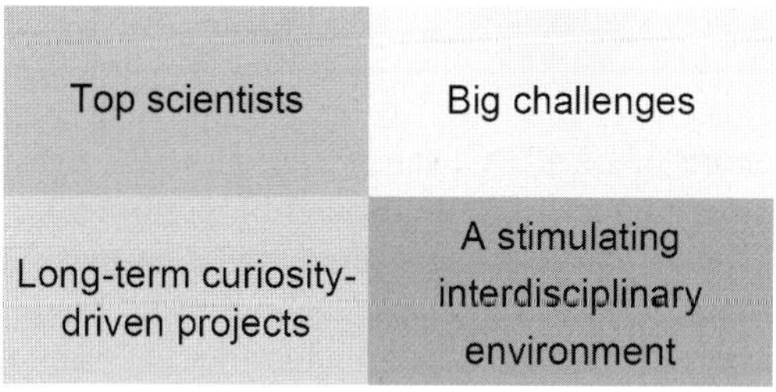

Figure 9.3. Components of the European research model.

In addition to EU-sponsored research, European BCI laboratories can take advantage of national support mechanisms as well as local, regional support. The level of national and regional support for BCI research was particularly strong in Germany, where the WTEC panel saw a highly sophisticated level of hierarchically organized support systems at a variety of levels that ranged from EU, to national, to regional, including an integration of government, academic, and industrial support. In general, the WTEC panel found a high level of commitment to research ranging from basic to applied, through to commercialization, both in the specific field of BCIs and in fundamental science and engineering fields relevant to BCIs.

Asia

BCI research in China, Japan, and the rest of Asia is in its infancy, so direct comparisons with BCI research programs in North America and Europe would be unrevealing. Nevertheless, BCI research in Asia must be considered in the context of a massive investment in the biological, engineering, and medical sciences by China and Southeast Asian countries. As a result of this investment, the overall scope and magnitude of BCI research in Asia is impressive. Many of the European manufacturers of BCI-related equipment and software informed the WTEC panel that their fastest growing markets were in China and Southeast Asia. Multiple BCI-related conferences in recent years have been sponsored in China. Moreover, in the latest international BCI competitions, more than half of the top-ten finishers have been from Asian institutions. Japan continues to forge new frontiers in robotics and is now beginning new BCI-directed research for brain-controlled robotics using the latest in mathematics and imaging technologies.

China

Although BCI research in China started only within the last ten years, it is already substantial in scope; there are many more BCI labs in China than those visited by the WTEC panel. However, in contrast to Europe, where EU programs have promoted and maintained strong associations between BCI labs, few such mechanisms exist in China; thus, the many BCI labs in China are still independent and have not benefited from the synergy of mutual interaction. Despite the "distributed" nature of most BCI labs in China, BCI algorithm development in China already leads the field, as evidenced by performance in international BCI competitions. WTEC panelists considered the robustness of BCI systems they witnessed in China, such as Dr. Gao's laboratory at Tsinghua University, to be impressive. Current BCI research in China is focused primarily on low-cost, low-technology solutions to BCI needs, likely a reflection of socioeconomic demands; given the size of the Chinese population and the still large percentage of that population living in low-income, rural areas, there is a need for large numbers of low-cost BCIs requiring minimal technical support.

The WTEC panel believes it was witnessing the beginnings of organized, collective BCI programs in China. This was particularly evident in Shanghai at the Shanghai Jiao-Tong University Institute of Laser Medicine and Biophotonics. The institute was a remarkable facility, complete with state-of-the-art infrastructure for neuroscience, engineering, and computing research dedicated to developing biomedical technologies, including BCIs. The facilities included *vivaria* and animal surgery suites for invasive brain studies. Working relationships with hospitals and patient populations for clinical studies were already established. The arrival of the WTEC panel was coordinated with an institute-sponsored symposium on BCIs held in a newly built, modern auditorium; approximately 75–100 faculty and students participated. The WTEC panel heard presentations from approximately a dozen faculty members whose laboratories were actively developing BCIs. The institute is only one part of a newly constructed, multibuilding campus consisting of multiple laboratories and facilities dedicated to scientific and medical research.

Likewise, at Tsinghua University in Beijing, the WTEC panel toured a brand new facility for biomedical engineering that was state-of-the-art and still being outfitted with new laboratories and equipment at least partially focused on BCIs. At Huazhong University of Science and Technology in Wuhan, the WTEC panel met with high-level administrators for a discussion about how the future success of universities depends on interdisciplinary research cutting across the physical sciences, engineering, and medicine. Administrators at Huazhong felt that BCI research represents a key example of such an interdisciplinary effort, and their plan was to invest in BCI research so that it could serve as a cutting-edge model.

It was evident to the WTEC panel that China was rapidly moving from an almost exclusive focus on noninvasive BCIs to invasive research platforms that will enable "systems-level" solutions evolving from fundamental studies of brain function. The laboratories at Tsinghua University were in the process of initiating invasive rat brain studies, including the design of silicon-based, multisite electrode arrays. Huazhong University in Wuhan was already developing and applying a novel, multisite, indwelling flexible electrode technology for epidural spinal cord stimulation. The technology was being used clinically to assist paralyzed human patients with spinal cord damage and for preclinical studies using a spinal cat preparation. In Shanghai at East China Normal University, the WTEC panel saw multisite electrophysiological recordings from the behaving mouse utilizing technologies and facilities that were as advanced as any in the world. Thus, in contrast to Europe and Japan where BCI research will remain noninvasive for the foreseeable future, China may become North America's partner in pushing the envelope with respect to invasive BCI paradigms and invasive fundamental research for BCI development.

Japan

BCI research in Japan should be evaluated within a context very different from that of China. The critical factors for understanding BCI research in Japan are (1) mature neuroscience and engineering research environments, (2) world-leading robotics programs (output of motor BCI systems), and (3) integrated academic-industrial research agendas and partnerships. Like China, however, Japan also is "discovering" BCI research in the sense that BCI-directed research represents a relatively small percentage of its total current research effort. But importantly, Japan appears to conduct BCI research as an extension of the challenge of understanding the brain and as an extension of its well-developed, world-leading robotics programs; that is, BCIs will become new "intelligent" controllers for robotics platforms. BCI research in Japan is almost exclusively noninvasive, despite the many experimentally based (invasive) neuroscience programs.

The decision to maintain primarily noninvasive BCI programs appears to the WTEC panel to be a deliberate decision motivated by estimates of the ultimate user base (users who do not require nervous system repair). Compared to most other countries visited by the WTEC panel, Japanese research and industrial entities have an enormous technology arsenal (combined fMRI, MEG, and near-infrared-spectrum [NIRS] resources) that they can utilize for noninvasive BCI research. Though some European sites compare favorably to Japan (e.g., Tübingen, Germany), Japan has more noninvasive resources than the North American sites reviewed by the WTEC panel.

In addition, Japan has a broad and advanced perspective on the ultimate use of BCIs in society—comparable to what the WTEC panel witnessed at the FIRST Institute in Berlin (BBCI project). In Japan, BCIs are not just for medical applications or for repairing nervous system damage. The WTEC panel's perspective of Japan's vision is one in which BCIs are integrated into the everyday life of "normal" individuals (e.g., enhancing desired movements, enhanced cognitive function, avoiding accidents). Commercial issues with respect to both medical and nonmedical applications of BCIs are already being considered and evaluated. There was, of course, consideration of neural prostheses for repair of the damaged nervous system—sensory, motor, and cognitive enhancement of neural function in the normal and the aged population. But other applications included hospital diagnosis, rehabilitation, and remote, home-based brain monitoring in healthcare; video games and sports training in entertainment; disaster rescue, bomb discovery and disposal in robotics; and electrical diagnosis, telecommunications, and pet robots in home electronics.

Ethical issues with respect to the use of BCIs, particularly in terms of "enhancing normal cognitive function," have already been elevated in Japan to a significant level of importance. The WTEC panel heard at the Advanced Technology Research Institute (ATR) in Kyoto that a group of academics, industrial representatives, and other non-research-oriented social leaders were meeting on a regular basis to discuss and consider the ethical and social implications of BCIs. Professor Kawato of the Computational Neurosciences Laboratory explained that this group was actively considering "Neuroethics" in terms of issues such as (1) the commercial benefits of "elective" enhancement of "normal" neural function, i.e., incorporating adaptive synaptic plasticity into BCIs; (2) military applications: the "super soldier" and the "substitute soldier;" the ever-increasing likelihood of "war at a distance"; (3) "ownership" of the mind: commercialization of cognitive-enhancing "downloads"; (4) public policy consequences of cognitive-enhancing technologies: social stratification; and (5) crime by BCI-controlled robots. This is a sophisticated range of considerations for BCIs given their current stage of development.

The consequence of these factors with respect to the organization and funding of BCI research in Japan is that, in contrast to China and North America, the major drivers of Japan's interest in BCIs are as (1) another set of tools for better understanding brain function, and particularly cognitive brain function; (2) a mechanism for developing "high-level reasoning" for robotics control; and (3) as a commercial opportunity for developing technologies that will better integrate individuals into their environments. Thus, industrial (e.g., NTT) and research institute (e.g., Riken, ATR) funding in Japan is far more developed than in either

China or North America (where funding occurs primarily through traditional government and academic channels). Japanese funding is more on the level of the research organizational schemes seen by the WTEC panel in Europe. In Europe and Japan there is a well-balanced array of mechanisms for realizing commercializable, next-generation products, including government/academic support for fundamental research, institute-based support that helps to transition BCI research output to prototypes or near-prototypes, and industry support.

FUNDING AND FUNDING MECHANISMS

United States and Canada

The primary sources for funding BCI research in the United States and Canada are the traditional ones through NIH (NINDS: Neuroprosthetics Program), NSF (particularly through the Engineering Research Center mechanism—see the Biomimetic Microelectronics Systems Center), the National Research Council of Canada, and the Neil Squire Foundation. In recent years, DARPA, through its Brain Machine Interface Program, the Human Assisted Neural Devices Program, and the Revolutionizing Prosthetics Program, has made major contributions to the advancement of neural prostheses and BCIs. Likewise, the Office of Naval Research has accelerated the growth of BCIs through its support of Adaptive Neural Systems and Biorobotics. Private sources have yet to make a major impact on BCI research in North America.

Europe

BCI research funding in Europe, as discussed above, is initiated from multiple origins, including primarily from EU programs, but also from national, state, and local sources. Although the United States has Small Business Innovative Research grants (SBIRs) and Small Technology Transfer Research grants (STTRs) as funding mechanisms that promote the transition from basic research to precommercialized prototypes, the range of such mechanisms in Europe is far greater and more creative.

Asia

The primary funding source for BCI research in China is the government. Funding entities include the Chinese Ministry of Science and Technology, NNSF China

(National Natural Science Foundation of China), and the China High-Tech Research and Development Program. Funding through these sources is not allowed to pay for faculty salaries. Support for graduate students is allowed, but that support is partial, not full; the panel understood that no tuition costs are paid from government support. The laboratories reimburse the university for space through an indirect charge. Funding for BCI research in Japan can occur through a variety of mechanisms that include governmental and industrial sources.

TRANSLATION-COMMERCIALIZATION

United States and Canada

The WTEC panel was very interested in the extent to which BCI research and BCI technologies had reached the stage of translation to industry and commercialization. In the United States, commercialization of BCIs is just beginning to occur, e.g., Cyberkinetics (Figure 9.4), which combines technology from Brown University and the University of Utah, for a BCI system that allows the user to move cursors on a computer screen using 2-D kinematic information extracted from motor cortical population single-unit recordings.

Also in the United States, the NSF ERCs actively promote the involvement of industry in academic research programs through Industrial Associates Boards that advise academics and researchers about the needs and opportunities in the industrial arena. Directed research is also strongly encouraged by the ERCs.

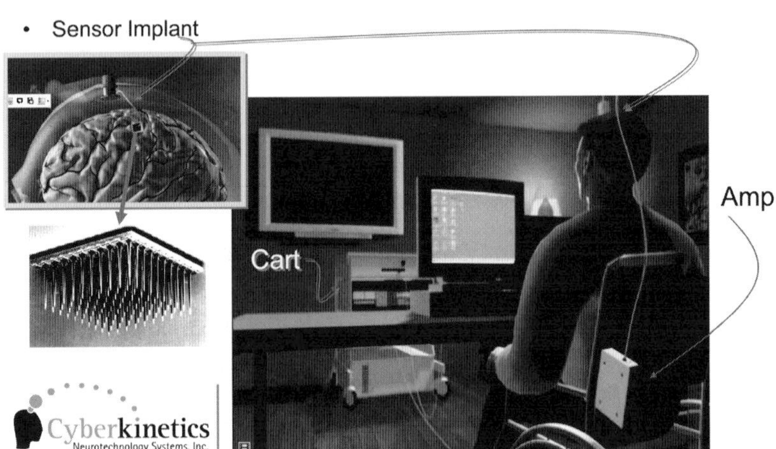

Figure 9.4. Cyberkinetics, a U.S. BCI company (Courtesy John Donoghue, Brown University).

Directed research is funded by industrial sources when academic and industrial interests converge to common goals (http://www.erc-assoc.org/centers.htm). Directed research arrangements usually require a set of agreements concerning intellectual property (IP).

Europe

In Europe, the WTEC panel found specific mechanisms for joint academic/scientific and industrial collaborations leading to the translation of BCI research, incorporation of BCI technology into small companies, and the creation of "spin-offs" from research efforts. For example, industrial entities can participate in EU-sponsored research as "just another project" that receives part of the research budget, i.e., a company can propose to partner with research members of an EU project to develop and shape a given technology to fit the research requirements of the global project. The only requirement is that each commercial entity must provide 50% of the costs of its project. Researchers benefit when industry is an integrated member of a large project because it maximizes research needs and available technology. It also benefits the company because it essentially guarantees a customer base; often, industry-related projects are producing technologies ultimately sold to other research-related projects. Scientific progress is achieved through a closer relationship between researchers and the sources of their technology, which allows a faster evolution of next generation technology. EU projects can require industrial involvement, so relevant businesses often are actively pursued. Example outcomes of the EU encouragement of industry participation include: (1) Multi Channel Systems GmbH (MCS) (http://www.multichannelsystems.com/), a leading worldwide supplier of multisite electrodes and multichannel recording/stimulation systems for brain slices/cultures and a partner in many EU projects; and (2) g.tec (http://www.gtec.at/), a worldwide supplier of multichannel EEG amplifiers that grew out of activities of the University of Graz BCI Laboratory and is also now a partner in many EU projects.

In Germany, the panel was introduced to institutional infrastructures that actively promote interactions between academia and industry. For example, the Fraunhofer Institute (Berlin) for Computer Architecture and Software Technology (see the site report in the appendix; http://www.bbci.de/) pursues the development of BCI research and BCI technology both for medical and commercial applications (e.g., gaming, auto industry). The Fraunhofer Institute in Berlin is one of four throughout Germany. The director of a given "research group" in the Fraunhofer FIRST Berlin holds an 80% appointment in Potsdam University (Berlin) and a 20% position in the Fraunhofer Institute. Support is derived from any source, but the university pathway allows funding for basic research, whereas the institute pathway provides an avenue for industrial support. At least 30% of the funding through the Fraunhofer Institute must be provided by industrial sources. So for example, the Intelligent Data Analysis Group (IDA) directed by Professor Dr. Klaus-Robert Müller engages in a wide range of theoretical research in machine learning and signal processing and develops new algorithms for real-world data analysis. The group also receives funding from the automobile industry to develop pop-up displays for the driver when periods of "cognitive overload" or "high-attention demand" occur—a form of "nonmedical" BCI. It also receives support from the gaming industry to develop brain-driven video games. Through active collaborations with the Charité University of Medicine Berlin, one of the premiere medical universities in Germany, the group is able to conduct experiments for clinical applications of BCIs. Through additional fundamental work on the neurophysiological underpinnings of BCI signals, the Charité group develops new experimental paradigms to point the IDA team to new directions of analytical development. This is an exciting state- and local-sanctioned infrastructure to support the highly interdisciplinary interactions at the fundamental, clinical, and industrial levels necessary for the development of BCIs.

This level of interaction between academia and industry is designed, of course, to develop new IP and eventually, new products. The WTEC panel heard that in these German (as well as European) systems, patent royalties are shared, spin-offs are promoted, and licenses revert to inventors. The information given to the WTEC panel was that no unique or strict rules for patenting and sharing of royalties are imposed; negotiations between relevant parties result in equitable sharing of IP and IP-generated returns. Distribution of royalties does not appear to be a major obstacle.

Another example found in Germany of an institutional infrastructure that promotes interactions between academia and industry is the Natural and Medical Sciences Institute (NMI) in Reutlingen (see site report in the appendix). The NMI is one of eleven institutes of applied research in the state of Baden-Württemberg alone, and conducts interdisciplinary applied research in the natural sciences and medicine. The NMI is internally organized according to multiple disciplines, which at the time of the panel's visit, contained thirteen "competence teams." Each competence team is responsible for generating its own projects and its own cash flow. The teams work with a network of clinics, universities, and other research institutes to develop new products for industry—in the case of the NMI, primarily biological and medicine systems. The internal organization is highly flexible and changes over time, evolving with the needs and opportunities of academia and industry. Although the NMI is independent of the University of Tübingen, there are close collaborations on research projects as graduate and undergraduate students from the university conduct their thesis research at the NMI. The WTEC panel saw a wealth of technologies and support staff that were available to academic researchers and small businesses so that cutting-edge approaches that ordinarily would be beyond the reach of small businesses can, in fact, be utilized to promote their success. One of the examples of this synergy between small businesses and the NMI was Multi Channel Systems, Inc. (see the site report in Appendix B). The NMI also serves as an incubator for small start-ups and spin-offs, one example of which is Retina Implant, Inc., which is developing a retinal prosthesis. During the two years preceding the WTEC panel's visit, the NMI collaborated with over 230 companies on research projects totaling €70 million. The WTEC panel was extremely impressed with this highly successful mechanism for generating new IP and new products for BCIs from basic research.

Industrial Collaborations

Although not detailed in this section, analogous mechanisms for promoting academic-industrial collaborations were found in other countries as well. The following are examples of industrial collaborations (nonexhaustive) for several of the universities and institutes visited in Germany and Switzerland:

Fraunhofer Institute (Berlin). Volkswagen, Daimler Chrysler, DuPont, Schering, ITSO, idalab, Overture/Yahoo, KPMG, IBM, Honda, Sony, Voice Trust, Microsoft, and financial companies.

NMI (Reutlingen). Abbot Laboratories, Bayer, BMI Biomedical Informatics, Boehringer Ingelheim Pharma, Carl Zeiss SMT AG, Accclab GmbH, BIBraun Aesculap AG, Brucker Daltonic, Altana Pharma, Biopharm, CellMed, Evotec Technologies, MAN, Mikrogen, Multi ChannelSystems, TETEC, Robert Bosch, and ZF Friedrichshafen.

University of Freiburg (Bernstein Center). Boehringer Ingelheim, GIF, Honda, and Multi Channel Systems.

EPFL (Swiss Federal Institute of Technology) (Lausanne). IBM, Silicon Graphics, and other sources.

Scuola Superiore Sant'Anna. Figure 9.5 shows examples of spinoff and startup companies for just the Scuola Superiore Sant'Anna (SSSA) in Italy, one of the foremost European institutions specializing in advanced robotics technologies.

For several reasons, the WTEC panelists felt that the EU model that includes funding for company members of a research team may offer some advantages compared to the U.S. model that promotes industry transition through SBIRs and STTRs. First, the EU model provides an integrated relationship between business and academic units where research and industry objectives evolve jointly. Second, the timeline to bring a product to market is shortened in the EU model because the U.S. model requires the research technology to precede the development of an industry prototype through SBIR/STTR; that prototype then becomes developed into a product through additional steps. Third, the EU system creates a partnership between academia and industry. In contrast, a largely antagonistic relationship between academia and industry exists and is promoted by the U.S. system. Thus, the EU approach is much more likely to lead to successful results.

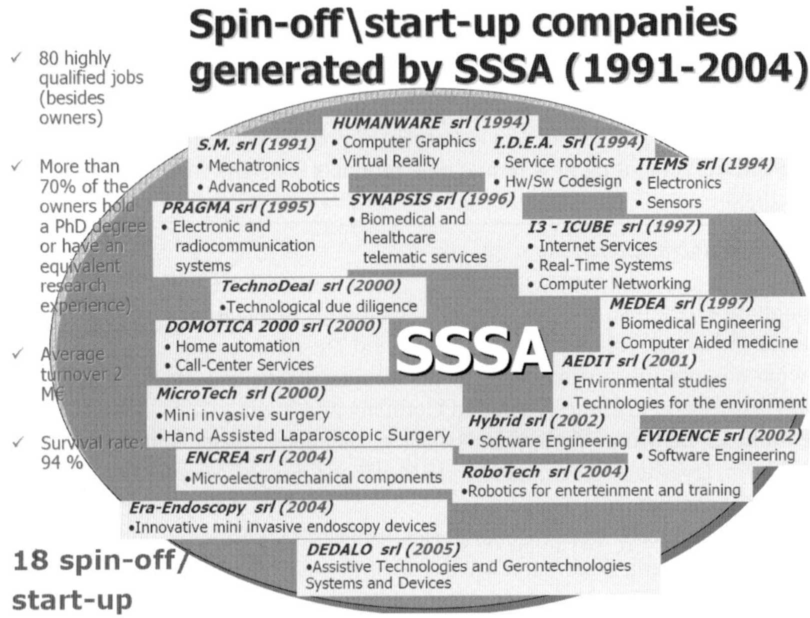

Figure 9.5. Scuola Superiore Sant'Anna (SSSA) spinoffs and startups.

It should be pointed out that such industrial collaboration was not universal throughout Europe (see the site reports in Appendix B). The WTEC panel visited BCI research sites in Oxford, England, and found a minimal number of industrial partnerships and commercialization. In Edinburgh, Scotland, the WTEC panel found no significant translational activities. In Tübingen, Germany, there was virtually no attempt to commercialize otherwise effective BCI systems. In the case of Tübingen, the lack of interest in commercialization appeared to be a deliberate decision to limit the focus to integration of BCI systems into patients' homes.

Asia

BCI research is in its beginning stages in China, and thus it is too early for significant industrial involvement or commercialization. Nonetheless, the WTEC panel saw evidence for multiple patents, particularly on the part of researchers developing devices, and there was clearly an increasing consciousness on the part of researchers for commercialization.

In Japan, BCI research already is becoming well integrated with large-scale industry, e.g., Nippon Telegraph and Telephone (NTT) and ATR, as the site reports in Appendix C indicate. The growth of industrial involvement in BCI research should increase in future years. This largely reflects the fact that BCI research is being propelled by a long and well-established academic-industry investment in robotics research. The WTEC panel heard repeatedly that much of the investment in BCI technology was being driven by a need for "smart," "cognitive" control of current and future robotics platforms important to Japanese industry. The major issue that arose in discussions with the WTEC panel was whether the rate of BCI growth would be higher in the "agile," small-sized companies than in the less dynamic, but well-funded, large-sized companies.

TRAINING-EDUCATION

Throughout Europe and Asia, the WTEC panel found that surprisingly little attention is paid to developing BCI-specific training programs at the undergraduate, graduate, or postdoctoral levels. Cross-disciplinary training occurs in an almost haphazard manner; with some noted exceptions, obtaining interdisciplinary training is largely the responsibility of the student. This is not to suggest that interdisciplinary training is not successful in European or Asian institutions—on the contrary, students are very well trained in multiple fields. It is simply that interdisciplinary training is not as formalized as it is in the United States, and this probably reflects: (1) the greater strength of traditional disciplinary boundaries in the European culture, (2) broadly based faculty salaries and student funding in Europe and Asia that limit specialized training programs, and (3) the relative "youth" of BCI research in Europe and Asia. For example, because of both the early stage of development of BCI programs in China and the explosive growth of educational institutions in general, efforts are focused primarily on forming foundational departments and programs (e.g., biomedical engineering); as a consequence, traditional disciplines have precedence. The United States clearly has more comprehensive, well-developed educational/training programs in BCI, with greater sensitivity to recruiting underrepresented minorities.

It was evident that many of the European institutions visited are now seeing the need for more formalized interdisciplinary training, particularly for the field of BCIs, and are moving in that direction. The WTEC panel saw that several European universities are fostering entrepreneurship training for their students to help promote translation of research into industry. Also, long-term industrial internships are common in Europe and Japan, allowing students to conduct their thesis research at collaborating companies.

REFERENCES

Berlin Brain Computer Interface (BBCI) website, http://www.bbci.de/ (accessed on July 17, 2007).

Engineering Research Centers Association. List of ERCs, http://www.erc-assoc.org/centers.htm (accessed on July 17, 2007).

European Commission. 2005. *Information society technologies: IST workprogramme for 2003 and 2004*. Available online at http://www.digital-ecosystems.org/doc/so.htm

European Robotics Search Network (EURON). n.d. *Where to find EURON members* (website and map): http://www.euron.org/members/index.html

g.tec (Guger Technologies) website, http://www.gtec.at/ (accessed on July 17, 2007).

Multi Channel Systems MCS GmbH website, http://www.multichannelsystems.com/ (accessed on July 17, 2007).

APPENDIX A. BIOGRAPHIES OF PANELISTS AND DELEGATION MEMBERS

Theodore W. Berger (Panel Chair)

Dr. Theodore W. Berger is the David Packard Professor of Engineering, Professor of Biomedical Engineering and Neuroscience, and Director of the Center for Neural Engineering at the University of Southern California. He received his Ph.D. from Harvard University in 1976; his thesis work received the James McKeen Cattell Award from the New York Academy of Sciences. He conducted postdoctoral research at the University of California, Irvine, from 1977–1978 and was an Alfred P. Sloan Foundation Fellow at the Salk Institute from 1978–1979. Dr. Berger joined the Departments of Neuroscience and Psychiatry at the University of Pittsburgh in 1979, being promoted to full professor in 1987. During that time, he received a McKnight Foundation Scholar Award, twice received an NIMH Research Scientist Development Award, and was elected a Fellow of the American Association for the Advancement of Science. Since 1992, he has been Professor of Biomedical Engineering and Neurobiology at the University of Southern California, and was appointed the David Packard Chair of Engineering in 2003.

While at USC, Dr. Berger has received an NIMH Senior Scientist Award, was elected a Fellow of the American Institute for Medical and Biological Engineering in 1998, received a Person of the Year "Impact Award" by the AARP in 2004 for his work on neural prostheses, was a National Academy of Sciences International Scientist Lecturer in 2003, and an IEEE Distinguished Lecturer in 2004–2005. Dr. Berger was elected a Senior Member of the IEEE in 2005, and received a "Great Minds, Great Ideas" award from the EE Times in the same year. Dr. Berger is currently chair of an NIH study section that evaluates grants related to clinical neurophysiological, medical devices, and neural prosthetics. Dr. Berger became Director of the Center for Neural Engineering in 1997, an organization that helps to unite USC faculty with cross-disciplinary interests in neuroscience, engineering, and medicine. He has published over 200 journal articles and book chapters, and is the coeditor of a book recently published by the MIT Press: *Toward Replacement Parts for the Brain: Implantable Biomimetic Electronics as Neural Prostheses.*

Dr. Berger's research focuses on electrophysiological and theoretical studies of hippocampal neurons and circuits for the purpose of developing neural prostheses and biological-based pattern recognizers.

John K. Chapin (Panelist)

Dr. John K. Chapin received his BS from Antioch College and his Ph.D. from the University of Rochester. After winning the Donald Lindsley Prize in 1980, he was appointed Assistant Professor of Cell Biology at the University of Texas Southwestern School of Medicine (1981), Associate Professor of Physiology at Hahnemann University in Philadelphia (1987), Professor of Neurobiology at the Medical College of Pennsylvania (1995), and Professor of Physiology at SUNY Downstate School of Medicine (2000).

In the 1980s Dr. Chapin developed the techniques for simultaneously recording from large numbers of single neurons in awake-behaving animals. Since then he has used this approach to extract both sensory and motor information from the brain in real time. In 1999 he was the first to demonstrate that neural information recorded from multisingle neurons in the motor cortex could be used to allow an animal to directly control a robot arm to obtain water. His current work involves stimulating through electrode arrays in the somatosensory system to provide feedback from such a robot directly to the brain.

Greg A. Gerhardt (Panelist)

Dr. Greg A. Gerhardt received his doctorate in chemistry with additional training in neuroscience from the University of Kansas in 1983. He did his postdoctoral training in Psychiatry and Pharmacology from 1983–1985 at the University of Colorado Health Sciences Center (UCHSC) in Denver. He rose to the rank of Professor (with tenure) in Psychiatry, Pharmacology, and the Neuroscience Training Program from 1985–1998 at UCHSC. He is currently a Commonwealth of Kentucky Research Challenge Trust Fund Professor in the Departments of Anatomy & Neurobiology, Neurology, Psychiatry, and Electrical Engineering, and Director of the Morris K. Udall Parkinson's Disease Research Center of Excellence at the University of Kentucky Chandler Medical Center in Lexington, Kentucky. This is 1 of 12 Parkinson's Disease centers in the United States funded by NINDS. He is also the Director of the Center for Microelectrode Technology (CenMeT), and he has been Editor-in-Chief (Americas and Australasia) of the *Journal of Neuroscience Methods* since 1999. He has received numerous awards, including a recent Level

II Research Scientist Development Award from NIMH (2000–2005), and he has published more than 220 original peer-reviewed papers, 50 book chapters, and 380 abstracts and conference proceedings.

Dr. Gerhardt's research focuses on Parkinson's disease and the repair of damaged dopamine neurons in the basal ganglia of the brain using growth factors such as glial cell line-derived neurotrophic factor (GDNF). In addition, his laboratory develops technologies to directly measure chemical communication in the brain.

Dennis J. McFarland (Panelist)

Dr. Dennis J. McFarland received his Ph.D. in Psychology from the University of Kentucky in 1978. Since then he has been a Research Scientist at the Wadsworth Center for Laboratories and Research at the New York State Department of Health. Dr. McFarland has experience with the development of recording, signal processing, and training of EEG signals, as well as methods and theory in Psychophysics. His current research interests are in developing a brain-computer interface and central auditory processing. Dr. McFarland has published over 80 articles in peer-reviewed journals as well as numerous book chapters, commentaries, and abstracts. He is currently an associate editor for the *IEEE Transactions on Neural Systems and Rehabilitation Engineering*.

José C. Principe (Panelist)

Dr. José C. Principe has been Distinguished Professor of Electrical and Biomedical Engineering at the University of Florida since 2002. He joined the University of Florida in 1987 after an eight-year appointment as Professor at the University of Aveiro in Portugal. Dr. Principe holds degrees in electrical engineering from the University of Porto (Bachelor), Portugal, University of Florida (Master and Ph.D.), USA, and a Laurea Honoris Causa degree from the Universita Mediterranea in Reggio Calabria, Italy. Dr. Principe is a Fellow of the IEEE and the AIMBE, past President of the International Neural Network Society, past Editor-in-Chief of the *Transactions of Biomedical Engineering*, and a former member of the Advisory

Science Board of the FDA. He holds five patents and has submitted applications for seven more. Dr. Principe was supervisory committee chair of 50 Ph.D. and 61 master's students, and he has authored over 400 refereed publications (3 books, 4 edited books, 14 book chapters, 116 journal papers, and presentations in 276 conference proceedings).

Dr. Principe's interests lie in nonlinear non-Gaussian optimal signal processing and modeling and in biomedical engineering. He created in 1991 the Computational NeuroEngineering Laboratory to synergistically focus the research in biological information processing models. He recently received the Gabor Award from the International Neural Network Society for his contributions.

Dawn M. Taylor (Panelist)

Dr. Dawn M. Taylor is an assistant professor of Biomedical Engineering at Case Western Reserve University and a research scientist with the Veterans Administration Cleveland Functional Electrical Stimulation (FES) Center of Excellence.

Dr. Taylor's primary research focus is on brain-machine interfaces designed to restore arm and hand function in people paralyzed below the neck. Dr. Taylor is developing ways to extract intended arm and hand movements in real time from neural activity recorded from intracortical microelectrodes as well as from field potentials recorded outside the brain. Her primary interest is in developing adaptive decoding functions that facilitate beneficial learning in the brain. She is applying her adaptive decoding methods to the control of the upper limb neuroprosthesis systems developed by her colleagues at the Cleveland FES Center. These systems restore arm and hand function by activating paralyzed muscles via low levels of electrical current applied to the peripheral nerves. Dr. Taylor and her colleagues are working to enable paralyzed individuals to once again move their arms and hands just by thinking about doing so.

Patrick A. Tresco (Panelist)

Dr. Patrick A. Tresco received an M.S. in pharmacology and toxicology from the University of Rhode Island and a Ph.D. in medical sciences from Brown

University. He is currently a professor in the Department of Bioengineering, Director of the Keck Center for Tissue Engineering, and Associate Dean for Research in the College of Engineering at the University of Utah. He is a scientific advisory board member of Acorda Therapeutics, Inc., Hawthorne, NY, and advisor to the Biomimetic MicroElectronic Systems Engineering Research Center at the University of Southern California. In addition, he has been a biomaterials consultant to such companies as Bard Access Systems, Fresenius, Microislet, Cytotherapeutics, Medtronic, and Smith Kline Beecham. Dr. Tresco is a regular peer reviewer for a number of top bioscience and engineering journals, as well as for the National Institutes of Health and the National Science Foundation. In addition, Dr. Tresco is a Fellow of the American Institute of Medical and Biological Engineering and was recently inducted into Tau Beta Pi as an eminent engineer.

Dr. Tresco is recognized for his work in various tissue engineering applications and for contributions to understanding how nervous tissue interacts with a broad range of implanted materials. He has published over 70 peer-reviewed publications and has over 150 presentations at top conferences in his field. He currently holds 16 issued and pending patents relating to this and other areas of biotechnology, and has taught graduate and undergraduate courses in cell and molecular biology, biomaterials science, and tissue engineering.

Walid V. Soussou (Associate Panelist)

Dr. Walid V. Soussou received his BS degree in biochemistry from Boston College in 1995 and a Ph.D. in Neuroscience at the University of Southern California (USC) in 2005. He has interned at Harvard Medical School and worked as a research technician at Boston University. Dr. Soussou is a consultant with Neural Consultants, a consortium of USC postgraduates in biomedical engineering and neuroscience specializing in neural prosthetics research. He is currently a postdoctoral fellow at the Burnham Institute for Medical Research. Dr. Soussou is a member of the Society for Neuroscience and is a recipient of the 2005 MIT Arab Student Organization's Science and Technology Graduate Student Award.

Semahat S. Demir (Lead Sponsor)

Dr. Semahat S. Demir received her BS degree in electronics engineering from Istanbul Technical University, M.S. degree in biomedical engineering from Boğaziçi (Bosphorus) University, and second M.S. degree and Ph.D. degrees in electrical and computer engineering from Rice University. She did her postdoctoral training at the Biomedical Engineering Department at The Johns Hopkins University. She has 18 years experience in academic research, ten years experience in teaching in academia, two years experience in the medical industry, and three years experience in research funding administration in the U.S. Federal Government. Dr. Demir is currently Program Director for Biomedical Engineering at NSF; chair of the Neurotechnology Group of NSF's Engineering Directorate; co-chair of the Bioengineering Consortium (BECON) Bridges Team; and the NSF representative on the National Science and Technology Council Subcommittee on Biometrics and Identity Management. Among many awards for excellence, she received the NSF Director's Award for Program Management Excellence and Director's Award for Collaborative Integration in 2006. Dr. Demir initiated and sponsored this WTEC study on Brain-Computer Interfaces.

Dr. Demir's own academic research integrates research, education, and training, with an emphasis on mathematical modeling and computer simulations in both cardiac electrophysiology and neuroscience. She is an internationally published expert and lecturer on the bioelectricity of subcellular, cellular, and multicellular systems and on the development of simulation-based teaching and learning resources, such as her interactive cell modeling resource, iCell, http://ssd1.bme.memphis.edu/icell/.

Hassan B. Ali (WTEC Project Manager)

Hassan B. Ali is a physicist with over 36 years experience in science and technology (S&T) with the U.S. Government and private industry as a researcher, project manager, and team leader. He retired from the Federal Government in 2002 and has since been working as an independent consultant (Director of International

Operations/Project Manager) for WTEC. He has more than a decade of expertise in international collaborations/assessments in science and technology, including extensive experience working with multidisciplinary groups in the Asia-Pacific region, Europe, and Latin America. In this capacity, he played a key role in establishing S&T collaborations between the U.S. Office of Naval Research and several countries in the Asia-Pacific region and Latin America. He is multilingual to varying degrees of proficiency in the languages English, Turkish, Italian, Japanese, German, French, Spanish, and Chinese. He has authored approximately 40 refereed papers and conference proceedings, more than 30 technical reports, and numerous abstracts. He has strong communication skills and has made formal presentations to audiences ranging from international conferences to the United States Senate Foreign Relations Committee.

Gerald Hane (WTEC Advance Contractor)

Dr. Gerald Hane received his Ph.D. from Harvard University (1992) and his BS and M.S. degrees from Stanford University (1980). A technology competitiveness specialist, he formerly was head of international strategy and affairs for the White House Office of Science and Technology Policy (1995–2001), worked for the Science Committee of the House of Representatives (1992–1995), and was a research engineer for Battelle Pacific Northwest Laboratory (1980–1988). Dr. Hane is founder and principal of Globalvation, a consulting firm specializing in the research of science and technology policy and management.

Grant Lewison (WTEC Advance Contractor)

Dr. Grant Lewison was trained as a mechanical engineer and experimental hydrodynamicist at the University of Cambridge and spent two years at the University of California, Berkeley, before joining the British civil service as a scientist. He worked on ship motions research for many years before switching in 1981 to science policy, in which capacity he worked for the British Department of Trade and Industry, the European Commission in Brussels, and a small consultancy firm in the UK. His own research has focused on bibliometrics. In 1993 Dr. Lewison joined the Wellcome Trust to design and manage its Research Outputs Database (ROD). Since then he has carried out many consultancy assignments in bibliometrics and written about 70 papers. At the end of 2000, the ROD was transferred to The City University on contract from the Trust, and he moved with it as visiting professor in the Information Science Department. The ROD project ended in 2003, and Dr. Lewison left City University at the end of 2005 to set up his own consultancy company in Richmond (UK), Evaluametrics, Ltd., which undertakes research evaluation through publication metrics. He was recently appointed as a Senior Research Fellow at University College, London. His particular interest is in presentation of research to the public through the mass media and policy documents, and its evaluation by these means.

APPENDIX B. SITE REPORTS—EUROPE

Site: **Aalborg University**
Department of Health Science and Technology
Fredrik Bajersvej 7D
DK 9220
Aalborg, Denmark
http://www.hst.aau.dk

Date Visited: May 29, 2006

WTEC Attendees: P. Tresco (report author), H. Ali, J. Chapin, S. Demir, J. Principe

Hosts: Prof. Kim Dremstrup Nielsen
Head of Department of Health Sceince and Technology
Tel: +45 96 35 88 11
Fax: +45 98 15 4
Email: kdn@hst.auc.dk
Prof. Thomas Sinkjaer
Director, Center for Sensory-Motor Interaction (SMI)
Tel: +45 9635 8824
Fax: +45 9815 4008
Email: ts@hst.aau.dk

BACKGROUND

Aalborg University, founded in 1974, is located in the north of Denmark. It has a unique educational and research mission that stresses teamwork and is organized around practical problems that are highly interdisciplinary in nature. The university emphasizes cooperation with business, organizations, and institutions. It has set internationalization as a high priority. Aalborg University is divided into three faculties: humanities; social sciences; and engineering, science, & medicine. It offers more than 60 different programs of study and has over 13,000 students.

RESEARCH AND DEVELOPMENT

Our visit began with a broad overview of the educational and research activities in the Department of Health Science and Technology presented by our host, Professor Kim Dremstrup Nielsen. Established in 2002, the department consists of 125 employees. It hosts a five-year master's program in biomedical engineering and health informatics (currently 200 students), a five-year master's program in medicine within industry (established in 2006 with 60 students in the first class), a two-year master's program in health informatics with 150 students under the open university, and a three-year doctoral program in biomedical science and engineering (51 Ph.D. students in 2006). Broadly speaking, the expertise within the department includes research in stem cells, motor control and rehabilitation, sensory systems and technology, and medical and health informatics. Specifically, the neural prosthetic research is focused on FES, BCI, electrode development, biomechanics, and rehabilitation. Other areas of interest include human brain mapping, pain and biomechanics research, EEG analysis, human performance, motor control, health information systems, surgery simulation, image analysis, and virtual reality. The group is very productive, having published over 780 peer-reviewed papers in a two-year span with a significant number of patents. About half of the research funding is extramural, split between national research agencies, the EU, and a variety of private sources. We saw several successful examples of the commercialization of university research including imaging technology, bioinstrumentation, and biomedical device technology. For example, the Neuro Rehabilitation Group recently developed a multichannel, implantable stimulator device used to correct foot drop. The device is called Actigait and is sold by Neurodan-Hans OttoBoch Gmbh. To facilitate ambulation, newer versions under development use multisite cuff electrodes to record sensory activity from skin stimulation of foot contact to trigger motor stimulation within the same nerve.

The strong practical research focus is complemented and strengthened by the educational programs at the (three-year) bachelor's level and (two-year) master's level in biomedical engineering and health informatics and a (three-year) Ph.D. training program in biomedical science and engineering. By design, the curriculum is strongly multidisciplinary, is focused on practical, problem-based learning, and stresses entrepreneurship. The academic unit trains over 50 Ph.D. students with an internal budget of approximately 30 million DKK (~$5 million in May 2006), with salaries included (2005). The educational model fosters translational research.

We heard presentations from several faculty members in the Department of Health Science and Technology. Professor Sinkjaer, the Director of the Center for Sensory Motor Interaction (35 researchers, 30 Ph.D. students), provided an overview of part of the research there, focused on developing new techniques to study human movement control, including the mechanics of the muscle tendon complex, small and large muscle afferents, and central control and modulation of movement. The center develops a variety of external and implantable electrode systems to record and stimulate the neuromuscular system used in rehabilitation. The group seeks to understand how to extract information from physiological signals and how to activate paralyzed muscles in a biological-based manner.

Next, we heard presentations from Professors Omar do Nascimento and Dario Farina. With expertise in biomedical signal detection, analysis, and processing, the two are working on, among other things, the next generation of brain-computer interfaces for reestablishing complex motor tasks in disabled patients. The research is focused on identifying the best possible signal features in the EEG to be used as command signals for external systems intended to restore more complex motor functions than are currently possible using two-choice commands. We heard of their efforts to increase detection specificity and speed of selection to develop better user interfaces for communication or control with particular focus on movement-related cortical potentials (MRCPs) (e.g., slow EEG signals that precede voluntary movements) associated with real and imaginary movement. Finally, Ph.D. students Ying Gu and Alvaro Cabrera discussed their research projects. Mr. Cabrera presented the status of the lab work within steady-state visual-evoked potential (SSVEP)-based BCI systems and preliminary work based on spatial navigation. Most notable was Ms. Gu's presentation, which examined methods to extract MRCPs from real and imaginary movements of amputees. She presented some intriguing but preliminary data.

Following informal discussion at lunch, we received tours of the labs including the pain and biomechanics research lab, EEG analysis lab, biomechanics and rehabilitation lab, human performance lab, and the electrode development lab. In a roundtable discussion concluding our visit, a couple of interesting issues arose. In answer to the question of what are the biggest challenges remaining, a number of responses were offered: (1) it is necessary to increase the number of control signals, or the degrees of freedom, to move beyond the present binary system of on and off; (2) it is necessary to understand the patient-to-patient variation in MRCPs to understand the role of plasticity; and (3) it is necessary to refine the hardware. The first application envisaged as part of rehabilitation therapy is a BCI device coupled to functional electrical stimulation as volitional control.

SUMMARY AND CONCLUSIONS

Striking features of this department were the high degree of integration of the labs, the congenial atmosphere among the investigators, their practical focus on human problems, and the high level of involvement of patients with the educational and research programs. There also was a strong emphasis on the dissemination of technology, including explicit training in entrepreneurship and new venture development. Collaborative interaction with industry and networking with other research institutions and organization, both within and outside of Denmark, were encouraged. At present, the lab has clear strengths in noninvasive BCI and related technologies. Although not formally presented during our visit, the group also has research activities in the design and development of invasive electrodes and associated animal experimentation that are directed by Professors Ken Yoshida and Professor Winnie Jensen. Research is being conducted to optimize the neural interface by understanding the influence and interactions of the biological factors with the implanted hardware.

REFERENCES

Aalborg University website, http://www.hst.aau.dk (accessed July 10, 2006).
Center for Sensory-Motor Interaction website, http://www.smi.auc.dk/ (accessed July 10, 2006).
Nascimento, O.F., website, http://omar.nascimento.dk/ (accessed July 10, 2006).
Yoshida, K., website, http://www.smi.hst.aau.dk/%7Eky/ (accessed July 10, 2006).

Site: **Berlin Brain-Computer Interface**
http://www.bbci.de

Fraunhofer-Institute for Computer Architecture and Software Technology
Intelligent Data Analysis Group
Kekuléstrasse 7
Berlin, Germany
http://www.first.fhg.de/

Charité University of Medicine Berlin
Campus Benjamin Franklin
Hindenburgdamm 30
Berlin, Germany
http://www.charite.de/international/

Date Visited: May 31, 2006

WTEC Attendees: T. Berger (report co-author), W. Soussou (report co-author), G. Gerhardt, D. McFarland, D. Taylor, G. Lewison

Hosts: Prof. Dr. Klaus-Robert Muller
Department Head of IDA
Tel: +49 30 6392 1860
Email: klaus@first.fraunhofer.de
Prof. Dr. Gabriel Curio
Neurologist
Tel: +49 30 8445 2276
Email: gabriel.curio@charite.de

BACKGROUND

Site Structure and Organization

The **Berlin Brain-Computer Interface** (BBCI) project is a collaboration between Fraunhofer Institut Rechnerarchitektur und Softwaretecknik's (FIRST) Intelligent Data Analysis (IDA) research group and the Neurophysics Group in the Department of Neurology at the Campus Benjamin Franklin of Charité University of Medicine with the cooperation of the Technical University of Berlin (TUB). Part of the BBCI research is performed in collaboration with the Bernstein Center for Computational Neuroscience in Berlin. The project is funded by the German Federal Ministry of Education and Research (BMBF).

The Fraunhofer FIRST Institute for Computer Architecture and Software Technology is a computer software institute that engages in contract research with the objective of developing its basic research into applied technology. The institute's work focuses on information technologies for intelligent data analysis, embedded and safety-relevant systems, and innovative human-computer interaction technologies. The Intelligent Data Analysis Group performs theoretical research in machine learning and signal processing and develops new algorithms geared for real-world data analysis.

The Fraunhofer-Gesellschaft "undertakes applied research of direct utility to private and public enterprise and of wide benefit to society. Its services are solicited by customers and contractual partners in industry, the service sector, and public administration. The Fraunhofer-Gesellschaft maintains roughly 80 research units, including 58 Fraunhofer Institutes, at over 40 different locations throughout Germany. A staff of some 12,500, predominantly qualified scientists and engineers, works with an annual research budget of over €1 billion. Of this sum, more than €900 million is generated through contract research" (Fraunhofer-Gesellschaft, 2005).

The Charité University of Medicine Berlin is the premiere medical university in Germany. The University conducts experiments for clinical applications and develops new experimental paradigms to point the IDA team to new directions of analysis development. The BBCI's Charité group provides expertise in neurophysiology and cognitive neuroscience. They are therefore able to branch out into face processing using MEG and EEG to study incidental and explicit learning in order to understand brain processing or for BCI applications. MEG may enable reconstruction of 3D sources. This Charité group also provides access to patients with movement disorders and deep brain stimulation MEAs in their basal ganglia. This enables the researchers to record from these electrodes and explore BCI applications for this group.

Professors Muller and Curio also hold appointments at the **Bernstein Center for Computational Neuroscience in Berlin**. This center is part of the German Network for Computational Neuroscience established by the BMBF with the goal of integrating advances in neurobiology, cognitive science, systems biology, and information technology to advance brain research. The other three centers are in Freiburg, Goettingen, and Munich. These centers are funded for five years with a commitment by their hosting universities to provide space, faculty positions in various departments, and a promise of continued support to match grant-based funding. The Bernstein Center Berlin is focused on the issues of precision and variability of neural signals. One of its main objectives is to understand the variability in EEG signals to enable real-time, single-trial control of a feedback BCI.

Educational Environment and Infrastructure

Three universities contribute students to IDA: Technical University of Berlin, Potsdam University, and Charité University of Medicine Berlin. Most students are math majors, some computer science, a few from Charité are in neurophysiology, and there are also some master's students from the Bernstein Computational Neuroscience Center. The group mixture was selected *ad hoc*; however, there are plans for a more formal selection process. Currently, there is no official BCI educational program. The FIRST seminar series brings in different experts, including some in BCI. Aside from a BCI journal club and group meetings, the seminar series is the only course at IDA.

Funding Sources and Commercialization

University salaries cover 80% of a professor's time, allowing 20% of the time for other work, e.g., Professor Muller's appointment as a director of IDA at Fraunhofer. Fraunhofer mandates that at least 30% of a professor's research funds come from industry in order for the institute to maintain the claim that it is funding research that supports industrial applications. Additional funding is provided by the German Ministry of Education and research or other grants. Although universities do not charge overhead on grants in Germany, Fraunhofer Institute does charge indirect costs, motivating the institute to pursue grants.

Fraunhofer and German universities encourage patenting, especially when working with companies. Universities want to retain intellectual property rights, but companies ask for it since they are investing the funds and universities are public. However, several patents are shared between universities and industry, with ownership assigned to the largest financial contributor and royalties shared with all involved. Additionally, Fraunhoffer supports spin-offs by providing licenses for inventors. Industries contract specific projects to IDA. For example, Schering supports a drug discovery project where characterization of solubility or liver degradation is predicted, based on similarities to other known drugs.

Industrial collaborations include VW, Daimler Chrysler, DuPont, Schering, ITSO, idalab, overture/yahoo, KPMG, IBM, Honda, Sony, Voice Trust, Microsoft, and financial companies.

Academic collaborations include Riken, UT, TIT, ANU, UCSC ETH, HUT, Inesc-ID, McMaster U, INPG, UHB, TUBS, HUB, TUB, FUB, and WIAS. The Charité PASCAL project involves several universities in a data analysis collaboration.

RESEARCH AND DEVELOPMENT

Short- and Long-Term Scientific Goals

A major directive of IDA is to learn from a small data set and extrapolate classification to unknown activity patterns. As such, the BBCI project is focused on real-time classification of single-trial EEG data from users engaged in overlearned motor imagery. They are examining different machine learning adaptive classification algorithms with various visual feedback programs for the user. Another major objective of the BBCI is to "let the machines learn" (Leitmotiv), and therefore they use healthy subjects untrained for BCI.

BBCI is using only noninvasive EEG interfaces and does not plan to implant or operate; its researchers believe patients will not want to sacrifice their last remaining healthy brain tissue in risky surgery. At the time the WTEC panel visited, the subjects were all healthy, but there were plans to start studies with movement-disabled patients. There is also an interest in higher frequency activity (>100 Hz) because there is evidence of use of 600 Hz wavelets observed in EEG signals (N20) for successful classification, with supportive biological evidence of potential usefulness derived from observations of tuned cortical cells bursting at similar frequencies.

The IDA group develops theoretical data analysis techniques, then also applies them to real-world data to complete the methodology development cycle and enhance the analysis for collaborators.

Due to collaborations among the TUB's computer science department, the Bernstein's Computational Neuroscience Center and Charité's hospital, the goals of the project are not only clinical rehabilitation, but include advances in computational neuroscience and understanding the brain.

Additional applications of IDA machine learning methods include protein/DNA analysis, drug discovery, intrusion detection, handwritten character recognition (OCR), financial time-series forecasting, consumer data privacy, and fraud detection; all of these involve learning the statistics of a class of activity in order to determine when a new similar pattern of activity arises.

Tools and Methods Used and Sources

After a thought is generated, if a movement is generated there is a reverberation of activity from the periphery. Thus, BBCI looks at premotor areas and preplanning time to predict volition for BCI control. The averaged *bereitschaftspotential* is a low amplitude (10 μV) signal that is generated 1,000 ms before the initiation of a movement.

One implementation of averaged *bereitschaftspotential* classification is the binary (right/left hand) decision coupled to an overlearned motor output for self-paced typewriting on a computer keyboard. Left- and right-hand keyboard presses can be differentiated before the movement from averaged EEG *bereitschaftspotential* from the premoter cortex of that hand. Reactive responses are sharper than those from spontaneous movement. However, there is a lot of variance in individual *bereitschaftspotentials* and trends are usually visible only after averaging, but not from single trials. There is also a large intersubject diversity in signals—even averaged ones. BCIs will therefore need individual calibrations. Another source of variability is a shift of distribution of activity between training and feedback sessions, which indicates that the system is nonstationary and requires adaptive classification algorithms.

PREPROCESSING

EEG signals are first windowed, and then filtered with Fourier transform to keep gamma frequencies (0.4–5 Hz). This generates a smooth curve from which three points from the 200 ms interval before movement onset are selected. These three points are fed into machine-learning algorithms, which are well suited to handle such multidimensional space with low sample numbers (Figure B.1).

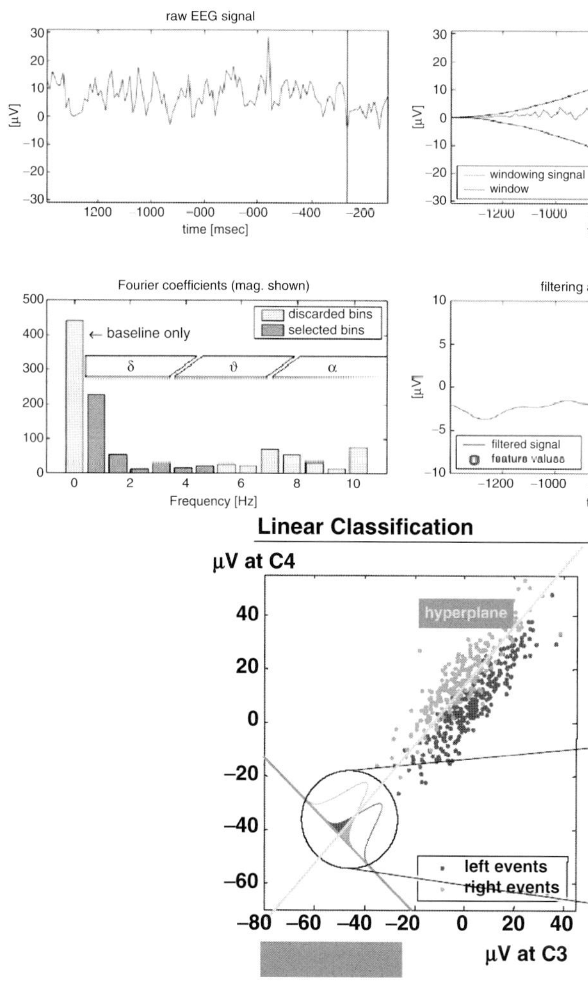

Figure B.1. Classification.

Another paradigm BBCI used involves the analysis of ERD/ERS-EEG data for imagined movements, to control an interface with a visual feedback. In this system, multiple features are extracted and used for classification: a FFT-based low-pass filter, a Band-pass 4–40 Hz to determine artifact removal coefficients, and a subject-specific band-pass filter 7–14 Hz with multiclass CSP for spatial localizations of data sources. ICA projections are often used for artifact and noise removal before feature extraction and selection (Figure B.2).

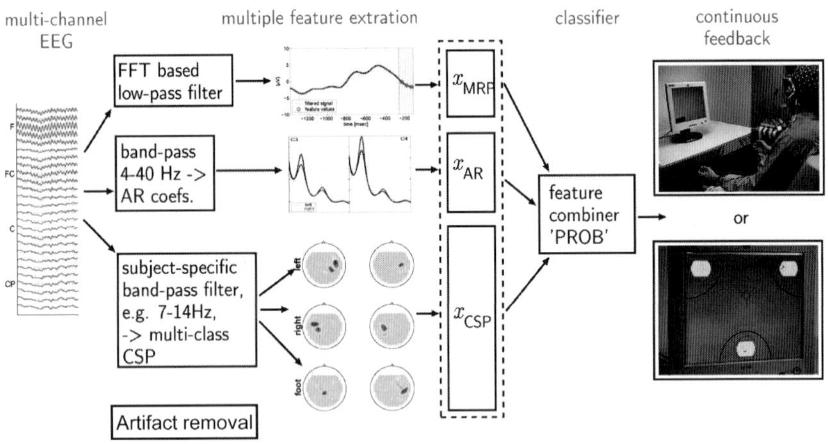

Figure B.2. BCI data extraction and classification paradigm with multiple features extraction.

This paradigm enabled the control of several interfaces including variants of Brain Pong games or typing with a virtual keyboard optimized with text prediction (Figure B.3), with bit rates up to 50 bits/minute, and typing speed of up to 8 characters/minute in noisy and stressful environments such as the CeBit 06 computer fair, and with untrained subjects with only 20 minutes of calibration time and 10 minutes of machine learning.

This system enables continuous control over the range of classification by weighing the confidence of the classification. In addition, this system is adaptive to changes in the user's signal over the course of the training and experiment.

PAST PERFORMANCE: PUBLIC RELATIONS AND PUBLICATIONS

The group has published over a dozen journal articles on BCI and has presented its findings at numerous conferences. A few IDA algorithms for signal processing and classification are already patented. In addition, the group sponsored and organized all BCI competitions, publicized, and reported on them (e.g., the third competition is described at http://ida.first.fraunhofer.de/projects/bci/ competition_ iii).

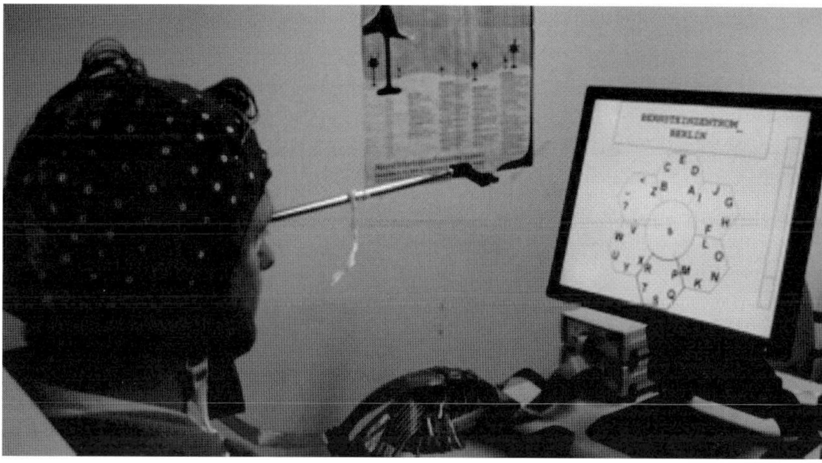

Figure B.3. BCI typing feedback interface with text prediction.

Media coverage, including several magazine and TV reports, varies from good and scientific to science fiction and hype. For example, before and after the CeBit06 conference there was extensive news coverage and an extended TV demonstration. The public is more interested in games or sci-fi applications than in typing or rehabilitation implementations of BCI.

SUMMARY AND CONCLUSIONS

Scientific and Technological Challenges and Solutions

The BBCI group aims to reduce the time for patient training in BCI applications and to enable faster neuroscience experimentation protocols by using signals from an overlearned movement. In order to have rapid, real-time control, the BBCI is focused on extracting single-trial EEG control signals. The burden of classification is therefore laid on machine learning with decision trees or support vector machines classifying multiclass signal features. These machines will be interacting with adaptive humans and will require adaptive algorithms. Moreover, the adaptive loop will need to be stable.

There is a need to demonstrate applicability for real-world, locked-in patients, which will be approached through the collaboration with Charité. Functional MRI shows that there are gravitational activity centers in the brain where there is activity that needs to be de-intermingled to get better resolution and control. What is possible with noninvasive approaches with respect to topographical mapping? Invasive work and animal BCIs are more advanced here than in the United States. Long-term (more than five-year) collaborations between invasive and noninvasive groups could allow investigation of the limitations of the approaches and the study of how signals are intermingled in cortex. In addition, the development of a toolbox for sensor fusion to bring various signals together (ECoG, EEG, MEG, NIRS, ensemble activity and chemical states) could further help understand brain activity and recording-method limitations.

BCI Competition III revealed a need for data sharing in order to test different algorithms with standardized datasets and error scoring. Someone needs to sponsor and support such a databank, which would not, however, preclude the need for real-time testing of developed algorithms and systems.

Competitive Advantages

The BBCI project is supported by the Bernstein Center for Computational Neuroscience emphasis on signal precision and variability and Fraunhofer's IDA strength in statistical and machine-learning methodologies. These include supervised learning approaches such as nonlinear classification, regression, and prediction, with support vector mMachines (SVM) (in which Dr. Muller was a pioneer), and kernel Fisher discriminant (KFD) analysis, and unsupervised learning methods such as support vector data description (SVDD), clustering, and nonlinear feature extraction for explorative data analysis. The group also has expertise in signal processing, which includes denoising and blind source separation (BSS, ICA). Additionally, group members are adept at nonstationary time series analysis, which is important for adaptive learning. These advanced machine-learning techniques are powerful tools for BCI applications, because they enable information extraction from a high-dimensional feature space, even with small training sets.

The BBCI's Charité group provides expertise in neurophysiology and cognitive neuroscience, as well as access to patients with movement disorders.

REFERENCES

Blankertz, B., G. Dornhege, C. Schafer, R. Krepki, J. Kohlmorgen, K.R. Muller, V. Kunzmann, F. Losch, and G. Curio. 2003. Boosting bit rates and error detection for the classification of fast-paced motor commands based on single-trial EEG analysis. *IEEE Trans. Neural Syst. Rehab. Eng.* 11:127–131.

Dornhege, G., B. Blankertz, G. Curio, and K.R. Muller. 2004. Boosting bit rates in noninvasive EEG single-trial classifications by feature combination and multiclass paradigms. *IEEE Trans. Biomed. Eng.* 51:993–1002.

Blankertz B., K.R. Muller, G. Curio, T.M. Vaughan, G. Schalk, J.R. Wolpaw, A. Schlogl, C. Neuper, G. Pfurtscheller, T. Hinterberger, M. Schroder, and N. Birbaumer. 2004. The BCI competition 2003: Progress and perspectives in detection and discrimination of EEG single trials. *IEEE Trans. Biomed. Eng.* 51:1044–1051.

Fraunhofer-Gesellschaft. 2005. *Annual report 2005*. Available online at http://www.fraunhofer.de/fhg/EN/publications/annualreports/jb_2005.jsp.

Site: **Commissariat à l'Énergie Atomique (CEA)**
CEA-Fontenay aux Roses
Boite Postale 6
F-92265 Fontenay aux Roses Cedex, France
http://www-list.cea.fr/

Date Visited: May 30, 2006

WTEC Attendees: J. Principe (report author), J. Chapin, S. Demir, P. Tresco, H. Ali

Hosts: Dr. Philippe Bidaud
Professor at University Paris 6 and
Director of the Robotics Laboratory of Paris
Tel: +33 1 46 54 78 91
Email: bidaud@robot.jussieu.fr

BACKGROUND

The Commissariat à l'Énergie Atomique (CEA) collaborates with the Collège de France (CdF) in areas of interest for brain-machine interfaces. Please see the site report for the Collège de France. At the CEA we were greeted by a large group of professors and students from the Institut de Systèmes Intelligents et de Robotique (ISIR) of the Université Pierre et Marie Curie, Paris 6 (Professor Phillip Bidaud); from the Interactive Systems Group of Software Intensive Technologies (Laboratoire d'Intégration des Systèmes et des Technologies, LIST) (Professor Rodolphe Gelin [rodolphe.gelin@cea.fr]); and from the AnimatLab (Professor Jean-Arcady Meyer [jean-arcadymeyer@lip6.fr]).

RESEARCH AND DEVELOPMENT

The WTEC team spent quite some time at CEA with the robotics group from both CEA and Université Pierre et Marie Curie (Paris 6), which is conducting state-of-the-art work on master-slave systems as well as on autonomous robots. The work on master-slave systems is sophisticated, using virtual reality environments and haptic interfaces, interactive robotics (to decrease the numbers of degrees of freedom), and many medical applications (assistive robotics, tele- and high-precision surgery, micro- and nanomanipulation—particularly impressive—and rehabilitation). Applications to the nuclear and services industries is also well advanced (interesting projects in assisted driving and navigation). Interesting use of robotics for education (or edutainment) stressing the use of imagery for mathematical and physical understanding is also being pursued.

In terms of autonomous systems, the research is being conducted at the Animat Lab from U. of Paris 6, and it is influenced by the College of France collaboration. We saw three projects of interest: the Psikharpax (an artificial rat), the Robur (artificial bird), and the Kodamat (intelligent bots). Each has unique features that are bioinspired. The most salient from our perspective is the use of reinforcement learning to teach Psikharpax its world and group behavior in Kodamat, and the use of genetic optimization to create the neural controller for Robur.

The CEA/U of Paris 6 team has a history of IP creation with several patents in the robotics area: manual control with tactile and/or kinesthetic feedback, articulated mechanical arm, control arm, transmission by screw, nut and cable attached to screw, articulated mechanism comprising a cable-driven reduction gear that can be used in a robot arm, telescopic arm, and control device with three parallel legs.

The CEA/U of Paris 6 team is also involved in spin-off company creation: ACTICM was created in 2000 by two LIST engineers. The company offers to industry 3D measurement and digitalization technology that combines the principles of photogrammetry with digital image processing. ACTICM is now operating in the region of Grenoble, with the support of Emertec and Anvar funds, and has a staff of around ten. The company continues to work in close collabor-ation with the LIST, particularly in the field of virtual reality research on two RNTL (French national software technology innovation and research network) projects relating to heightened reality and three-dimensional environment reconstruction.

HAPTION (http://www.haption.com/index.php?lang=eng&p=0), which was created in September 2001, designs, produces, and markets computerized force feedback peripherals, also referred to as "haptic interfaces." As a LIST spin-off, HAPTION has benefited from the granting of technology through a license and by the training of its staff in the techniques of design and construction of force feedback systems.

NEWPHENIX (http://www.new-phenix.com/) was formed in early 2004 and is one of the first editors in the world to supply a range of fully-operational products combining use of text and images in a truly cross-lingual context. Using cutting-edge technology originating from the LIST, this startup commercializes a range of products based on an engine with a set of multimedia information modeling and analysis functions (text and image) specially designed to enable automatic use of its content.

SUMMARY AND CONCLUSIONS

This group (CdF and CEA/U of Paris 6) is already orchestrating a substantial portion of the European research and development in bioinspired robotics. The group is academically strong, with many publications and books in the area of brain recognition of spatial orientation, robotics, knowledge engineering, virtual reality, and sensory interfaces. Although the collaboration with the CEA is still at an early stage, and it is limited to specific topics, it has the ingredients and the potential to develop into a full-fledged neuro-robotics effort. Professor Berthoz is an eminent scientist and a visionary. The LIST and ISIR comprise a strong group of roboticists and signal processing/computer scientists, with excellent facilities and enthusiastic students. The group is well articulated, with many of the strong players in Europe as is evident in the Neuroprobes project. Funding for the CdF comes from national funding and EEC projects. Funding for LIST is around €3.3 million (equivalent to ~$4.2 million in May 2006) for the robotics component, but only a small percentage is for biomedical applications.

Site: **CNRS/Collège de France Physiology of Perception and Action Laboratory**
11, Place Marcelin Berthelot
75231 Paris, Cedex 05, France
http://www.college-de-france.fr/chaires/chaire8/EN/

Date Visited: May 30, 2006

WTEC Attendees: J. Principe (report author), J. Chapin, S. Demir, P. Tresco, H. Ali

Hosts: Dr. Alain Berthoz
Director of the CNRS-CdF Laboratoire de Physiologie de la Perception et de l'Action (LPPA), and Professor, Collège de France (CdF)
Tel: +33 (0) 1 44 27 16 29
Email: alain.berthoz@college-de-france.fr
Dr. Sidney Wiener
Research Director and Adjunct Director, LPPA
Email: sidney.wiener@college-de-france.fr

BACKGROUND

The Physiology of Perception and Action Laboratory (Laboratoire de Physiologie de la Perception et de l'Action, LPPA) is a joint undertaking of France's National Center for Scientific Research (Centre National de la Recherche Scientifique, CNRS) and the Collège de France (CdF).

The LPPA group is divided into two complementary specialties: neuroscience and robotics. The neuroscience component is devoted to the study of the neural bases of four major cognitive sensorimotor-motor functions:

- Saccades and eye movements
- Generation of locomotor trajectories
- Cognitive strategies for spatial memory
- Perception and expression of emotions.

These functions are studied by means of various methods: imagery, recording of movements, and use of virtual reality in both healthy subjects and patients. A space exploration component studies the effect of microgravity on these sensori-motor functions. Mathematical models of the biological functions are developed with two goals: (1) model validation through simulation in robotic platforms developed in cooperation with signal-processing specialists and robotics engineers (at the AnimatLab of LIP6 at University Pierre and Marie Curie, Paris 6, and

École Normale Supérieure); and (2) inspire biological principles for autonomous robotic design. This symbiosis, called neurorobotics, seems very important.

Work on BCI was only briefly mentioned. This is primarily associated with the EU Sixth Framework Neuroprobes ("Development of Multifunctional Microprobe Arrays for Cerebral Applications") project (http://www.neuroprobes.org/) in which CNRS-CdF-LPPA participates, and which aims to develop a new integrated tool that combines multiple functions to allow multichannel electrical recording and stimulation as well as chemical sensing and stimulation.

Unfortunately, WTEC panelists did not have the opportunity to visit the College of France (CdF), but we did have the opportunity to meet with the director of the LPPA, Professor Alain Berthoz, and Dr. Sidney Wiener, research director and adjunct laboratory director of the LPPA and an expert in the neural bases of spatial cognition.

RESEARCH ACTIVITIES

Neurorobotics is the major topic of concern in regard to BCI research at the CdF. The aims of neurorobotics are both scientific and technological. On the scientific front, the aim is to test in an "animat," the plausibility of the neuroscience modeling of the nervous system and of the mechanisms that contribute to its adaptive capacities. The animat is an artificial system that is confronted with situations similar to those faced by a real animal. On the technology side, the goal is to develop adaptive autonomous robots that choose goals and actions to ensure "survivability" and achieve their mission without the assistance of human operators. This is articulated through two integrated projects of the EC Sixth Framework CogSys (Cognitive Systems) program in which LPPA is a partner laboratory. One is called ICEA (Integrating Cognition Emotion and Autonomy; http://www2.his.se/icea/); the other is BACS (Bayesian Approach to Cognitive Systems; http://www.bacs.ethz.ch/).

SUMMARY AND CONCLUSIONS

This group is orchestrating a substantial portion of the European research and development in bioinspired robotics. The group is academically strong, with many publications and books in the area of brain recognition of spatial orientation, robotics, knowledge engineering, virtual reality, and sensory interfaces. Although the collaboration with the CEA (see earlier site report) is still at an early stage and limited to specific topics, it has the ingredients and the potential to develop into a full-fledged neurorobotics effort. Professor Berthoz is an eminent scientist and a visionary. The group is well articulated, with many of the strong players in Europe, as is evident in the Neuroprobes project. Support for the LPPA comes from national funding and EEC projects.

One of the key points of this group is its central role in collaborative efforts at the EC level in many interdisciplinary projects gravitating around the science and technologies necessary for brain-computer interfaces. Listed below are members of the Nuroprobes project, besides LPPA, to indicate the extent of the collaborative efforts:

Neuroprobes Project

- Dr. Herc Neves (Coordinator), Interuniversitair Micro-Elektronica Centrum (IMEC; Belgium)
- Joerg Kohnle, Hahn-Schickard-Gesellschaft, Institut für Mikro und Informationstechnik (HSG-IMIT; Germany)
- Patrick Ruther and Dr. Oliver Paul, Institute of Microsystem Technology (IMTEK; University of Freiburg Germany)
- Professor Sven Oscarsson, Mälardalen University, and Dr. Karin D. Caldwell, Department/Centre for Surface Biotechnology, Uppsala University (Sweden)
- Professor Nicolaas F. de Rooij, Institute of Microtechnology(IMT), University of Neuchâtel (Switzerland)
- Dr. Guy A. Orban, Laboratorium voor Neuro-en Psychofysiologie, Katholieke Universiteit Leuven (Belgium)
- Professor Trevor Robbins, Department of Experimental Psychology, University of Cambridge (United Kingdom)
- Professor Giacomo Rizzolatti, School of Medicine, University of Parma (Italy)
- Dr. Istvan Ulbert, Institute of Psychology, Hungarian Academy of Sciences (Hungary)
- Professor Eduardo Fernandez, University Miguel Hernández de Elche (Spain)
- Dr. Youri V. Ponomarev, Philips Innovative Technology Solutions (Belgium)
- Mr. Micha Mulder, Micronit Microfluidics (The Netherlands)
- Dr. Carl Van Himbeeck and Dr. ir Ben Kloeck, Cochlear Technology Centre (United Kingdom).

REFERENCES

Albertin, S.V., A.B. Mulder, E. Tabuchi, M.B. Zugaro, and S.I. Wiener. 2000. Lesions of the medial shell of the nucleus accumbens impair rats in finding larger rewards, but spare reward-seeking behavior. *Behav. Brain. Res.* 117:173–183.

Berthoz A., I. Israel, P. Georges-Francois, R. Grasso, and T. Tsuzuku.1995. Spatial memory of body linear displacement: what is being stored? *Science* 2699:95–98.

Burguiere E., A. Arleo, M. Hojjati, Y. Elgersma, C.I. De Zeeuw, A. Berthoz, and L. Rondi-Reig. 2005. Spatial navigation impairment in mice lacking cerebellar LTD: a motor adaptation deficit? *Nat. Neurosci.* 8:1292–1294.

Girard, B. and A. Berthoz. 2005. From brainstem to cortex: Computational models of saccade generation circuitry. *Prog. Neurobiol.* 77:215–251.

McIntyre, J., M. Zago, A. Berthoz, and F. Lacquaniti. 2001. Does the brain model Newton's laws? *Nat. Neurosci.* 4:693–694.

Viaud-Delmon, I., Y.P. Ivanenko, A. Berthoz, and R. Jouvent. 1998. Sex, lies, and virtual reality. *Nat. Neurosci.* 1:15–16.

Watroba, L., P. Buser, and C. Milleret. 2001. Impairment of binocular vision in the adult cat induces plastic changes in the callosal cortical map. *Eur. J. Neurosci.* 14:1021–1029.

Wexler, M., F. Panerai, I. Lamouret, and J. Droulez. 2001. Self motion and the perception of stationary objects. *Nature* 409:85–88.

Wiener, S.I. 1996. Spatial, behavioral, and sensory correlates of hippocampal CA1 complex spike cell activity: implications for information processing functions. *Prog. Neurobiol.* 49:335–361.

Zugaro, M.B., A. Arleo, A. Berthoz, and S.I. Wiener. 2003. Rapid spatial reorientation and head direction cells. *J. Neurosci.* 23:3478–3482.

Site: **European Commission, Research Directorate General**
Unit F2 (Major Diseases)
CDMA, Office 2/5
B-1049 Brussels, Belgium

Date Visited: June 7, 2006

WTEC Attendees: Semahat S. Demir (report author)

Hosts: Dr. Philippe Cupers
Scientific Officer
Tel: + 32 2 2998796
Fax: +32 2 2955365
Email: philippe.cupers@ec.europa.eu

OVERVIEW OF EU FUNDING MECHANISMS

Dr. Cupers described EU's different funding instruments of Integrated Projects: five years, 10–20 partners; STReP (Strategic Targeted Research Projects): three years, €3,000,000, 5–8 partners; NoE (Network of Excellence): broader for research partners; SSA (Specific Support Action): €50,000–300,000 for a few partners to prepare workshop, white paper, etc.; CA (Coordinating Action): €1,000,000–2,000,000, 5–20 partners). The EU 6th Framework Programme (2002–2006) has funded mainly consortia.

Dr. Cupers also described the review and evaluation process. The evaluations are performed as peer reviews. Evaluation is completed by independent experts. IP proposals are reviewed by 7–9 reviewers, and small proposals are reviewed by 4-5 reviewers.

The review criteria depend upon

- Relevance of the topic
- Quality of consortium management and human resources
- Quality of scientific work
- Impact of the project
- Mobilization of the resources.

The review panels must address ethical issues. If the experts cannot review the ethical issues, the proposal may be forwarded to the ethical issues panel. For example, there are differences in stem cell research in EU countries; Sweden allows embryonic stem cell research whereas Germany does not. Ethical approval of each country is required in partnerships.

Consortium funding requires a minimum of three different partners from three different countries. Partners can be from universities, government labs, and industry.

Spin-off companies can also be developed. The consortia must also address intellectual property and licensing issues in their consortium agreement.

The program/scientific officer chooses the reviewers, sends proposals for reviews, and compiles the scores of the criteria. The annual reports of the grants are also reviewed by the officer and the panels.

Five percent of research funded in EU countries is by the European Commission. Ninety-five percent of research funding is in the individual country.

BRAIN SCIENCE

The Brain Science grants have been funded by EU's Research and Technology Development (RTD) and Information Society (INFSO) Directorate. Some examples of brain science related to brain-computer interfaces are listed below:

- Overview of Information Society Technologies (IST) projects related to neuroscience, http://cordis.europa.eu/ist/ (search "neuroscience").
- EU Sixth Framework Programme (FP6), http://cordis.europa.eu/fp6/.
- The Future and Emerging Technologies (FET) website for the BIO-i3 call on Neuroinformatics, http://www.cordis.eu/ist/fet/bioit.htm; this links to the DAISY (PI: Kennedy) and FACETS (Investigators Meier and Markram) integrated projects; http://daisy.ini.unizh.ch/; http://www.facets-project.org.
- Info on the Neurobotics integrated project (development of prostheses linked to the CNS/PNS) can be found at: http://www.neurobotics.info/.
- Some neuron-related STRePs have also been funded under the FET Open call: Golden Brain (http://icadc.cordis.lu/fepcgi/srchidadb?ACTION=D&CALLER=PROJ_IST&QM_EP_RCN_A=72301, coordinator IMEC, Belgium) which deals with electronic interfaces to neurons
- Mental Augmentation through Determination of Intended Action (MAIA), Non Invasive Brain Interaction with Robots (http://www.maia-project.org/, coordinator IDIAP, Switzerland) is researching a noninvasive BCI.

SUMMARY AND CONCLUSIONS

The EU 7th Framework program (2007–2013) will have brain research as a priority, according to Dr. Cupers. The focus might be more in neuroinformatics, neural networking, and databasing of the brain. OECD (Organisation for Economic and Commercial Development) in Paris is also placing neuroinformatics as one of the future areas. The EC supports but has not signed the memorandum of understanding. The EU 7th Framework Programme will also fund investigator-initiated grants; the EU 6th Framework included more contract-based funding. Research for 6th Framework (2002–2006) had a budget of €17.5 billion for four years. This budget funded many disciplines (sciences, engineering, fishery, health, etc). The request for 7th Framework is €6.5 billion per year for seven years.

Site: **Graz University of Technology**
Laboratory of Brain-Computer Interfaces
Institute for Knowledge and Discovery
8010 Graz, Austria
http://bci.tugraz.at

Date Visited: June 2, 2006

WTEC Attendees: D. McFarland (report author), T. Berger, G. Gerhardt, D. Taylor, G. Lewison

Host: Gert Pfurtscheller
Tel: +43 316 873 5300
Fax: +43 316 873 5349
Email: pfurtscheller@tugraz.at

BACKGROUND

The Laboratory of Brain-Computer Interfaces (BCI) is in the Institute for Knowledge Discovery at the Graz University. It involves cooperative research by three teams specializing in biomedical engineering (headed by G. Pfurtscheller), psychology (C. Neuper), and computer science (A. Schlogl). The Graz group has published articles about BCI since 1994. During this time the lab has published 140 articles, as listed in PubMed, of which at least 58 are directly concerned with BCI. Current funded projects related to BCI include Presenccia, Eye-to-IT, and Direct Brain Interface.

The Presenccia project consists of three parts. These are developing strategies and designs for asynchronous (uncued) BCI, development of wireless BCI hardware, and a feasibility study. The eye-to-IT project will develop methods for online classification of EEG to support prompting during foreign language translation. The ECoG project is funded as a BRP from the NIH in conjunction with Simon Levine at the University of Michigan, School of Public Health.

RESEARCH AND DEVELOPMENT ACTIVITIES

BCI systems are designed to obtain useful control signals directly from the brain. Applications include spelling devices, neuroprosthesis, biofeedback therapy (e.g., for reduction of seizures, stroke therapy, and attention problems), person identification, and navigation in virtual environments.

There are several potential control features that can be derived from the surface EEG. These include the P300 potential, slow cortical potentials, steady-state potentials, and event-related desynchronizations and synchronizations (ERD/ERS).

The Graz BCI group specializes in the use of ERD/ERS. As the pioneers in this area (Pfurtscheller and Aranibar, 1979), group members have developed an extensive knowledge base of these phenomena. ERDs are a decrease in a spectral peak (amplitude modulation) that occurs in response to some event. Basic EEG rhythms are identified by location on the scalp and reactivity. ERDs occurring in response to motor imagery over central regions are particularly relevant for BCI research. The alpha-band rhythm associated with motor function is known as the mu rhythm. ERDs also occur in response to other events, such as the desynchronization of the posterior alpha rhythm with visual stimulation.

Recently, Pfurtscheller and Lopes de Silva have described phenomena of focal ERD and surround ERS. As an example, foot movement causes desynchronization at central midline sites over areas associated with representation of the feet in sensory-motor cortex. Furthermore, more lateral areas associated with representation of the hands show enhanced synchronization. Generally, alpha-band activity is interpreted as an idling rhythm of underlying cortex and desynchronization in this range (10–13 Hz) is associated with cortical activity. Thus, the focal ERD and surrounding ERS phenomena is interpreted as a result of activation of foot areas and inhibition of surrounding areas associated with other body parts.

A key feature of the mu rhythm is that motor imagery produces ERD/ERS in a manner analogous to actual movement. This allows the use of these signals in individuals with motor-control dysfunction. The Graz group has recently described an important distinction between kinesthetic and visual motor imagery. Subjects were given instructions to imagine how movement feels, or to imagine how it looks. Results indicate that mu-rhythm ERD/ERS is much more robust with kinesthetic imagery.

These basic research findings of the Graz group provide a rational basis for the design of BCI systems. The group has also been a leader in development of signal processing methods. Group members are working on navigation through a virtual environment by means of features extracted from the surface EEG. Since navigation is not cued but rather asynchronous, the classification problem is particularly difficult. Separate classifiers are trained for detection of presence and detection of the direction of movement. Preprocessing consisted of either independent components analysis or common spatial patterns. Features consisted of band power, phase information, or autoregressive parameters. Results reported in a recent publication (Pfurtscheller et al., 2006) showed that subjects were able to successfully navigate the virtual environment. This basic paradigm was shown to the WTEC panel as a live demonstration in which a BCI user successfully navigated a virtual maze (see Figure B.4).

The Graz group is involved in other applications including reduction of seizures through biofeedback and the use of EEG feedback in virtual reality as a therapy in stroke rehabilitation. These projects are based on collaborative arrangements with the Medical University Graz (B. Urlesberger), Clinic Judendorf-Straßengel (P. Grieshofer), and the Orthopedic University Hospital II Heidelberg (R. Rupp).

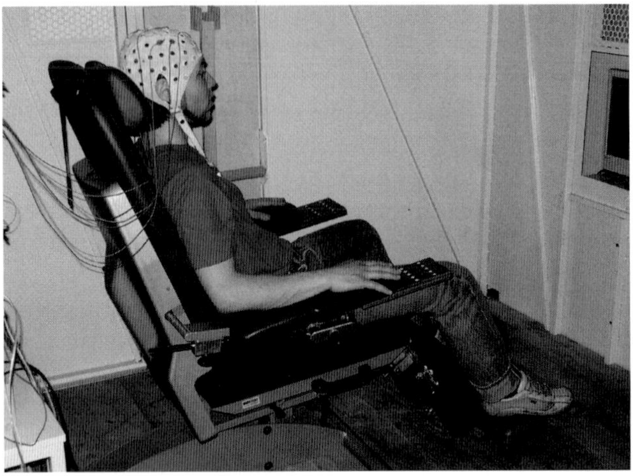

Figure B.4. Live demonstration of navigation in a virtual maze. Navigation was based upon asynchronously-detected. event-related synchronizations and desynchronizations.

The Graz group has also recently shown that feedback in the form of images of a moving hand is particularly effective for a BCI. Another live demonstration showed a user manipulating a mechanical arm with four degrees of freedom based on the steady-state visual potential. This is shown in Figure B.5.

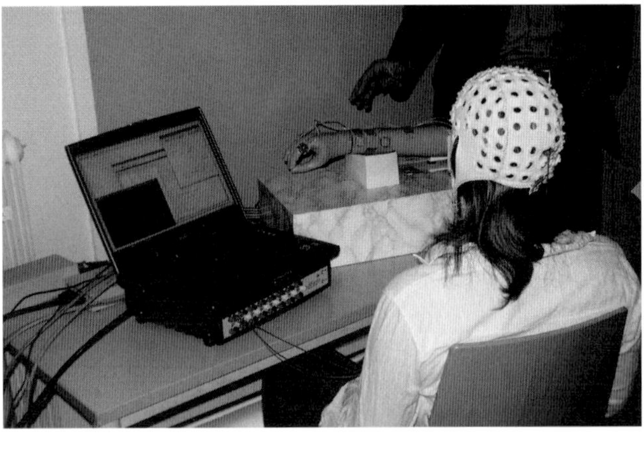

Figure B.5. Live demonstration of a mechanical arm controlled in four degrees of freedom by steady-state visual potentials. Each of the lights oscillates at a different characteristic frequency. Attending to a given light enhances EEG at that frequency and harmonics. When the BCI detects this signal, it causes the arm to move about the associated axis.

CONCLUSIONS

The Graz group has an active research program mainly concerned with the development of noninvasive methods for EEG-based communication and control. It is one of the leaders in this field.

REFERENCES

Muller-Putz, G.R., R. Scherer, C. Brauneis, and G. Pfurtscheller. 2005. Steady-state visual evoked potential (SSVEP)-based communication: impact of harmonic frequency components. *Journal of Neural Engineering* 2:123–130.

Neuper C., R. Scherer, M. Reiner, and G. Pfurtscheller. 2005. Imagery of motor actions: differential effects of kinesthetic and visual-motor mode of imagery in single-trial EEG. *Brain Research: Cognitive Brain Research* 25:668–677.

Pfurtscheller, G., and A. Aranibar. 1979. Evaluation of event-related desynchronization (ERD) preceeding and following voluntary self-paced movement. *Electroencephalography and Clinical Neurophysiology* 46:138–146.

Pfurtscheller, G., and F.H. Lopes Da Silva. 1999. Event-related EEG/MEG synchronization and desynchronization: basic principles. *Clinical Neurophysiology* 110:1842–1857.

Pfurtscheller, G., R. Leeb, C. Keinrath, D. Friedman, C. Neuper, C. Guger, and M. Slater. 2006. Walking from thought. *Brain Research* 1071:145–152.

Pfurtscheller, G., G.R. Muller-Putz, A. Schlogl, B. Graimann, R. Scherer, R. Leeb, C. Brunner, C. Keinrath, R. Lee, G. Townsend, C. Vidaurre, and C. Neuper. 2006. 15 Years of BCI research at Graz University of Technology: Current projects. *IEEE Transactions on Neural Systems and Rehabilitation Research* 14:205–210.

Site: **Guger Technologies OEG, "g.tec"**
Herbersteinstrasse 60
8020 Graz, Austria
http://www.gtec.at

Date Visited: May 29, 2006

WTEC Attendees: G. Gerhardt (report author), T. Berger, D. McFarland, D. Taylor, W. Soussou, G. Lewison.

Hosts: Dr. Christoph Guger and Dr. Gunter Edlinger
Tel: +43 316 675 106
Fax: +43 316 675 106 39
Email: office@gtec.at

BACKGROUND

The private company g.tec was founded in 1999 by Dr. Christoph Guger and Dr. Günter Edlinger, who were students in the Graz University of Technology with Professor Gert Pfurtscheller. This company is composed of an interdisciplinary team of engineers, computer scientists, and psychologists. Products developed by g.tec include, but are not limited to, sensors and electrodes, computer-controlled recording systems, EEG/ECoG/ECG/EMG/EOG signal amplifiers, data acquisition software, real-time data acquisition processing and analysis neuro-feedback, offline signal analysis, and neurostimulation. Currently, g.tec exports products into more than 40 countries around the world. The primary customers are universities, hospitals, R & D departments, and industry. Guger Technologies has a scientific pipeline with Universities in UK, Spain, and Austria. They jointly sponsor Ph.D. students and research fellows. Students who have finished their degrees and who have successfully worked in this field in related studies are encouraged to join g.tec. In addition, g.tec works in conjunction with Universities and research centers on workshops involving Brain-Computer Interface technology.

RESEARCH AND DEVELOPMENT ACTIVITIES

Internationally known, g.tec is a world supplier of high-quality hardware and software devices for brain-computer interface applications involving EEG, ECoG, EMG, EOG, and ECG amplifier technology. The company develops and distributes technologies for laboratory and mobile applications. Its hardware technologies involving analog data acquisition are some of the finest in the field, and its low-noise amplifier technology ranging from 8–64 channels are used by a large number of laboratories in the field of brain-computer interface technology throughout the world. The company's software is largely based on C++ applications involving MATLAB® and Simulink® products from MathWorks, Inc.

Electrodes and Electrode Technologies

For a variety of Brain-Computer Interface technologies, g.tec is a source for arguably one of the best head caps used in the field involving wet electrode recordings. To name a few other products, g.tec also provides ECG/EMG/EOG electrodes, cables, consumables, and various sensors for measuring respiration, pulse, galvanic skin response, breathing sounds, snoring sounds, swallowing sounds, and temperature movement acceleration. The company's unique head cap (EEG electrodes) design allows for some of the best signal-to-noise ratio achievable from wet electrode technology. In particular, the electrode cap design requires some extra time for attachment of electrodes, but achieves excellent signal-to-noise characteristics. This highly versatile design can be employed with other g.tec products and amplifiers, as well as instrumentation produced by other suppliers.

Amplifier Technology

For amplifiers, g.tec is a source for some of the best analog amplifier technology for noninvasive and invasive BCI applications. G.tec's researchers have developed a standalone data acquisition analog amplifier referred to as g®.BSamp, or biosignal amplifier, and an advanced software-supported g®.USBamp (USB biosignal amplifier, Figure B.6) for simultaneous recordings from 16 to 64 electrodes. The company's novel mobile laboratory EEG system (Figure 2.16 in Chapter 2), known as g®.MOBIlab, is a battery-powered system for four EEG/EOG channels, ECG/EMG channels, and two analog inputs that can be utilized for other sensors. The system operates for up to approximately one week with a single set of batteries. This interface to g.tec's portable biosignal acquisition and analysis system is a perfect tool for recording multimodal biosignal data on a standard PC, pocket PC, or notebook. This is a unique mobile technology that has been adopted for use in the United States for computer animation.

Figure B.6. USB biosignal amplifier from g.tec.

Software

Another major avenue in the g.tec line is development of software technologies for data acquisition, real-time data processing, signal analysis, stimulation, and conducting Brain-Computer Interface technologies. As a MathWorks® partner, g.tec provides software developed in conjunction with MATLAB® and Simulink® drivers. These are software packages that are fully integrated with g.tec's amplifier technologies and allow for recording processing and data analysis of a wide range of EEG-signal activity. Clearly, g.tec is one of the major developers of integrated systems for Brain-Computer Interface technology involving EEG and related surface recording and other noninvasive and invasive recording technologies.

Future Hardware Development

Along with other hardware manufacturers, g.tec has been working to make its g.USBamp hardware systems compatible for potential human use. In conjunction with universities in Europe and the United States, g.tec has been collaborating to integrate its system with the ECoG work on epilepsy recordings being carried out at Graz University of Technology for combined BCI/ECoG electrophysiological investigations in patients with advanced epilepsy. This is an interesting and novel area of study combining both surface recording and ECoG-type measurements.

One of the interesting areas of development by g.tec is an amplifier system that will be able to carry out EEG/ECoG depth electrode recordings and eventually the instrumentation certification for patient use. This may expand g.tec into the field of multiple-single-unit recording for more invasive recordings such as ECoG and multiple-single-unit electrode recordings. The company is also working on a wireless EEG cap technology for possible coupling with the g.MOBIlab technology.

This company has a series of international cooperations and projects involving development for BCI 2000, real-time data processing in Windows, and the development of high-altitude medicine EEG technology. It has a variety of partnerships with MathWorks®; University College London; and several institutes and initiatives in Austria, Spain, Germany, the United States (including NASA Langley), Israel, and others. However, g.tec does not have an extensive focus on human applications. In addition, it is not developing any new dry-type EEG electrodes, which are sorely needed in the field. Finally, g.tec is not developing any preamplifier technology for electrode design.

SUMMARY AND CONCLUSIONS

Austria's g.tec is one of the major developers of systems for noninvasive and invasive brain-computer interface technologies. The company has excellent products, including amplifiers for recordings of up to 64 channels of EEG/ECOG/ECG/

EMG/EOG and related signals. Its totally portable systems are state of the art and represent some of the most unique products on the market for freely moving measures of BCI signals. The company has a strong software development program, which is extensively linked with MATLAB® and Simulink® drivers. It is clearly the source for most laboratories of high-quality amplifiers and software modules to control brain-computer interface technology involving both noninvasive measures such as EEG signals and invasive measures such as ECoG signals.

Site: **Max Planck Institute of Biochemistry**
Am Klopferspitz 18
D-82152 Martinsried, Germany
http://www.biochem.mpg.de/en/

Date Visited: June 2, 2006

WTEC Attendees: G. Gerhardt (report author), D. McFarland, D. Taylor, T. Berger, W. Soussou, G. Lewison

Hosts: Prof. Dr. Peter Fromherz
Department of Membrane and Neurophysics
Tel: +49 89 85 78 x2820
Email: fromherz@biochem.mpg.de
http://www.biochem.mpg.de/en/rd/fromherz/
Dr. Roland Thewes
Guest from Infineon
Email: roland.thewes@infineon.com

BACKGROUND

Peter Fromherz directs a group of researchers in the Max Planck Institute of Biochemistry Department of Membrane and Neurophysics who are primarily physicists, but the group also includes biochemists and neurobiologists. Professor Fromherz is also professor for experimental biophysics in the Physics Department of the Technical University Munich. Max Planck Institute directors hold dual appointment at universities to accommodate students. Because here are few courses at the Institute, many students must find a university advisor who will allow them to do research at the institute.

Dr. Fromherz's lab has about 15 students and postdocs, with about 10 setups for neuron-chip coupling and electrophysiology. The lab is mostly independent from its neighboring labs. The main focus is development of silicon/neuron interface technology. This has encompassed development of semiconductor-based microarray chips for recording and stimulation of neurons in culture. In particular, his group has focused on neuron interfaces that involve neuron-to-chip separation of 50 nanometers from cell bodies of neurons relative to the silicon interfaces. Through these interfaces, researchers can carry out either voltage recordings with transistors or electrical stimulation of cells with capacitors. This unique group has developed a remarkable series of semiconductor-based chip technologies that allow for up to 16,000 recording sites adjacent to organotypic slice cultures of the hippocampus. In addition, they have recorded from isolated neurons and ensembles of neurons from snails and rats *in vitro*. This group is clearly an exceptional research group that has carried out studies with silicon/neuron interfaces that are unique and truly state of the art.

FUNDING SOURCES AND COMMERCIALIZATION

Program directors at Max-Planck Institute have ample space and facilities in addition to large amounts of funding to use as they wish. They are free to pursue goals that might otherwise be hard to finance and are free from grant-writing duties. The directors are evaluated in two-year periods by an international advisory board of high-rank peers. If positively evaluated, there is no risk of decrease or loss of support.

Even though the Max Planck Institute has a center for patenting and licensing, Dr. Fromherz does not rush for patents or commercialization.

Industrial Collaborations

Infineon was a subsidiary of Siemens that manufactured CMOS chips, but its research branch was closed after company reorganization two years ago. Before, there was collaboration with Dr. Roland Thewes of Infineon in the area of field-effect transistors (FET) arrays. Infineon and other companies have sought larger output products than the market for FET arrays would provide. At this time, pharmaceutical companies are not too interested in FETs for drug screening.

Academic Collaborations

A few groups interested in the technology have sought collaborations with Dr. Fromherz.

RESEARCH AND DEVELOPMENT ACTIVITIES

Short- and Long-Term Scientific Goals

The major goal of Dr. Fromherz's lab is to study the basic biophysical processes in the interface of brain tissue and semiconductor chips. The lab focuses more on developing neuron-silicon interfaces than on BCI applications. The effort ranges from investigating the structure of the contact to developing devices for neuro-prosthetics and neurocomputing. Arrays of FETs and of capacitive stimulators are being developed that can be used for neuroscience research to study neuronal microcircuitry or to investigate drug effects with recombinant channels. A major competition of the FET array is high-speed imaging of voltage sensitive dyes. There are no plans for constructing *in vivo* arrays or FET-EEG sensors.

Tools, Methods Used, and Sources

Reliable recording and stimulation of nerve cells and brain tissue requires knowledge of the microscopic structure and the electrical features of the contact. Novel fluorescence techniques were developed such as fluorescence interference contrast (FLIC) microscopy and fluorescence Stark phase microscopy. The Fromherz group found that cell and chips are separated by a 50 nm layer of bulk electrolyte that is in exchange with the culture medium. That layer gives rise to an electrical resistance R on the order of one megaohm that determines the strength of interfacing between cells and transistors or capacitors. The resistance is also a source of thermal noise with a voltage power density $Sv = 4k_BTR$ (Nyquist).

Transistors are used to study ion transport across cell membranes. Recombinant Na and K channels were tested. Comparison with patch-clamp recording shows that cell adhesion does not damage the functional properties of the ion channels. Applying voltage ramps to a capacitor allow activation of ion channels by extracellular stimulation. For these experiments, the silicon chips were insulated with TiO_2 to get higher capacitance and better stability without DC current across the electrode.

Transistor recording and capacitor stimulation are also used to observe and activate ion channels in cultured neurons for snails and rats to induce and record electrical excitation under conditions without electrochemical perturbations and without electroporation of the membrane. This clearly has outstanding implications from the viewpoint of neural network studies *in vitro* that are not achieved with other electrode technologies. An example of a neuron from a rat brain on a linear array field-effect transistor is seen in Figure B.7 below.

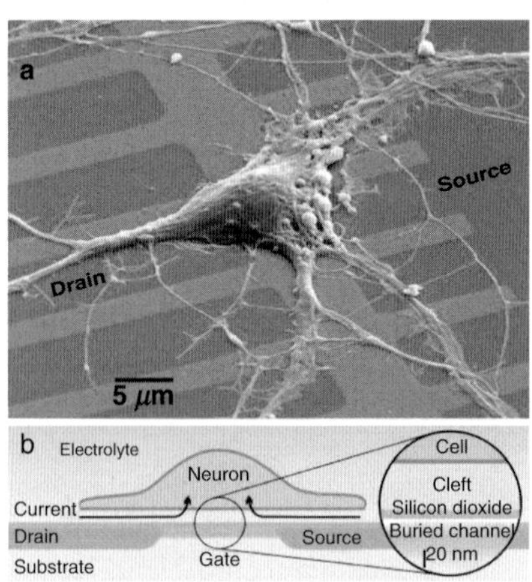

Figure B.7. Rat neuron on electrolyte-oxide-silicon (EOS) field effect transistor. (a) Electron micrographs (colorized) of a hippocampal neuron on a silicon chip array; (b) schematic cross-section of a neuron on a buried-channel field-effect transistor with blow-up (drawn to scale) of the contact area.

Cultured hippocampal slices are stimulated through capacitors and their responses recorded with FET. Increasing stimulation intensity yielded stepwise increases in responses, suggesting that single neurons in the slice may be coupled to the capacitors.

CMOS chips with a 16,000 multitransistor array and with a multicapacitor array were manufactured at Infineon. They work just like the simple silicon chips. The TiO_2 surface provides high stability and allows culturing of dissociated neurons and even brain slices without corrosion. The chips are used to stimulate snail neurons by capacitive interactions and to record excitation with transistors. Distribution of channels can be mapped when there are many transistors below each neuron.

High-Resolution Multitransistor Array Recording of Electrical Field Potentials in Cultured Rat Brain Slices

The team had recently recorded electrical activity in cultured hippocampal slices using a multitransistor array (MTA), with over 16,000 recording sites. Time-resolved imaging was achieved with a resolution of 7.8 μm on an area of 1 mm^2 at a sampling rate of 2,000 samples/second. Individual transistor signals were caused by local-evoked field potentials and agreed with micropipette measurements in amplitude and shape. The spatial recordings provided time-resolved images of evoked field potentials and allow detection of functional correlations over relatively large distances in the slice such as 1×1 mm. This is the highest-resolution recording that is known to date in a hippocampal slice preparation. Such technology may have some potential applicability to not only *in vitro* studies as documented, but to adaptation for potential development of high-resolution *in vivo* arrays for multiple single-unit ("multisingle") electrophysiological recording and field potential measures.

Electrodes and Electrode Technologies

The MTA devices were fabricated in coordination between Dr. Roland Thewes and the laboratory of Professor Peter Fromherz. The microcircuit manufacturing facility at the Max Planck Institute with 10/100 and 100/1,000 cleanroom capability is impressive. This is a remarkable program that allows for education of students that is exceptional for their professional growth, development of pilot semiconductor devices, and training for post-graduate employment. The unique capabilities of the MTA devices developed by Professor Fromherz and his group are unparalleled in the knowledge of the WTEC visiting team. At the present time, there are no commercialization considerations in the near or distant future for this technology.

SUMMARY AND CONCLUSIONS

Scientific and Technological Challenges and Solutions

There is variability in the recorded signal from neurons depending on the surface coupling and axon's location. Some attempts at patterning cells on the surface were abandoned in favor of higher density arrays. Acute slices have a dead cell layer that increases the distance between the FET and the live cells and could hinder signal recording. Cultured slices are better; however, there are promising preliminary results with acute slices. There is significant heating from the FET arrays; thus, there are no plans for developing *in vivo* recording implementations. Perhaps lowering the FET count could help. There are a large number of wires connecting the arrays requiring a need for optical or flip-chip bonding and wireless signal transmission. Because noise of sensors increases with lower size of the sensors, it is not possible to record dendritic signals (also due to poor coupling). However, this method is perfectly well suited for spike detection.

Competitive Advantages Compared to the World

Electrophysiology usually involves coupling between neurons or cells through silver chloride electrodes that get worn through electrical stimulation (regeneration is not controlled). Other metal electrodes are also difficult to control due to electrochemical properties (i.e., corrosion) of metals in ionic solution. The transistor and capacitor electrodes make purely capacitive interactions with the cells and do not affect their environment.

Professor Fromherz and his laboratory embody a unique research group developing state-of-the-art silicon/neuron interface technology that is without parallel. The 16,000 MTA arrays present a look at a resolution of performance that to the knowledge of the WTEC panel has not been achieved by any other laboratory. It does not appear that Professor Fromherz is interested in potential industrial partners or the further development of this technology for commercial applications. This is a unique technology that should be studied further for potential use in BCI technologies.

REFERENCES

Fromherz, P. 2005. The neuron-semiconductor interface. In *Bioelectronics*. Eds. I. Willner and E. Katz. Weinheim: Wiley-VCH, pp 339–394.

Gleixner, R. and P. Fromherz. 2006. The extracellular electrical resistivity in cell adhesion. *Biophysical Journal* 7:495–505.

Hinner, M.J., G. Hubener, and P. Fromherz. 2006. Genetic targeting of individual cells with voltage-sensitive dye by enzymatic activation of membrane binding. *ChemBioChem* 7:495-505.

Hutzler, M., A. Lambacher, B. Eversmann, M. Jenkner, R. Thewes, and P. Fromherz. 2006. High resolution, multitransistor array recordings of electrical field potentials in cultured brain slices. *Journal of Neurophysiology* 96:1638–1645

Schmidtner M. and P. Fromherz. 2006. Functional Na+ channels in cell adhesion probed by transistor recording. *Biophysical Journal* 90:183–189.

Voelker, M., and P. Fromherz. 2006. Nyquist noise of cell adhesion detected in a neuron-silicon transistor. *Physical Review Letters* 96:228102.

Wallrapp, F. and P. Fromherz. 2006. TiO_2 and HfO_2 in electrolyte-oxide-silicon configuration for applications in bioelectronics. *Journal of Applied Physics* 99:114103.

Site: **Natural and Medical Sciences Institute at the University of Tübingen and Retina Implant GmbH**
Markwiesenstrasse 55
72770 Reutlingen, Germany
http://www.nmi1.de//en/index.php

Date Visited: May 30, 2006

WTEC Attendees: T. Berger (report co-author), W. Soussou (report co-author), G. Gerhardt, D. McFarland, D. Taylor, G. Lewison

Hosts: Enzio Müller
Managing Director
Tel: +49 7121 51530 10
Email: mueller@nmi.de
Hugo Hämmerle
Deputy Managing Director
Tel: +49 07121 51530 45
Email: haemmerle@nmi.de

BACKGROUND

The Natural and Medical Sciences Institute at the University of Tübingen (NMI) in the Tübingen-Reutlingen Technology Park was founded in 1985 and is one of 11 institutes of applied research in the state of Baden-Württemberg. NMI conducts interdisciplinary applied research in the fields of natural science and medicine. Thirteen competence teams with 48 physicists, physical chemists, biophysicists, biochemists, biologists, and engineers collaborate with a network of clinics, universities, and other research institutes with the purpose of developing new products for industrial customers. NMI has over 100 employees divided among its 13 competency teams. Each team is responsible for its own cash flow; however, NMI's matrix organization allows it to engage efficiently in interdisciplinary research projects by drawing on strengths from all its members. The organization is controlled by a Management by Objectives system in which goals are defined within the framework of planned strategies. A transparent cash flow system allows each team leader to assess the team's performance according to indicators such as margin contribution. This new and developing organizational system is proving its usefulness in increased efficiency and growth at the NMI. Retina Implant AG is the latest NMI spin-off company. It is still housed at the NMI facility and collaborates closely with its researchers who developed some of its core technologies.

Educational Environment and Infrastructure

NMI is independent of the University of Tübingen. However, there are several close university collaborations on research projects. Also, 13 graduate and 10 undergraduate students from the university conduct their thesis research at NMI. NMI organizes or sponsors several workshops such as the biannual international MEA meetings. NMI employees also serve on the organizing committees of numerous international congresses.

Funding Sources and Commercialization

NMI receives €1.5 million of its annual budget of €9 million from the state of Baden-Würtemberg. Accounting for 35% of its budget, NMI has several grants from the Ministry of Education and Research, from the state of Baden-Wuerttemberg, and the EU. About 45% of the budget is obtained from industrial contracts. An additional source of funding for the NMI is derived from being a certified testing center for state and federal CE accreditations of medical products.

NMI has been a successful incubator for small startup and spinoff companies, with ten technology enterprises already spun-off from NMI activity in the last ten years. These companies can to use NMI knowledge, personnel, equipment, and infrastructure for research and development of new products. Additionally, NMI can provide up to 50% of the startup funds and own up to 49% of the company. The cities of Tübingen and Reutlingen, along with the chamber of commerce and BioRegio STERN, BIOPRO Baden-Wuerttemberg, and Attempto Service also foster an encouraging environment for startups. Commercialization and licensing of patents is handled by the NMI Technology Transfer (NMI TT) GmbH, with some intellectual property (IP) support being given to spin-off companies to ensure their financial survival. The NMI TT is a subsidiary of the NMI that takes advantage of its knowledge and resources and its nonprofit status to provide non-R&D products and services to customers. The NMI TT provides custom peptide synthesis and bioanalytics as well as various coatings for implantable medical products. In addition, it offers consulting and information brokerage such as the Internet platform biochipnet.com and patent licensing on NMI IP.

Multi Channel Systems (MCS) GmbH is a mature NMI spin-off, and Retina Implant GmbH was the latest start-up at the time of the WTEC visit. The latter was still housed at the NMI facility and collaborated closely with the researchers who developed some of its core technologies. It collaborated with several institutes, including the Ophthalmology Clinics of the University of Tübingen and Regensburg, the University of Stuttgart-Hohenheim, the Institute for Microelectronics in Stuttgart, and Multi Channel Systems.

The mission of Retina Implant GmbH is to return sight to people blinded by retinitis pigmentosa and age-related macular degeneration who lose their rod and

(a)

(b)

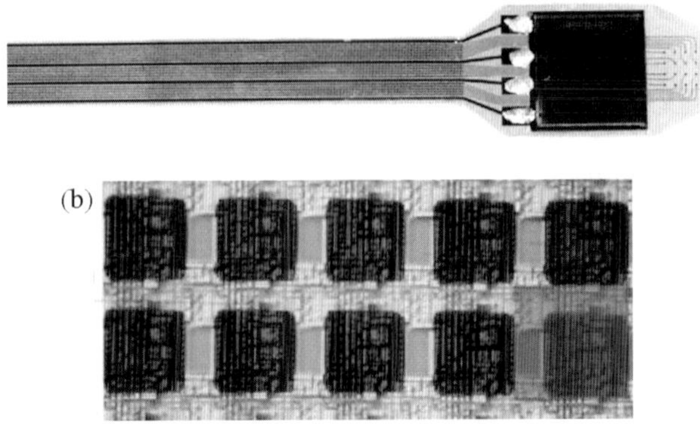

Figure B.8. (a) Implant of a Retina Implant GmbH CMOS electrode with flexible ribbon leads and MEA electrode array; (b) close-up of the CMOS stimulation chip.

cone cells. Retina Implant develops and implants a sub-retinal CMOS array that converts incoming light into electrical patterns to stimulate surviving bipolar cells (Figure B.8).

The Retina Implant electrodes, leads, and CMOS microchip array passivation layers are produced by NMI researchers. The 40×40 pixel array with 70 μm^2 pixel size converts incoming light to electrically stimulate bipolar cells. The electrodes are encapsulated with polyimide, which had proved to have less water absorption than silicon and had sustained 24 months of implantation without any sign of damage. These electrodes were being tested in clinical trials at the University of Tübingen. The wireless power transmitters for the implant were built at MCS.

Industrial Collaborations. In the two years 2004–2006, NMI collaborated with more than 230 small- and medium-sized companies on 37 research projects at a total cost of €70 million. Of this total, industrial partners contributed €28 million and public grants provided €42 million. The NMI coordinates the competence cluster "Biochip Technologies Baden-Würtemberg." Some of NMI's industrial partners and customers include Abbot Laboratories, Bayer, BMI Biomedical Informatics, Boehringer Ingelheim Pharma, Carl Zeiss SMT AG, Accelab GmbH, BIBraun Aesculap AG, Brucker Daltonic, Altana Pharma, Biopharm, CellMed, Evotec Technologies, MAN, Mikrogen, Multi Channel Systems, TETEC, Robert Bosch, and ZF Friedrichshafen.

Academic Collaborations. Although NMI has several national and international partners, it prefers local collaborations that benefit small and medium enterprises in the state of Baden-Würtemberg. Among other collaborations, NMI has close ties to the University of Tübingen, the Max Plank Institute with its focus on neuroscience and biotech, and the University of Freiburg.

Regulatory and Political Environment

The state of Baden-Würtemberg is number one in patent applications per inhabitant in Germany; it is heavily invested in medical technology commercialization. There are 69 colleges and universities in this state alone.

EU grants are found to be too restrictive, and the requirement for international collaboration is not necessarily conducive to good research. The large distances and lack of professional associations between groups also make communication and progress difficult.

To obtain permission to test Retina Implant's electrodes in human patients, applications were filed by Retina Implant GmbH with the University of Tübingen and local government.

RESEARCH AND DEVELOPMENT

Short- and Long-Term Scientific Goals

The mission of NMI is to produce innovative research in medical science that can be translated into the industrial sector in the state of Baden-Würtemberg. NMI has three core competencies: (1) pharma- and biotechnology, which is a growing field and includes applications in functional genomics, test systems for drug discovery, and bioanalytics and peptide synthesis; (2) biomedical technology that emphasizes applications of electrophysiology, neurotechnology and micromedicine, and regenerative medicine and biomaterials for the development of medical products; and (3) surface and interface technology, which constitutes NMI's applied research at the junction between life and material science and includes research in microsystem and nanotechnology, functional surfaces and layers, bonding, adhesion, and tribologic systems.

Retina Implant develops and implants a sub-retinal CMOS array that converts incoming light into electrical patterns to stimulate surviving bipolar cells.

Tools and Methods Used and Scientific and Technological Challenges and Solutions

NMI has an impressive array of state-of-the-art technologies in its arsenal, which includes a silicon manufacturing cleanroom with PVD and PECVD plants capable of nanomanufacture, all types of microscopy (light, Scanning Electron [REM, ESEM], atomic force, transmission electron, and scanning tunnel), spectroscopy (photoelectron, secondary ion and secondary neutral particle mass, Raman, optical, Fourier-transformed infrared, electron energy loss, and energy dispersive x-ray), and a Zeiss CrossBeam®.

The pharmacology and biotechnology core of NMI conducts R&D and provides services in functional genomics, which include molecular biology tools such as gene transfer, peptide synthesis, bioanalytics, protein microarrays, and cell biochips for the screening of cell-matrix interactions. This core also conducts *in vitro* electrophysiology to test the effects of drugs on cells and ion channels, and has developed Multi Channel Systems' (MCS) planar MEA system and Roboocyte patch-clamp robot, as well as Cytocentrics' automated patch-clamp instrument. NMI's test systems for the biological assays competence group has also co-operated to develop the MCS QT-Screen instrument that is a high-throughput device that measures drug effects on QT-interval prolongation in heart cells, an important drug safety assay as well as the MCS Roboocyte patch clamp robot.

The biomedical technology core of NMI is bringing modern biotechnology to medical applications. In the fields of neurotechnology and micromedicine, NMI researchers explore and develop materials and processes for the manufacture of microimplants that are biocompatible with their biological environment and whose surfaces can be either inert or used for electrical and chemical stimulation. The processes for reliable and long-lasting surface coating and cleaning are developed and the ensuing mechanical and electrochemical properties are tested at NMI (Figure B.9).

In collaboration with industrial partners, NMI has developed NanoVirDetect and NanoBioPore, two microsystems for diagnostic separation and detection of biological micro- and nanoparticles. Furthermore, a collaboration with Zeiss developed the BioFIB (Bio Focused Ion Beam) CrossBeam, a laser SEM Microscope that allows 3D reconstruction of interfaces between tissue and metals or polymers for the analysis of interface interactions.

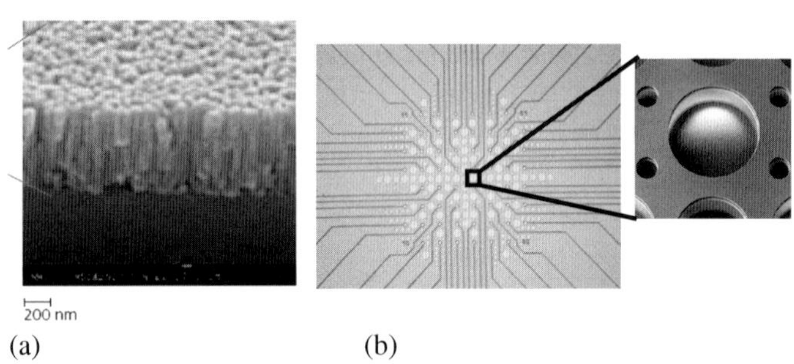

Figure B.9. (a) The nanostructure of the titanium nitride MEA electrodes shows the columnar structure of the layer which provides larger current discharge capacity; (b) porous MEA with expanded view of pore structure.

A BMBF project is designing neuroimplantable biohybrid chips that package genetically modified cells in chips that can stimulate them to release opioids for the treatment of pain and paraplegia. A similar hybrid cell and surface approach is being undertaken in collaboration with TETEC AG to develop regenerative articular cartilage from encapsulated bone marrow stem cells. Biomaterials for such regenerative medicine designed to either last or dissolve in the body are being developed and are also applied to nerve regeneration or control of cell behavior and 3D structural arrangement *in vivo* or *in vitro*.

The surface and interface technology core combines many competencies of the NMI to produce and assess microsystems and nanotechnology. The NMI cleanroom is able to manufacture small series of microchips for medical products or prototype nanostructures for biotechnology applications. Some of these applications include the manufacture of MEAs for MCS, nanoporous electrodes as chemosensors, and microfluidic and dielectrophoretic lab-on-a-chip. Manufactured surfaces can be encapsulated, coated, or treated with several methodologies including diamond-like carbon, passivation, or laser printing on stainless steel to enhance biostability, and control wetting or water permeability. These surfaces can be examined or manipulated with NMI-developed tools such as the nanoworkbench, nano-cryo-SIMS, and BioFIB, which enable examination of interface interactions at the nanoscale. These methodologies are also used in bonding and tribologic investigations at the NMI.

Past Performance: Patents, PR, and Publications

Research funding has been growing at an average of 22% per year from 1999 till 2005, with most growth from federal and EU grants and industrial collaborations.

NMI has systematically pursued patent applications since 1998, and has since registered 48 patent families.

NMI publishes and presents its research findings in scientific journals and conferences. In addition, visibility is promoted through the Internet, print media and press releases, film reports and interviews, exhibitor fairs, organization or sponsorship of workshops (such as the MEA workshops), scientific publications and presentations, teaching and mentoring of university students, and through membership in competence clusters. Finally, its greatest success is the tremendous satisfaction of its industrial customers as assessed by an internal survey.

DISCUSSION AND CONCLUSIONS

Competitive Advantages Compared to the World

NMI possesses a combination of competencies in biology, physics, biophysics, biochemistry, physical engineering, and coatings/analytics technology that enable it to execute first-rate research at the junction of life and material sciences. Its matrix infrastructure enables smooth collaboration between its core competencies, while providing a unified structure to industrial partners seeking its services and resources. The emphasis on applied research and industrial contracts enhances and expedites the technology translation of NMI research projects.

NMI is a DIN EN ISO/IEC 17025 accredited testing laboratory for state and federal accreditations, and its quality management fulfills ISO 9001:2000 requirements. Product testing and accreditation provides an additional source of income, and the proactive Quality Management standard ensures high-quality work and products at NMI.

REFERENCES

Egert, U. 2005. Perforated microelectrode arrays optimize oxygen availability and signal-to-noise ratio in brain slice recordings. Presented at Mikrosystemtechnologie Kongress, Freiburg.

Egert, U., B. Schlosshauer, S. Fennrich, W. Nisch, M. Fejtl, T. Knott, T. Muller, and H. Hammerle. 1998. A novel organotypic long-term culture of the rat hippocampus on substrate-integrated multielectrode arrays. *Brain Res. Brain Res. Protoc.* 2:229–242.

Janders, M., U. Egert, M. Stelzle, and W. Nisch. 1996. Novel thin film titanium nitride micro-electrodes with excellent charge transfer capability for cell stimulation and sensing applications. Presentation at the 18th Annual International Conference of the IEEE Engineering in Medicine and Biology Society, Amsterdam.

NMI Annual Report 2005: http://www.nmi.de/NMI-download/Publikationen/NMI-GB-E-Internet-R1.pdf.

Retina Implant website, http://www.eye-chip.com

Zrenner, E., A. Stett, S. Weiss, R.B. Aramant, E. Guenther, K. Kohler, K.D. Miliczek, M.J. Seiler, and H. Haemmerle. 1999. Can subretinal microphotodiodes successfully replace degenerated photoreceptors? *Vision Res.* 39:2555–2567.

Site: **Multi Channel Systems (MCS)**
Aspenhaustrasse 21
72770 Reutlingen, Germany
http://www.multichannelsystems.com

Date Visited: May 31, 2006

WTEC Attendees: T. Berger (report co-author), W. Soussou (report co-author), G. Gerhardt, D. McFarland, D. Taylor, G. Lewison

Hosts: Karl-Heinz Boven, CEO
Tel: +49 7121 909250
Email: boven@multichannelsystems.com
Andreas Moeller, President
Tel: +49 7121 909250
Email: moeller@multichannelsystems.com

BACKGROUND

Founded in 1996, MCS is a privately-owned company focusing on the development of advanced electrophysiological equipment for neuroscience and pharmacological research. The company's first product, a multielectrode array (MEA) recording system, was launched in 1997. It was followed in 2002 by an automated, two-electrode voltage clamp system, Roboocyte, and a QT-Screen system in 2004.

MCS was housed at the NMI to commercialize its MEA technology until 2003, when it moved into its own building. It n has 16 employees, of which 7 hold Ph.Ds. in physics, two in biology, one in biochemistry, and two are senior management. MCS outsources its sales to international distributors in 19 countries.

Educational Environment and Infrastructure

MCS has been sponsoring a biannual international scientific workshop on MEA technology and applications in Reutlingen since 1998, as well as annual short symposia during the American Society for Neuroscience meetings.

Funding Sources and Commercialization

MCS is a privately-owned company that receives much of its research funding from grants and collaborations. Many EU research projects require universities to collaborate with small companies, and MCS is therefore often pursued as a partner. MCS has to match any research funds it receives from the government with investment of its own. As a company, it can use some of those governmental research grants to pay for a percentage of its overhead costs. This is in contrast to universities, which usually obtain additional funds to pay for their overhead (though most universities do not charge research overhead). MCS is a partner in research and provides work time as well as electronics to support its collaborators. Most collaborative projects limit the development of products to the prototype stage over the course of two years; the company alone is responsible for development and expense past that time period.

Patents are owned by the largest financial contributor to a collaborative project but are usually free to use by all partners. Royalties from licensing are generally divided accordingly as well.

Industrial Collaborations. Several MCS products were developed with industrial collaborations, such as the Roboocyte collaboration with Bayer AG Leverkusen, the Synchroslice with Lohmann Research Equipment (Castrop-Rauxel, Germany), the small animal heart activity recording and stimulation system with Millar Instruments (Houston, Texas, USA), and the wireless power supply with Retina Implant (Reutlingen, Germany).

Academic Collaborations. MCS has collaborated on several projects funded by the German Ministry of Education and Research (BMBF). These include projects to develop MEAs, multimicrowellplates, and biochips. Several are funded by the state of Baden-Württemberg, such as the Roboocyte project and a high-density MEA project. Further, MCS is a partner in at least two EU projects for fluorescent biochips and ionchannel screening. In addition, MCS is a member of the Innovations in Electrophysiology Innovation Partnership (e.IP), which includes NMI, NPI (electronic instruments for the life sciences), Cytocentrics (an NMI spinoff to automate high-throughput patchclamping), and the Department of Neurobiology and Biophysics at the University of Freiburg. The aim of e.IP is to develop cutting-edge technology and electronic and electrophysiological products and services for customers and project partners.

RESEARCH AND DEVELOPMENT

Short- and Long-Term Scientific Goals

MCS develops automated multichannel measurement devices that provide fast secondary functional screening for basic research and pharmaceutical applications.

MCS is constantly supporting and improving its hardware and software product lines. The company is developing new planar MEA electrode layouts—some with holes to enable perfusion from the bottom. It is also planning 1000-electrode MEAs. Roboocyte is also in its second version, and the QT-Screen has recently been launched. They are collaborating with several companies to release new products such as the Synchroslice and heart activity recording and stimulating system, and the wireless power for Retina Implant products. The company plans to harness its electronics expertise to collaborate with others to produce quality electrophysiological equipment. For the future, MCS is considering maintaining its own sales department.

Tools, Methods Used, and Sources

MCS's flagship product, the MEA system, is shown in Figure B.10a. It consists of planar MEAs that have TiN columnar electrodes that lower the recording impedance while enabling good current stimulation. These arrays can be used with a variety of tissue preparations, from dissociated neural cultures to acute brain slices and even isolated retinas. The system's amplifiers are close to the electrodes, and the A/D conversion is now also on the same circuit card. The system supplies a USB plug-and-play interface and enables control of multiple systems from one computer. The data acquisition software provides additional filters and simple analytical tools with more advanced functions available as a free MATLAB toolbox developed through collaboration with the University of Freiburg. An *in vivo* recording system is now also available to connect to third-party, penetrating-shaft MEAs. In addition, flexible arrays are being developed for epicortical recording. Programmable stimulation generators (Figure B.10b) with up to 8 channels are now in their second generation and selling very well.

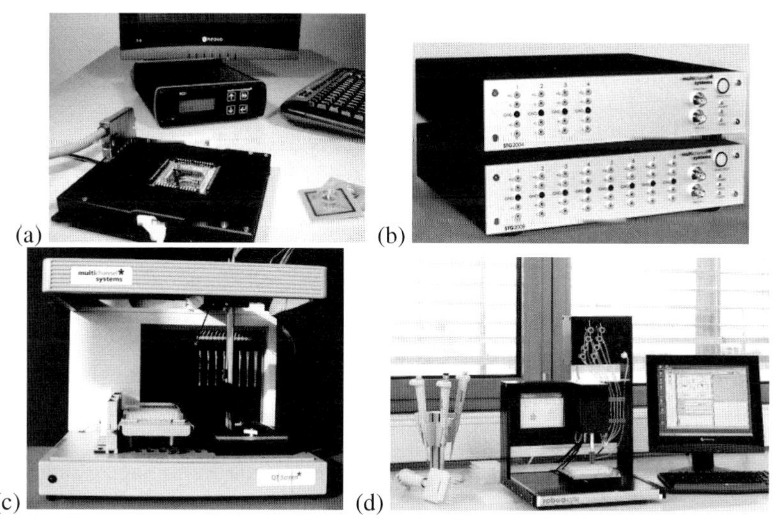

Figure B.10. (a) MEA System; (b) stimulation generators; (c) QT-Screen; (d) Roboocyte.

A new product designed for drug profiling and safety testing was launched in 2005. QT-Screen (Figure B.10c) examines drug effects on cardiac myocyte repolarization and QT-prolongation in 96 well plates, allowing high-throughput investigations of up to 100 compounds per day.

MCS produces several products that have been developed in collaboration with other companies, for example, an automated two-electrode voltage clamp system called Roboocyte (Figure B.10d). This system was developed in collaboration with Bayer AG Leverkusen and is now on the market. In collaboration with Millar Instruments (Houston, Texas, USA), MCS is producing an OEM product for rat and mice heart activity recording and stimulation. Moreover, a collaboration with Lohmann Research Equipment (Castrop-Rauxel, Germany) is the Synchroslice, a slice-electrophysiology system for recording and stimulation from eight slices simultaneously for rapid neuropharmacologal and toxicologal experimentation.

MCS is also leveraging its strength in electronic circuitry development to collaborate with Retina Implant by providing that company with a wireless power supply and telemetric control.

Past Performance: Patents, PR, and Publications

MCS has exclusive rights to several NMI patents and has several patents of its own. However, company management is generally more interested in rapid development of new technology rather than focusing on patenting company products. The feeling is that the company is too small to protect its patents in legal proceedings.

MCS representatives and distributors attend numerous scientific conferences to promote MCS products. Numerous publications have been written by scientists worldwide who use their products.

SUMMARY AND CONCLUSIONS

Scientific and Technological Challenges and Solutions

MCS has contributed to the advancement of MEAs. Conformal MEAs are now being produced by MCS to optimize the location of recording electrodes given the current limitation of channels. Perforated MEAs (shown in Figure B.11) have been developed to enhance the perfusion of tissue slices on the arrays. In order to penetrate the dead tissue layer and stimulate and record from the live tissue, sharp-tipped MEAs (Figure B.12) are being produced by a small company at EPFL in Switzerland, Ayanda Biosystems SA.

In addition, to allow stimulation from any electrode during recording, rapid blanking circuitry has been developed and included in the newest systems.

MCS plans to develop 1,000+ electrode arrays for multiunit recordings. The challenges for this project involve miniaturization and interconnections. Another long-term project involves the development of carbon nanoelectrodes that grow from the array into the tissue.

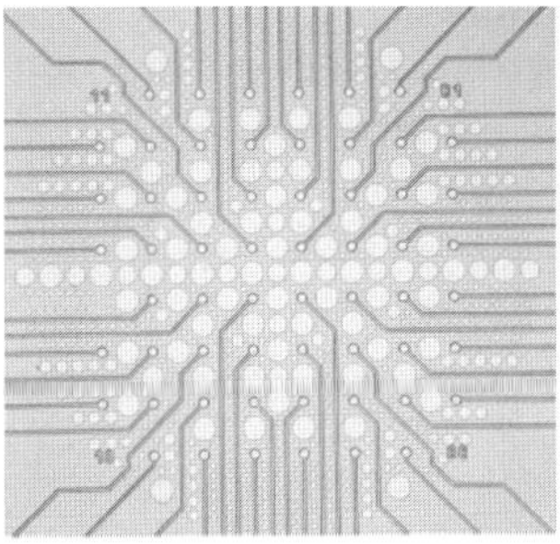

Figure B.11. Perforated MEA.

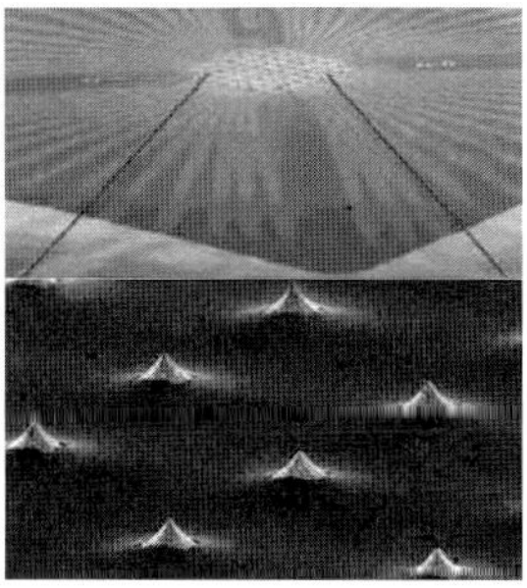

Figure B.12. 3D MEA.

Competitive Advantages Compared to the World

MCS is a small company with an excellent electrophysiological recording-systems product line including the leading MEA system on the market. MCS leverages its expertise in electronics for electrophysiological applications to collaborate with research institutions such as the NMI and partner with other companies on novel products. Outstanding customer relations enable MCS to update its products according to changing research needs.

REFERENCES

Egert U. 2005. Perforated microelectrode arrays optimize oxygen availability and signal-to-noise ratio in brain slice recordings. Presented at Mikrosystemtechnologie Kongress, Freiburg.

Egert U., B. Schlosshauer, S. Fennrich, W. Nisch, M. Fejtl, T. Knott, T. Muller, and H. Hammerle. 1998. A novel organotypic long-term culture of the rat hippocampus on substrate-integrated multielectrode arrays. *Brain Res. Brain Res. Protoc.* 2:229–242.

Egert U., T. Knott, C. Schwarz, M. Nawrot, A. Brandt, S. Rotter, and M. Diesmann. 2002. MEA tools: An open source toolbox for the analysis of multielectrode data with MATLAB. *J. Neurosci. Methods* 117:33–42.

Heuschkel M.O., M. Fejtl, M. Raggenbass, D. Bertrand, and P. Renaud. 2002. A three-dimensional multielectrode array for multisite stimulation and recording in acute brain slices, *J. Neurosci. Methods* 114:135–148.

Meyer T., K.H. Boven, E. Gunther, and M. Fejtl. 2004. Micro-electrode arrays in cardiac safety pharmacology: A novel tool to study QT interval prolongation. *Drug Saf.* 27:763–772.

Schnizler K., M. Kuster, C. Methfessel, and M. Fejtl. 2003. The Roboocyte: Automated cDNA/mRNA injection and subsequent TEVC recording on Xenopus oocytes in 96-well microtiter plates. *Recept. Channels* 9:41–48.

Stett A., W. Barth, S. Weiss, H. Haemmerle, and E. Zrenner. 2000. Electrical multisite stimulation of the isolated chicken retina. *Vision Res.* 40:1785–1795.

Site: **The Santa Lucia Foundation**
Laboratory of Neuroelectrical Imaging and Brain-Computer Interface
Via Ardeatina 306
00179 Rome, Italy
http://www.hsantalucia.it
http://www.neilab-fsl.it/

Date Visited: June 5, 2006

WTEC Attendees: J. Chapin (report author), J. Principe, P. Tresco, S. Demir, G. Lewison

Hosts: Fabio Babiloni, Ph.D.
Associate Professor
Tel: +39 328 769 7914
Email: Fabio.Babiloni@uniroma1.it

BACKGROUND

The Santa Lucia Foundation is a "Scientific Institute for Hospitalization and Treatment of National Importance and High-level Specialisation in Neuromotor Rehabilitation (IRCCS)" affiliated with the University of Rome "*La Sapienza*" and "*Tor Vergata*." Its research and clinical programs are devoted to neurorehabilitation and developmental strategies. The general director is Dr. Luigi Amadio, and the scientific director is Professor Carlo Caltagirone. Our hosts gave us a tour of the hospital's excellent physical therapy facilities and then collected us in a seminar room for a series of PowerPoint presentations. The head of the neurophysiopathology research line, Professor Maria Grazia Marciani, introduced the multidisciplinary research team: Dr. Mattia in the Neuroscience area, and Professor Babiloni and Drs. Cincotti and Bianchi in the Bioengineering area. The team also includes five Ph.D. students and several undergraduates.

RESEARCH AND DEVELOPMENT ACTIVITIES

Fabio Babiloni, whose Ph.D. is in Neural and Computational Engineering, introduced the scope and overall aims of research in brain-computer interfaces carried out at the Neuroelectrical Imaging and Brain-Computer Interface (NEI and BCI) lab at the Santa Lucia Foundation. The major aim is to develop computer-controlled environments for disabled patients who could utilize EEG recordings for assisted control. In attempting to improve the resolution of EEG signals for BCI control, Professor Babiloni has developed an approach that does not require complex neural networks or nonlinear filters for classification. Instead, this group focuses on increasing the spatial resolution of EEG recordings by using MRI images of the head compartments and the brain to define the individual anatomical characteristics, which are then used to pinpoint regions of interest (RoIs) of the modeled cortical mantle. This enables accurate recognition of the cortical sources of the scalp recorded EEG signals by applying the linear inverse solutions to measure the neural activity in a given RoI (Figure B.13). This combined MRI and EEG approach is being used to reveal the locations of cortical activity relating to execution of a task. This cortical activity can then be used to control a computer cursor or other device.

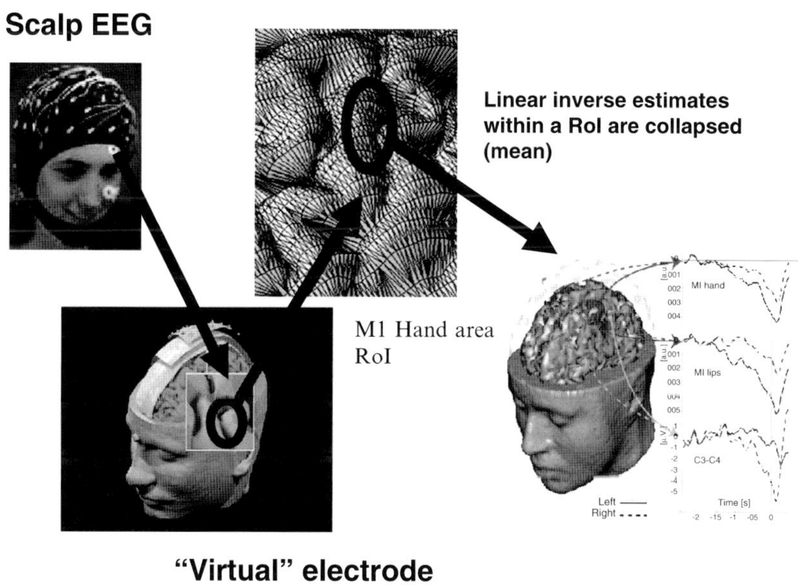

Figure B.13. Scalp to cortical EEG using RoIs drawn on the modeled cortical surface.

A further approach has been developed to understand how different cortical areas cooperate with each other during a particular task. The Granger causality (implemented through the partial directed coherence estimator) is being used to estimate the direction and strength of information flows through the brain during such tasks (Figure B.14). The assumption of causality is based on measurable time delays between neural activities recorded above different cortical areas in different EEG frequency bands. Computational models are also being tested to improve the resolution of spatial filtering online during BCI recordings for subject training. Overall, this group now aims to improve the detection of imagined mental activity in the brain in one or two dimensions by use of the above methods to analyze data obtained from arrays of 32 or 64 EEG electrodes on the scalp.

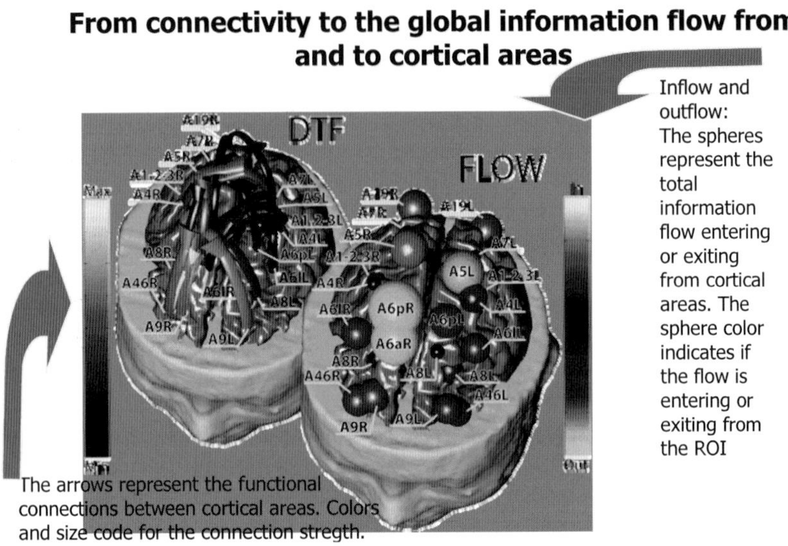

Figure B.14. Global information flow by using Granger causality and partial directed coherence.

Dr. Luigi Bianchi is developing a general-purpose software package (BF++) for BCI coding in biofeedback systems. He aims to develop a collection of C++ programming tools for the development of brain- and human-computer interfaces (BCI and HCI). The goals are to develop cross-platform tools that work in Windows, Linux, and embedded environments. BF++ applies object-oriented programming using ANSI C++ to develop cross-platform and cross-compiler tools that work in real time. It includes a low-level application programming interface (API) for easy migration and dissemination.

BF++ also incorporates an NPX file format (NeuroPhysiological signals in eXtensible Markup Language) that is based on the XML markup language widely used for online documents and interchangeable data systems. This format provides an NPX viewer for analyzing and evaluating EEG data, "toys" such as metrics, tuning, optimization and simulation, as well as file and data utilities. One important use of this system is for real-time removal of electrical artifacts from the fMRI machine and the heart. Based on the ability to delete these artifacts, this group now hopes to implement BCI platforms for P300 and mu-rhythm experiments during fMRI imaging.

Finally, a longer-term goal is to adapt the use of BCIs for use in the low-gravity environment of space, specifically the ALTEA (Anomalous Long-Term Effects in Astronauts) project on the International Space Station. The general project (disorders of motor control and cardio-respiratory systems) and the specific subproject (neuro-prosthesis and brain-machine interface) will be developed under the funding of the

Italian Space Agency (ASI). The specific subproject studies the possibility of implementing some BCIs in zero-gravity environments.

Dr. Febo Cincotti focuses on the use of biofeedback for training subjects to intentionally modify their brain signals in a way that allows EEG recordings to detect and classify their intent. This requires user training, appropriate feature extraction of the EEG signals, and computer training on appropriate feedback strategy. Expertise on this approach has been contributed into a project called MAIA (Mental Augmentation through Determination of Intended Action; see also the EPFL site report for further information on MAIA), involving multiple European universities: IDIAP (coordinator), Katholieke Universiteit Leuven, University Hospital of Geneva, Fondazione Santa Lucia-Rome, and Helsinki University of Technology. This consortium hopes to achieve: (1) noninvasive estimation of brain activity; (2) an adaptive, shared autonomy between user and robot; (3) use of haptic feedback to facilitate training and accuracy; (4) recognition of cognitive states of brain; and (5) online adaptation to maintain tuning of the BCI interface.

Finally, the Italian charity Telethon ONLUS Foundation has funded the ASPICE project (Assistive System for Patient's Increase of Communication, ambient control and mobility in absence of muscular Effort, ASPICE,) developed at the BCI laboratories at IRCCS Santa Lucia (Figure B.15). Such a project is aimed at the development of a technological aid that allows neuromotor-disabled users to improve their mobility and communication within the surrounding environment by remotely controlling a set of home-installed appliances, including a Sony AIBO mobile robot.

Dr. Donatella Mattia is a neuroscientist-neurologist who is working with this group to bring BCI technology to actual clinical practice. She works with spinal-cord-injured (SCI) patients to determine their ability to generate motor cortical activity in the absence of physiological actuation (i.e., muscle activity). The findings suggest that the cortical motor areas in these patients remain functional even when these areas have been disconnected (by the SCI) from the body parts that they normally would command. Thus, these deprived motor areas remain functionally available to be used by an appropriately configured BCI. Additional studies have used the Granger's causality approach to examine the functional connectivity between different brain areas during arm movements of a manipulandum to a target against a viscous force field. Appropriately, a naïve subject exhibited strong functional connectivity between the visual and motor/premotor cortices, suggesting use of visual feedback to guide target reaching. In contrast, an experienced subject showed stronger functional connectivity between the supplementary motor area (SMA, thought to be involved in self-paced movements) and the motor cortex. Finally, Dr. Mattia is heavily involved in the ASPICE program that will allow neurologically damaged patients to use BCI to control electronic and robotic devices.

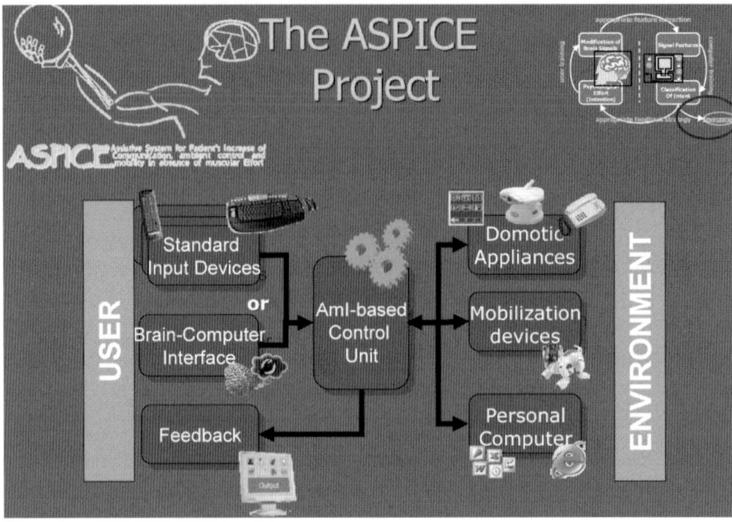

Figure B.15. The ASPICE Project. Since the ultimate goal is to improve the quality of life of patients, BCI was integrated with other more common communication aids for the motor disabled and used to send commands to a modular and expandable computerized system for domotics and robotic control.

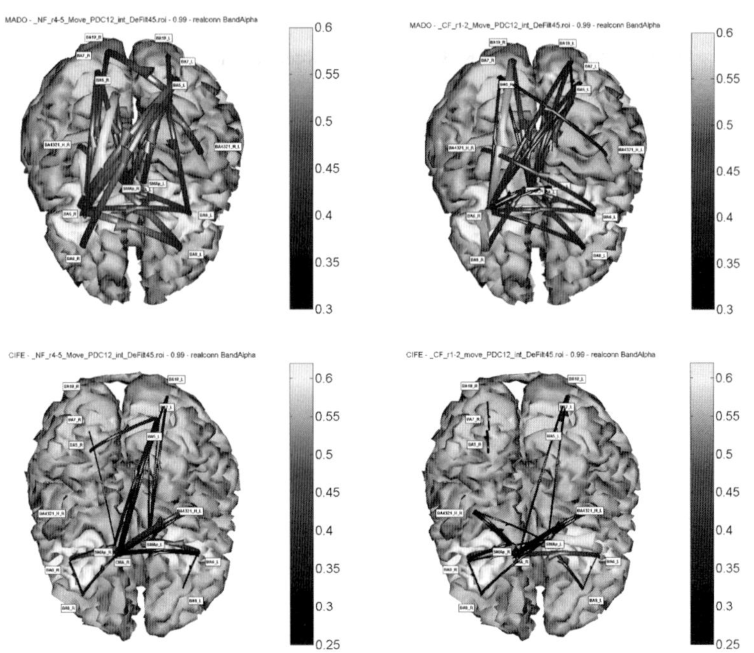

Figure B.16. Connectivity patterns under "force field." The images represent the cortical areas of different subjects. The colored zones are the RoIs considered; the lines depict the flow of causality between RoIs during the experimental task analyzed in a particular frequency band.

Funding

Funding in Europe is multi-institutional, which works well for multidisciplinary projects such as BCI. This group receives funds from private, national sources such as the Telethon foundation and from governments (Ministry of Foreign Affairs' relations with Chinese universities, three years; and the Ministry of Health, two years). The EU also has funded two research programs for three years each. Finally, a U.S. NIH grant (with the Wadsworth Center) involves software maintenance work.

Commercialization

There was negotiation taking place with the IBM foundation, and the Italian EEG equipment firm EBNeuro had provided hardware furniture.

REFERENCES

Aloise, F., F. Cincotti, F. Babiloni, M.G. Marciani, D. Morelli, S. Paolucci, G. Oriolo, A. Cherubini, F. Sciarra, F. Magnola, A. Melpignano, F. Davide, and D. Mattia. 2006. ASPICE: An interface system for independent life. *Proceedings of 4th International Conference on Smart Homes and Health Telematics (ICOST)*.

Babiloni, F., F. Cincotti, C. Babiloni, F. Carducci, A. Basilisco, P.M. Rossini, D. Mattia, L. Astolfi, L. Ding, Y. Ni, K. Cheng, K. Christine, J. Sweeney, and B. He. 2005. Estimation of the cortical functional connectivity with the multimodal integration of high-resolution EEG and fMRI data by directed transfer function. *Neuroimage* 24(1):118–131.

Cincotti, F., F. Babiloni, M. Mattiocco, L. Astolfi, S. Bufalari, M.G. Marciani, and D. Mattia. 2005. Laboratory of functional neuroelectrical imaging and brain computer interfacing at Fondazione Santa Lucia. *Cogn. Process*. 6:75–83.

Mattia, D., F. Cincotti, M. Mattiocco, G. Scivoletto, M.G. Marciani, and F. Babiloni. 2005. Motor-related cortical dynamics to intact movements in tetraplegics as revealed by high-resolution EEG. *Hum. Brain Mapp.* 27(6):510–519.

Site: **Polo Sant'Anna Valdera**
Scuola Superiore Sant'Anna
Viale Rinaldo Piaggio, 34
56025 Pontedera (Pisa), Italy
http://www.sssup.it/sssup/index.jsp?lang=en

Date Visited: June 1, 2006

WTEC Attendees: P. Tresco (report author), J. Chapin, J. Principe, S. Demir, H. Ali

Host: Paolo Dario, Ph.D.
Professor of Biomedical Robotics
Tel: +(39) 050 883420 400
Fax: +(39) 050 883497
Email: dario@sarts.sssup.it

BACKGROUND

The Polo Sant'Anna Valdera (PSV) is the Sant'Anna School of Advanced Studies' research park located in Tuscany and directed by Paolo Dario. The PSV was established in 1994 as part of a larger plan for strengthening the Tuscan territory's research and development network. It contains a large number of laboratories that combine education and research with training in entrepreneurship. These labs cover such areas of research as robotics, microengineering, surgical and rehabilitation robotics, computer science, engineering, virtual reality, biotechnology, economics, lab management, and longevity research.

Our visit focused on the Advanced Robotics Technology and Systems or ARTS Lab and the Center for Research in Microengineering (CRIM). The ARTS Lab is coordinated by Professor Maria Chiara Carrozza, and the CRIM Lab by the leadership of Professor Dario. These well-funded research groups are supported by several cleanrooms, CAD labs for electronic and mechanical design, an ultra-precision machine shop, micro-injection molding equipment, rapid prototyping facilities, computational facilities, and wet labs that support *in vitro* studies. In addition, space is available for teaching, offices, administrative staff, and lectures. The group has grown considerably since its inception in the early 1990s and now consists of approximately 90 people who are involved in the steady-state training of about 15 Ph.D. students per year. The group has a strong focus on translational research, including intellectual property and new venture development. An incubator for spinoffs had been recently established, and 18 start-up companies had been launched based on PSV or related technologies. In addition, PSV has developed a strong relationship with Waseda University that involves educational exchange and research collaboration in humanoid robotics.

RESEARCH AND DEVELOPMENT

Our visit began with a broad overview of the philosophy behind the educational and research activities of the Scuola Superiore Sant'Anna and the Polo Sant'Anna Valdera provided by our host, Professor Paolo Dario. This was followed by presentations of several assistant professors who work and direct the activities in the ARTS and CRIM labs, including Silvestro Micera, Cecilia Laschi, and Oliver Tonet. We were treated to an informal lab tour with demonstrations of technology provided by students and staff. The educational programs combine the study of life sciences with robotics and engineering. The long term vision is to interface biology with robotics. The present focus is on neurorobotics or the fusion of neuroscience with robotic engineering. An extensive collaborative network has been established among Dario's group in Pontedera and some of the best roboticists, engineers, and neuroscientists in Europe, Japan, and the United States. The thrust of the research is to use biological models to drive the design of biomimetic robotics and then use the robots as physical platforms for validating biological models. It is not surprising with this philosophy that most projects are biologically inspired or biologically based. Fusion of fundamental neuroscience with robotics has created a new training paradigm and new technologies that are producing a new breed of Ph.D. candidates with hybrid training and a focus on biomedical applications oriented toward innovation and new therapeutic products.

The ARTS and CRIM labs are supported by considerable numbers of staff and graduate students, and relatively new, state-of-the-art infrastructure. The labs focus on developing robotic components and systems for use as surgical aids, in rehabilitation, or for use in assistive technologies to aid the elderly. As an example of a diagnostic and therapeutic technology developed at the CRIM lab, the group presented research on a legged capsule for navigation in the GI tract. The ARTS lab research activities include such areas as biomechatronics, neurorobotics, biomorphic control, and robotics used for rehabilitation and assistive care.

Significant effort has been invested at the ARTS lab into developing artificial limbs, especially the upper extremities such as the human hand, which has reached a very high level of achievement. Using a biomechatronic approach to duplicate the natural hand, the group has produced Cyberhand (Figure B.17), an elegant robotic surrogate that uses an underactuated design with multiple degrees of freedom and incorporates tactile biomimetic sensor feedback. Shaking this hand provides an understanding of the considerable achievement of Dario's and Maria Chiara Carrozza's team. The group is part of several larger projects in Europe and in the United States. For example, group members participate in a DARPA-sponsored project called Revolutionizing Prosthetics that seeks to animate the Cyberhand through implanted neural interfaces that interact with an amputee's efferent and afferent nerves to provide intentional control and sensory feedback.

Another highlight of the visit was the presentation of humanoid robotic technology that contains an anthropomorphic head and a retina-like visual sensory system combined with an anthropomorphic arm and biomechantronic hand that is capable of communicating emotional states.

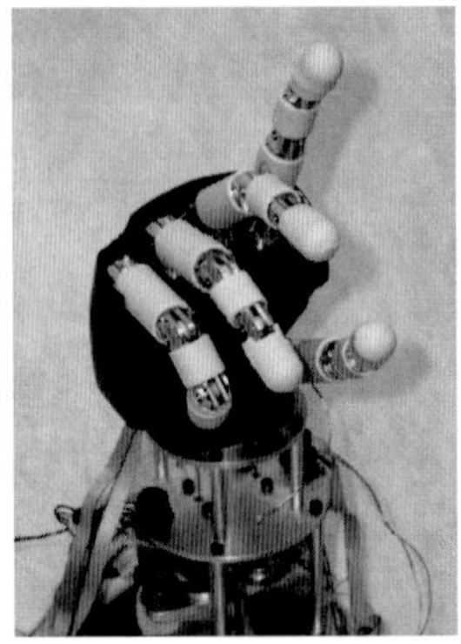

Figure B.17. Cyberhand.

Silvestro Micera reviewed the institution's activities in the area of implantable interfaces. We were told of several electrode technologies, including sieve electrodes and thin-film, longitudinal, intrafascicular electrodes, that were being developed with collaborating labs to animate the Cyberhand. The group is intent on selecting the "best" available neural interface for control of the bionic hand. Dr. Cecilia Laschi presented her work on natural interfaces, and Dr. Tonet described a nascent BCI research plan that has just begun.

SUMMARY AND CONCLUSIONS

A striking feature of this visit was the strong vision that was communicated of combining the study of biology with engineering, with a concurrent focus on discovery and invention, especially in the realm of integrating robotic components into the human body (or bionics). The group has a strong appreciation for biology that inspires its engineering practice. Unlike the engineering approach that followed Newton's contributions and the industrial age, this group is engineering using a biology-centric, or inside-out, approach. This approach has added intelligence to the robotic or neuroprosthetic components by limiting the number of control signals needed from the nervous system. Strong input by notable neuroscientists, such as Alain Berthoz of the College of France, and other collaborators should ensure the contribution of this group into the foreseeable future. Although its present direct efforts in BCI are modest, its long-term potential as a contributor in this area is significant. As with other sites that we visited throughout Europe, there was a strong emphasis on the dissemination of technology, including explicit graduate training in entrepreneurship and a focus on regional economic impact. Collaborative interaction with industry and networking with other research institutions and organizations within and outside of their labs were clearly encouraged.

Site: **Swiss Federal Institute of Technology, Lausanne**
Ecole Polytechnique Fédérale de Lausanne (EPFL)
EPFL, 1015 Lausanne, Switzerland
http://www.epfl.ch/

Date Visited: May 31, 2006

WTEC Attendees: J. Chapin (report author), J. Principe, P. Tresco, S. Demir, H. Ali

Hosts: Prof. Touradj Ebrahimi, EPFL
Tel: +41 21 693 2606
Email: Touradj.Ebrahimi@epfl.ch
Prof. José del R. Millán, IDIAP
Email: jose.millan@idiap.ch
Prof. Henry Markram
Email: henry.markram@epfl.ch

BACKGROUND

The EPFL is a highly regarded polytechnic institution that has recently developed a major presence in biology, BCI, and computational neurobiology. The program included presentations by groups from EPFL and the affiliated semiprivate research institute, the *IDIAP Research Institute* in Martigny, Switzerland (http://www.idiap.ch).

Professor José del R. Millán of the IDIAP Research Institute is also a professor at EPFL, where he teaches a graduate course on Brain-Computer Interaction. He described his goal of enhancing brain-robot interaction based on noninvasive brain recordings. He pointed out that this is a big challenge because fast decisionmaking is critical. Thus, he has created a large consortium to address the challenging task. The consortium includes HUG (Geneva University Hospital), the MAIA project (Mental Augmentation through Determination of Intended Action) for mental control of robots (coordinated by IDIAP), IM2.BMI (a Swiss NCCR, National Center of Competence in Research, also coordinated by Millán), and BACS (Bayesian modeling of brain functions), coordinated by ETHZ (Swiss Federal Institute of Technology in Zurich). Professor Millán also briefly mentioned two other BCI prototypes he has developed, a virtual keyboard and a brain game. His BCI research has received wide scientific and media coverage worldwide.

Professor Millán has pioneered the use of noninvasive brain recordings to control movement of robots and prosthetic devices. He has developed an asynchronous protocol for EEG analysis and married it to machine learning techniques and artificial-intelligence robotics. In this he has used principles of "adaptive shared autonomy" to enhance the BCI's functionality. This is necessary because EEG recordings alone cannot provide bit rates sufficient to control robots in real time. Therefore the system has been adapted to allow robotic algorithms to handle low-level tasks such as navigating a maze without running into walls. This allows the human user to focus on purely intentional tasks, such as determination of general direction.

This was tested by requiring subjects to mentally move a wheelchair icon on a computer screen through a 20 m virtual corridor (Figure B.18). The time to completion decreased from approximately 600 seconds to about 300 seconds when a higher level of robotic intelligence was applied. Online learning to enhance the detection of distinct spatial patterns in the alpha band was used to enhance the classifier in real time as the subject mentally controlled the robot. In order to classify "correct" versus "error" trials, field potentials were recorded over the midline regions superior to the anterior cingulate cortex. In four subjects these classifications were accurate for approximately 80% of trials.

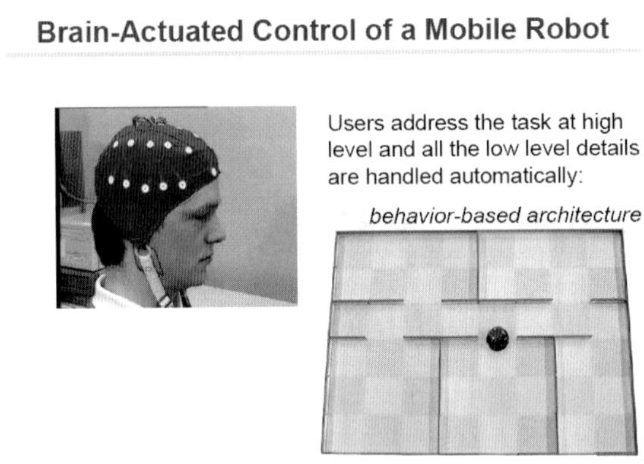

Figure B.18. Brain-actuated control of a mobile robot.

Another approach was to estimate local field potentials from scalp recordings. Use of these estimated LFP recordings in subjects with 111 electrodes produced the lowest error rates (Figure B.19). Finally, tactile and vestibular stimuli were used as natural feedback to signal the position of the computer cursor (Figure B.20). This freed the subject to use his visual sense to monitor the process that he was controlling.

Funding

Funding for the EPFL is through the Swiss National Science Foundation and the EPFL, which provides infrastructure and 25% of Ph.D. salaries. Support in Europe is multi-institutional, which works well for multidisciplinary projects such as BCI.

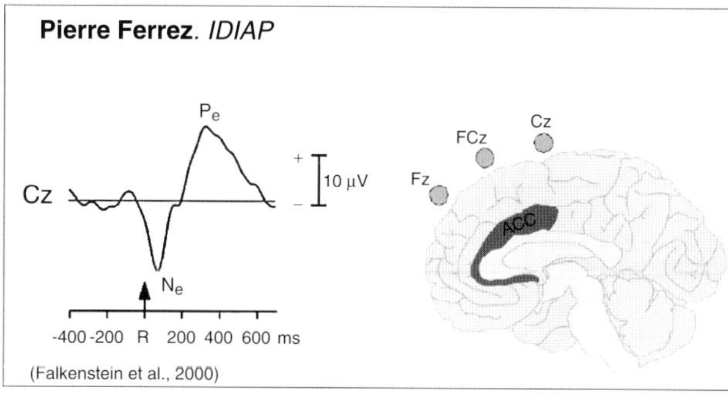

Figure B.19. "You Got Me Wrong!"— recognition of cognitive states.

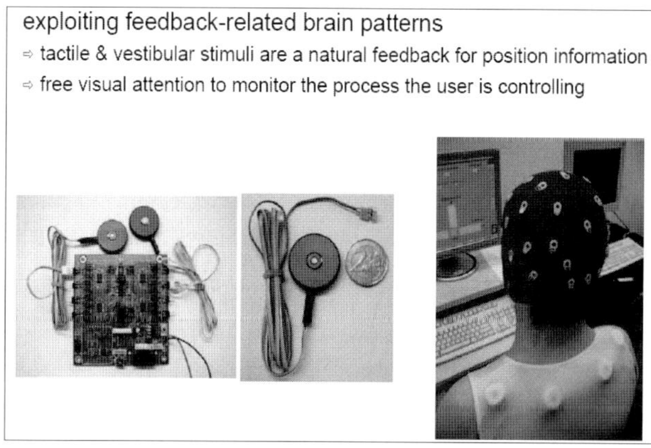

Figure B.20. "Tell Me More!"— multimodal feedback.

REFERENCES

EPFL. The Blue Brain Project website, http://bluebrainproject.epfl.ch (accessed July 10, 2006).

Ferrez, P.W., and J.d.R. Millán. 2007. Error-related EEG potentials in brain-computer interfaces. In *Towards brain-computer interfacing*. Eds. Dornhege G., J.d.R. Millán, T. Hinterberger, D. McFarland, and K.-R. Müller. Cambridge: MIT Press.

Markram, H. 2005. The Blue Brain Project. *Nat. Rev.* 7:153–160.

Millán J.d.R., F. Renkens, J. Mourino, and W. Gerstner. 2004. Noninvasive brain-actuated control of a mobile robot by human EEG. *IEEE Trans. Biomed. Eng.* 51(6):1026–1033.

Studer, D., U. Hoffmann, and T. Koenig. 2006. EEG dependency multichannel matching pursuit to sparse topographic decomposition. *J. Neurosci. Methods* 153(2):261–275.

Traub, R.D., and R.K. Wong. 1982. Cellular mechanisms of neuronal synchronization in epilepsy. *Science* 216:745–747.

Vesin, J.-M., U. Hoffmann, and T. Ebrahimi. 2006. Human-brain interface: State-of-the-art signal processing and machine learning. In *Wiley encyclopedia of biomedical engineering*. Ed. M. Akay. Hoboken, NJ: Wiley-Interscience.

Site: **University of Edinburgh**
Integrated Micro and Nano Systems Laboratory
Mayfield Road
Edinburgh EH9 3JL Scotland, UK
http://www.see.ed.ac.uk

Date Visited: June 2, 2006

WTEC Attendees: J. Principe (report author), J. Chapin, S. Demir, P. Tresco, H. Ali

Host: Dr. Alan Murray
Professor and Department Head
Tel: +0131 650 5589
Email: A.F.Murray@ee.ed.ac.uk

BACKGROUND

The main lab for BCI research is the Neural Network Group headed by Professor Alan Murray. His interests are neural networks, mixed-mode VLSI hardware, and more recently, neuromorphic VLSI and the interface between silicon and biology. The University of Edinburgh is very large and has three schools with interests related to BCIs: the Schools of Engineering and Electronics, Informatics, and Biomedical Sciences. The Neural Network Group is part of the Institute for Integrated Micro and Nano Systems (IMNS) in the School of Engineering and Electronics. We met with Dr. Murray and his two Ph.D. students, Keith Baldwin and Evangelos Delivopoulos. We did not see or contact any of the other collaborators.

RESEARCH AND DEVELOPMENT

The short-term goals include successful silicon-neuron interaction for cellular recordings and moving from *in vitro* to *in vivo* work. The long-term goals of the lab are centered in the design of neuromorphic learning systems. We discussed two of the current projects: patch clamping in silicon and 2-D patterning of cells. Patch clamping in silicon would open up the possibility of direct connection between neurons and silicon devices, and will allow the control of multiple patch clamps at the same time, which is time consuming today and requires expensive instrumentation. The method builds a 10x2 mm-deep channel on a silicon substrate leading to a 80×500 mm chamber. The cell is "sucked" into the chamber and the channel seals the lumen. Several different surface treatments have been tried and the best roughness was achieved from a nitride and Boron treatment. The group has not been successful so far in patch clamping, achieving input impedances in the 200 MΩ range instead of the expected 1 GΩ.

The patterning of cell work was conceived together with the cell clamping idea to guide the spatial growth of neurons over the holes to help position the cells for patch clamping in an effort to automatically monitor the activity of neural networks. The novelty was to use photoresist-patterned tracks instead of biological materials. Glia locked to the patterns (not neurons directly), but only in very small concentration (affinity to other cells was much higher). Parylene coating (instead of photoresist) worked much better, in particular when it was aged for two months. Several successful patterns were shown.

We had the opportunity to visit the wet laboratory in the School of Biomedical Sciences and observe the apparatus to position the cells over the silicon device and attempt the patch clamping.

SUMMARY AND CONCLUSIONS

The University of Edinburgh is a large university covering multiple disciplines that can collaborate to create extensive multidisciplinary research. The organization is different from U.S. institutions, and biomedical engineering is not included in the College of Engineering. There is no obvious organizational link between the engineering groups and those in biology or the hospital. Professor Murray's group strives for academic excellence. He is well known for his activities in neuromorphic systems and computation. In spite of the breadth of areas present at the University of Edinburgh, our impression is that interdisciplinary collaborations are still limited to specific topics, and there was no evidence of large collaborative projects. The university currently seems to lack translational research to clinical problems. Funding sources are the UK government and European projects.

Site: **University of Freiburg**
Brain-Machine Interfacing Initiative
Schänzlestrasse 1
D-79104 Freiburg, Germany
http://www.brainworks.uni-freiburg.de

Date Visited: June 1, 2006

WTEC Attendees: D. Taylor (report author), T. Berger, G. Gerhart, D. McFarland, W. Soussou, G. Lewison

Hosts: Dr. Ad Aertsen
Professor in Neurobiology and Biophysics
Institute of Biology III
Tel: +49 761 203 2718
Fax: +49 761 203 2860
Email: aertsen@biologie.uni-freiburg.de.
Dr. Tonio Ball
Faculty, Bernstein Center for Computational Neuroscience
Dr. Carsten Mehring
Assistant Professor
Animal Physiology and Neurobiology
Institute of Biology I
Dr. Jörn Rickert, PostDoc
Benstein Center for Computational Neuroscience
Dr. Simone Cardoso de Oliveira
Teaching and Training Coordinator
Bernstein Center for Computational Neuroscience

BACKGROUND

The Brain-Machine Interfacing Initiative at the University of Freiburg establishes a multidisciplinary collaboration within and across institutions to develop innovative approaches for connecting the human brain to computers or prosthetic devices. This initiative brings together members of the Bernstein Center for Computational Neuroscience Freiburg, the WIN-Kolleg of the Heidelberg Academy of Sciences and Humanities, and the METACOMP project funded by the German Federal Ministry of Education and Research (BMBF) within the framework of the German-Israel Project Cooperation (DIP).

Principal investigators include Ad Aertsen, Tonio Ball, Carsten Mehring, Jörn Rickert, Martin Nawrot, Stefan Rotter, Andreas Schulze-Bonhage, Simone Cardoso de Oliveira, and Kaus Vogt. Outside collaborators include Moshe Abeles, Eilon Vaadia, Niels Birbaumer, Klaus Pawelzik, Rony Paz, Nikolaus Weiskopf, and Alexa Riehle.

RESEARCH AND DEVELOPMENT ACTIVITIES

BMI-related activities reported during this site visit focused on two complementary research areas: (1) decoding of arm movement parameters from epicortical field potentials in humans, and (2) decoding of arm movement parameters from both unit activity and local field potentials recorded from intracortical microelectrodes implanted bilaterally in rhesus macaques.

Epicortical Field Potentials in Humans

This group is conducting research in human epicortical field potentials at the Epilepsy Center of the University Hospital, Freiburg. Participants in this study have arrays of thin, flat disk electrodes surgically implanted on the brain surface for the purpose of identifying seizure-generating areas prior to resection surgery for the treatment of intractable epilepsy.

For this research project, neural data were collected from five patients with subdural grid electrodes over the arm and hand areas of the precentral gyrus. These neural data were synchronized with arm movement data collected on videotape as the subjects made self-paced, center-out arm movements to four or eight targets, as well as when the subjects made additional continuous arm movements. In two of the patients, high-density electrode grids were used with 7.1 mm interelectrode spacing versus the standard 10 mm interelectrode spacing. Where the electrodes were located in relation to the functional and structural anatomy of each subject's cortex was accurately determined by a combination of cortical stimulation of each electrode site, brain imaging, and 3D reconstruction techniques.

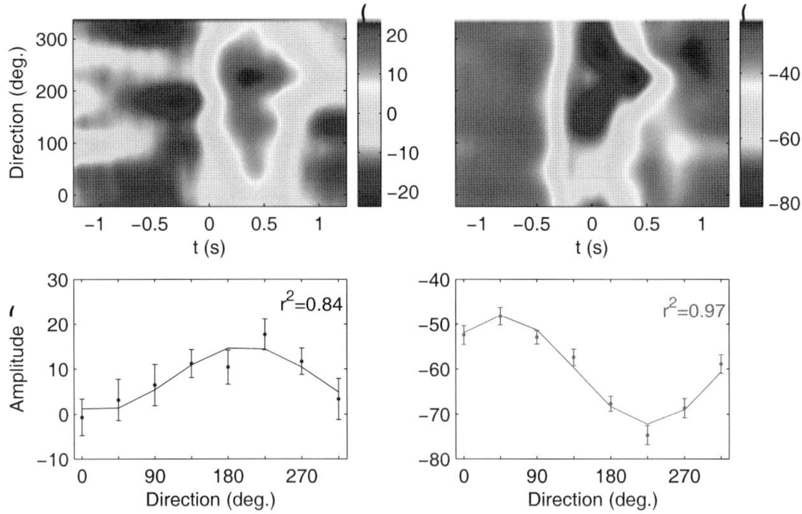

Figure B.21. Directional tuning in the time domain of two human epicortical field potentials recorded during center-out movements to eight targets.

The recorded neural signals were analyzed offline in both the time (Figure B.21) and frequency (Figure B.22) domains in conjunction with the videotaped arm movement data. Both time and frequency domain data exhibited directional tuning (i.e., amount of modulation was dependent on movement direction).

The movement modulation characteristics in the frequency domain tended to fall into three specific bands of frequencies (<4 Hz, 6–30 Hz, 34–128 Hz), shown in Figure B.22 Thus, those three bands were analyzed individually for their movement-related content. Useful directional information was highest in the less-than-4 Hz band, and lowest in the 6–30 Hz band, with all three bands providing useful movement information.

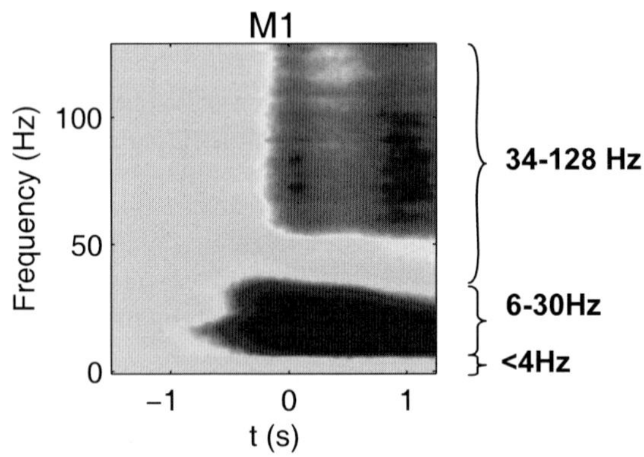

Figure B.22. Example of modulation across the frequency domain in epicortical field potentials starting prior to movement onset (t=0).

Comparisons of information from high-density grids versus standard density grids showed that additional *new* information regarding movement parameters can be gleaned by increasing the density of the electrodes covering the same cortical areas. This suggests that increasing the electrode density over current standards may be an effective way to increase movement information from epicortical potentials for brain-machine interface applications.

In the near future, this team will start real-time cursor control experiments with the ECoG patients in the Epilepsy Center. It has arranged its equipment to tap into the Epilepsy Center's ECoG signal lines, feed a copy of those signals to its own g.tec amplifiers, decode the signals in real time, and display the decoded movement signals back to the subject as cursor movements on a computer screen. Although the technical hurdles for setting up this combined system were easy to overcome, the regulatory hurdles took about a year to resolve. A common set of electrical safety standards needed to be derived when linking the two clinically approved systems together. This group reported receiving significant help from g.tec in working through this process. (Note: g.tec is a small neural recording hardware company that caters to the BCI community).

Intracortical Microelectrode Recordings from Rhesus Macaques

At our site visit, the Freiburg team reported on its *post-hoc* analysis of intracortical data recorded in Dr. Eilon Vaadia's lab at Hebrew University. Dr. Vaadia's experimental setup consisted of two horizontally moveable manipulanda (one for each hand) that controlled the movements of cursors to targets displayed on a screen. The analysis presented was from neural data collected during planar center-out movements to eight different target directions made by each hand separately. During each recording session, four movable tungsten microelectrodes were acutely positioned within the arm area of motor cortex in chronic recording chambers over each hemisphere. This allowed for data collection from eight new penetration sites during each recording session.

In these studies, local field potentials (LFPs) and unit activity were recorded simultaneously from each electrode. This allowed for a direct comparison of movement-related parameters extracted from well-isolated, single-unit activity (SUA), multiunit activity (MUA), and LFPs recorded from the same electrodes. As in the human ECoG studies, the local field potentials were analyzed both in the time and in the frequency domains. This analysis showed that the power in different frequency bands had unique modulation patterns throughout the time course of the movement, and many of these bands were further modulated by movement direction. The LFP time-domain signal was also modulated throughout the movement, and these movement modulations also often varied by target direction (i.e., both frequency and time domain LFPs were directionally tuned).

This LFP directional tuning enabled classification of intended target with the same accuracy as the directional tuning of the firing rates from single units or multiunit clusters recorded on the same electrodes. In many cases, the preferred directions of the time domain LFP signals were different from the preferred directions of the associated unit activity recorded on the same electrode, although a weak but significant correlation between the tuning of LFPs and single units from the same electrodes was present. When LFPs were combined with single or multiunit activity recorded from the same electrodes, target prediction accuracy improved over either one alone.

Various classification algorithms were used to predict to which of the eight targets the animals had been moving. Continuous movement decoding was also used to predict the actual evolving hand paths. By combining neural data recorded on different days, target prediction accuracy can be estimated for larger numbers of electrodes than the eight that were recorded at any one time. These estimates suggest about a 95% prediction accuracy could be achieved in an eight-target task if both unit and LFP activity from 48 electrodes were combined.

Unit activity often conveyed movement information about both ipsilateral and contralateral arm movements. However, on average, unit activity tended to be more strongly predictive of movements in the contralateral arm versus movements of the ipsilateral arm. LFPs tended to be more even in their ability to predict both contralateral and ipsilateral arm movements.

Predicted movement trajectories were recreated from single-trial recordings of LFPs, SUAs and MUAs using signals from only eight electrodes at a time. Both single unit and LFPs could predict hand position with similarly high accuracies (average correlation coefficients between actual and predicted trajectories approaching 0.7). Example predicted trajectories are shown in Figure B.23. Combining LFPs with single-unit activity further improved the correlation between predicted and actual hand movements.

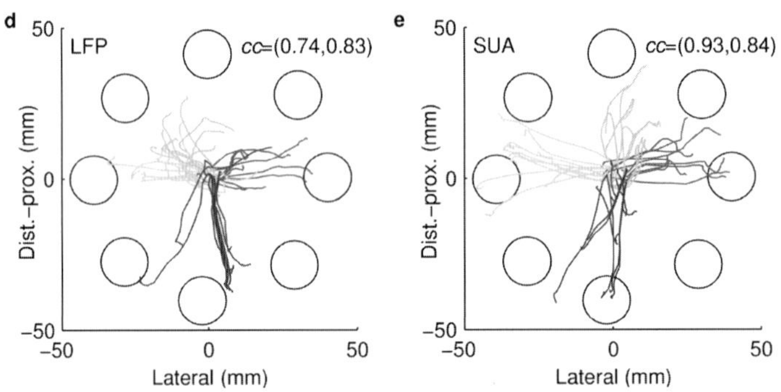

Figure B.23. Movement trajectories predicted from single-trial recordings of LFPs vs. SUAs from only eight intracortical microelectrodes.

SUMMARY AND CONCLUSIONS

The prediction accuracies reported by this group for both discrete target classification and continuous movement predictions from intracortical signals in monkeys are among the best currently reported in the literature using such small numbers of single-trial input channels. The group's analysis showed that useful movement information can be recorded both ipsilateral and contralateral to the moving limb and that both LFPs and unit activity can be combined to enhance the accuracy of the decoded information.

By having both the human and monkey subjects do similar center-out movement tasks, this group has been able to compare the directional information from typical LFP channels in monkeys with the directional information captured by ECoGs in humans. On average, each LFP channel exhibits more directional information than each ECoG channel (roughly twice as much). This group's evaluation of information content using high-density versus standard ECoG grids also suggests that higher density ECoG grids are worth further investigation for neuroprosthetic applications.

REFERENCES

Mehring, C., J. Rickert, E. Vaadia, S. Cardoso de Oliveira, A. Aertsen, and S. Rotter. 2003. Inference of hand movements from local field potentials in monkey motor cortex. *Nat. Neurosci.* 6:1253–1254.

Merhing, C., M. Nawrot, S. Cardoso de Oliveira, E. Vaadia, A. Schulze-Bonhage, A. Aertsen, and T. Ball. 2004. Comparing information about arm movement direction in single channels of local and epicortical field potentials from monkey and human motor cortex. *J. Physiol.* Paris 98:498–506.

Rickert, J., S. Cardoso de Oliveira, E. Vaadia, A. Aertsen, S. Rotter, and C. Mehring. 2005. Encoding of movement direction in different frequency ranges of motor cortical local field potentials. *J. Neurosci.* 25(39):8815–8824.

University of Freiburg BMI Initiative. (n.d.). Brain-machine interfacing initiative research overview. (PowerPoint presentation.)

Site: **University of Oxford**
Machine Learning and Pattern Analysis Research Group
Wellington Square, Oxford OX1 2JD, UK
http://robots.ox.ac.uk/~parg/
http://cswww.ac.uk/Research/BCIs/

Date Visited: June 5, 2006

WTEC Attendees: D. Taylor (report author), T. Berger, G. Gerhart, W. Soussou, M. DeHaemer

Hosts: Prof. Steve Roberts, Ph.D.
Tel: +44 (0)1865 273152
Fax: +44 (0)1865
Email: sjrob@robots.ox.ac.uk
Dr. Fransisco Sepulveda, Ph.D.
Tel: +44 (0)1206 874151
Fax: +44 (0)1206 872788
Email: fsepulv@essex.ac.uk

BACKGROUND

Professor Steve Roberts is the principal researcher of the Pattern Analysis and Machine Learning Research Group, a subgroup of the Robotics Research Group in the Department of Engineering Science at the University of Oxford. The focus of this group is on using probabilistic reasoning applied to problems in engineering and the life sciences. This group uses the tools of statistical inference, particularly Bayesian statistics, to deal rationally with uncertainty in a number of domains ranging from biology and biomedical engineering to image and signal processing. It has been applying machine learning approaches to EEG signals in brain-computer interface applications since 1996. Past and present team members conducting BCI-related research in this lab include Duncan Lowne, Chris Haw, Pete Sykacek, Will Penny, Maria Stokes, and Mike Gibbs. BCI-related research constitutes only about 20% of the total research activity. The primary research areas are machine learning, signal and image processing, and complex and adaptive systems.

Professor Roberts' group does not currently perform clinical work directly. However, it is collaborating with a University of Essex research team that is working to directly apply BCI technologies in various clinical applications. The Essex BCI research team was established in 2003; it is headed by Dr. Fransisco Sepulveda from the Department of Computer Science at the University of Essex. Other team members include Drs. R. Poli, J.Q. Gan, H. Lakany, R. Palanlapppan, and five Ph.D. students. The primary source of research funding for these groups is the Engineering and Physical Sciences Research Council (EPSRC). This is the UK Government's leading funding agency for research and training in engineering and the physical sciences. The Oxford Group receives additional funding from private foundations such as the Wellcome Trust, and the Essex group had recently received £273,000.00 (~$510,000 at the time of the WTEC visit) in equipment and infrastructure funding to establish a new state-of-the-art BCI lab (SRIF3 grant).

RESEARCH AND DEVELOPMENT ACTIVITIES

Oxford Research Philosophy and Activities

Although the Oxford Group is not directly involved with clinical testing of BCIs, its focus is on developing statistically rigorous signal decoding methods that will make EEG-based BCIs practical for real-world applications. Therefore, its primary goals are to develop a highly accurate real-time brain interface that can be used without training by anyone and requires only a very small number of EEG electrodes. This group has a small, in-house EEG lab where its researchers do their own data collection on able-bodied subjects. To keep their work relevant and practical, they often limit the signals used for decoding to a single pair of bipolar recording electrodes over the sensory motor areas (e.g., a C3 or C4 recording location) or other cortical areas. They have found that they can often decode intended movement or brain state just as well with a single bipolar recording channel as most labs do with a full cap of electrodes.

One problem of implementing BCIs in useful applications is that many of the standard EEG characteristics used for classification are different from person to person and may even be absent in some individuals. The Oxford group is investigating alternative EEG characteristics, such as resonance and signal complexity measures, as a means of making its BCI classification methods generalized to a larger number of users. This group has shown that alternative signal characteristics, such as decoherence, can be more user-invariant than the typical change in EEG power spectra currently used by many EEG-based BCIs.

This group's focus has been primarily on robust classification of discrete states from noisy EEG signals. Its researchers hypothesize that the brain undergoes state transitions that can be detected from the observed EEGs (hidden Markov model). By analyzing EEG data within a Bayesian statistical framework, they can optimize net information transfer rate regarding the hidden state classification. Their philosophy is to use probability distributions from data recorded across electrodes and over time to define and update an optimal classification function.

Because biological systems are constantly changing, the classification decision boundaries must be regularly modified as the recorded EEG signals change over time. Members of this team have applied extended Kalman filtering techniques to find optimal, closed-form solutions that maximize classification of these nonstationary signals. The methods they have employed are effective at maintaining good classification rates from nonstationary signals by modifying the classification boundaries as the class distributions shift. Their methods enable this modification to happen even when little or no feedback is available regarding what the correct classifications should have been. This ability of the system to adapt the classification function without complete knowledge of past prediction accuracy is a significant benefit of this extended Kalman filtering technique. It is particularly useful for adapting classifiers in real-world, asynchronous applications where information about what the person was really trying to do is simply not available.

One reason this group has focused on the development of adaptive classification functions stems from an interest in modeling the adaptive symbiotic machine-human learning processes where both partners adapt to each other's actions simultaneously. Although many BCI research groups have found decoding accuracies increase when their users are provided feedback regarding their decoded brain signals (i.e., closed-loop control), the Oxford group found the opposite to be true. In one Oxford study, the information transfer rate was higher on average when the user did *not* have any real-time feedback of the classification results. However, once the user was provided feedback, the average information transfer rate consistently decreased for each subject by an average of 0.21 bits/second. This suggests that the users intuitively modified their EEG output in a way that was unintentionally disruptive to the classification functions. However, when the classification algorithm was regularly adapted to account for the changes made by the user (updated about once every 10 seconds), the classification accuracy significantly improved and the information transfer rate increased by 0.42 bits/ second. This illustrates the importance of understanding this interplay between the adaptive user and the machine learning algorithms when implementing these BCI devices in real time with feedback.

One area that Oxford team members see as a potential use of their BCI decoding technology is in the field of BCI *assisted* devices. One of their goals is to detect the intent to move before actual movement onset. This signal could then be used to help initiate movement in individuals with a wide range of movement disorders, e.g., Parkinsons, stroke, brain injury, etc., as well as to initiate other BCI-assisted devices.

BCI Research at Essex University

The Essex research team is working more directly toward developing clinical applications of BCI technology in association with the Royal Hospital for Rehabilitation. This interdisciplinary team includes engineers, psychologists, and other researchers that are able to develop and test complete BCI systems to meet the needs of disabled individuals. They have assembled a 70 m^2 wheelchair-accessible research facility that includes three EM-noise-reduced experimental recording areas. They are equipping the facility with multiple EEG/EMG recording systems, including a 64- and a 128-channel EEG system with active electrodes that can record a broad range of frequencies from the scalp, including high gamma band activity (Biosemi). They are also expanding their equipment list to include a 24-channel near-infrared recording system, prosthetic hands, and a virtual reality system.

Basic research activities encompass identification of novel features for classifying movement intent, comparing the use of different types of motor imagery tasks, and asynchronous detection of movement intent. Work by this team has shown that nonmotor-brain areas may be just as useful for intent classification as motor areas. The team has also shown that activity in the high gamma band (80 Hz measured via active Biosemi surface EEGs) can be useful in detecting motor intentions. Work on higher-order statistical analysis has also proven to be fruitful in mental classification tasks.

Essex team members are working on EEG-based mouse control systems, including one that uses visual-evoked potentials and has achieved an information transfer rate of 5 bits/minute. They are also developing spelling programs and have devoted efforts to optimizing visual and auditory stimulus parameters (e.g., letter size, color) in programs such as the P300 speller. This group is also working to control of a five-degree-of-freedom prosthetic hand and is developing an asynchronous wheelchair controller.

The Essex group has funding from the EPSRC for a collaborative project with the Oxford group, Adaptive Asynchronous Brain-Actuated Control. This project aims to develop a novel adaptive and asynchronous BCI system for brain-actuated control of intelligent systems and robots. Total funds are £442,401.00 (~$830,000); funds for Essex are £261,939.00 (~$490,000). The methods developed as part of this grant are to be assessed through extensive experimentation with real-time brain-actuated control of an intelligent wheelchair and a robotic arm. The Essex group also has additional funding from the EPSRC for the project, Mining for Novel Signatures in Multi-Channel EEG for Brain-Computer Interfaces (£122,984.00 or ~$230,000).

SUMMARY/CONCLUSIONS

Although members of the Oxford Group do not conduct clinical testing themselves, the very practical and applied focus of their BCI efforts has resulted in the development of robust classification methods that are highly appropriate for use in real-world applications (i.e., require few electrodes, adapt with the user even without feedback of error information, and can use robust aspects of the EEG that are invariant across users). The collaboration between the Oxford and Essex research teams is an excellent example of how to move good ideas from theory into clinical practice. The Essex group's ability to implement and evaluate BCI applications with the disabled individuals will help ensure that these research and development efforts will proceed along clinically relevant lines. Ten years ago,

Dr. Roberts' team was virtually the only group in the UK performing BCI-related research. Now, more than 30 labs in the UK are moving into BCI research areas. The interest of the research community, along with the willingness of the UK government to fund these types of projects, suggest that growth in this area is likely to continue in years to come.

REFERENCES

Curran, E., P. Sykacek, S. Roberts, W. Penny, M. Stokes, I. Jonsrude, and A. Owen. 2004. Cognitive tasks for driving a brain-computer interfacing system: a pilot study. *IEEE Trans. Neural Syst. Rehab. Eng.* 12(1):48–54.

Lowne, D., C. Haw, and S. Roberts. 2006. An adaptive, sparse-feedback EEG classifier for self-paced BCI. BCI 2006 Workshop, Powerpoint Overview Presentations of Research at Oxford and Essex, June 5.

Roberts, S. and W. Penny. 2000. Real-time brain-computer interfacing: a preliminary study using Bayesian learning. *Med. Biol. Eng. Comput.* 38(1):56–61.

Roberts, S., W. Penny, and I. Rezek. 1998. Temporal and spatial complexity measures for EEG-based brain-computer interfacing. *Med. Biol. Eng. Comput.* 37(1):93-99.

Sykacek, P. and S. Roberts. 2002. Bayesian time series classification. In *Advances in neural information processing systems* 15. Cambridge, MA: MIT Press, pp 937–944.

Sykacek, P., S. Roberts, and M. Stokes. 2004. Adaptive BCI based on variational Bayesian Kalman filtering: an empirical evaluation. *IEEE Trans. Biomed. Eng.* 51(5):719–729.

Sepulveda, F., M. Meckes, and B.A. Conway. 2004. Cluster separation index suggests usefulness of nonmotor EEG channels in detecting wrist movement direction intention. *Proceedings of the 2004 IEEE Conference on Cybernetic and Intelligent Systems.*

Zhou S.M., J.Q. Gan, and F. Sepulveda. 2004. Using higher-order statistics from EEG signals for developing brain-computer interface (BCI) systems. *Proceedings of the 2004 UK Workshop on Computational Intelligence. Loughborough University.*

Site: **University of Tübingen**
Institute of Medical Psychology and Behavioral Neurobiology
Gartenstrasse 29
72074 Tübingen, Germany
http://www.mp.uni-tuebingen.de/mp/

Date Visited: May 31, 2006

WTEC Attendees: D. McFarland (report author), T. Berger, G. Gerhardt, D. Taylor, G. Lewison

Hosts: Andrea Kübler, Ph.D.
Tel: +49 7071 29 75 997
Email: andrea.kuebler@uni-tuebingen.de
Femke Nijboer
Tel: +49 7071 29 74 222
Email: femke.nijboer@uni-tuebingen.de

BACKGROUND

Niels Birbaumer has been training healthy subjects to control slow cortical potentials (SCPs) for the last 30 years. This involves the use of neurofeedback to produce voluntary control of SCPs and the study of behavioral correlates of this control. More recently, these methods have been applied to practical problems such as control of epileptic seizures, attention problems, and BCI. PubMed lists 286 publications for the Birbaumer group, of which at least 32 are directly related to BCI research.

RESEARCH AND DEVELOPMENT ACTIVITIES

BCI research at Tübingen began about 10–12 years ago and is based upon the premise that it is all about the patients. Few healthy subjects are studied. The SCP-based BCI was originally called the "thought translation device." The Tübingen group demonstrated that paralyzed patients can use SCPs to choose letters from a menu at rates up to three letters/min. (Birbaumer et al., 1999).

In 2000 the group developed BCI2000 in collaboration with the Wadsworth group in Albany, NY (Schalk et al., 2004). Since that time, it has used sensorimotor (SMR) and P300-based BCI systems in addition to the SCP-based BCI. All of these methods are noninvasive. Its researchers have recently shown that the SMR-based system can be used by patients with ALS, despite the fact that they have degeneration in motor areas (Kubler et al., 2005). More recently, the group has emphasized the P300-based matrix speller, originally developed by Farwell and Donchin (1988), since it is faster and does not require training.

The Tübingen group, in conjunction with its Wadsworth collaborators, is currently focusing on providing BCI systems to subjects who can benefit from the devices. This work is supported in part by a Bioengineering Research Partnership (BRP) grant from NIBIB and NINDS. The Tübingen part of this project involves working with individuals who have only residual movement that is difficult to control and who have decided to be on a respirator. BCI units will be placed in patients' homes so that patients can use these units on a regular basis. This requires developing a reasonably priced system that does not require frequent intervention by experts.

The Tübingen group provided a live demonstration of visual SMR-based cursor control, an auditory SMR-based task, and spelling with the P300-based matrix. The setup used BCI2000 and is shown in Figure B.24. A single user who did not have extensive training performed all three tasks. The user moved a cursor vertically on a video monitor to intercept targets with the visual SMR-based task. The user modulated bongo and harp sounds with the auditory SMR-based task. The P300-based matrix task presented the user with a matrix of flashing letters. The user selected these in sequence to spell the phrase "*Cogito, ergo sum.*"

Figure B.24. The WTEC panel talks with a user who demonstrated several of the Tübingen BCI systems.

The Tübingen group showed the WTEC panel several other projects, including a MEG-based system using SMR training to aid in stroke rehabilitation. This is based on brain-initiated movement therapy using MEG-triggered, pneumatically controlled orthosis that mechanically opens and closes the paretic hand. This group is also developing a real-time fMRI-based neurofeedback system that is being evaluated as a treatment for antisocial personality disorder. The system is based on the concept of training patients to regulate the activity of brain areas associated with the psychopathology. The hypothesis is that such training will lead to modification of symptoms.

SUMMARY AND CONCLUSIONS

The Tübingen group has an active research program mainly concerned with development of noninvasive methods for EEG-based communication and control. The group emphasizes working with patients in their homes. There is a need for more groups willing to work with disabled subjects, because these are the individuals most likely to benefit from BCI research in the short term. The Tübingen group is also exploring other uses of learned regulation of brain states that have potential benefits for individuals with epilepsy, movement disorders, and psychopathology. It is a recognized leader in these fields.

REFERENCES

Birbaumer, N., N. Ghanayim, T. Hinterberger, I. Iversen, B. Kotchoubey, and A. Kubler. 1999. A spelling device for the paralyzed. *Nature* 25.223–232.

Farwell, L.A. and E. Donchin. 1988. Talking off the top of your head: toward a mental prosthesis utilizing event-related potentials. *Electroencephalogr. Clin. Neurophysiol.* 70:510–523.

Kubler, A., F. Nijboer, J. Mellinger, T.M. Vaughan, H. Pawelzik, G. Schalk, D.J. McFarland, N. Birbaumer, and J.R. Wolpaw. 2005. Patients with ALS can use sensorimotor rhythms to operate a brain-computer interface. *Neurology* 64:1775–1777.

Schalk, G., D.J. McFarland, T. Hinterberger, N. Birbaumer, and J.R. Wolpaw. 2004. BCI2000: A general-purpose brain-computer (BCI) system. *IEEE Trans. Biomed. Eng.* 51:1034–1043.

APPENDIX C. SITE REPORTS—ASIA

Site: **Advanced Telecommunications Research Institute International**
Computational Neuroscience Laboratories (CNS)
2-2-2 Hikaridai "Keihanna Science City"
Seika-cho, Souraku-gun
Kyoto 619-0288 Japan
http://www.cns.atr.jp/
http://www.cns.atr.jp/~cns-naist/index_e.html

Date Visited: October 27, 2006

WTEC Attendees: T. Berger (report author), G. Gerhardt (report co-author), G. Hane

Hosts: Prof. Mitsuo Kawato, Director
ATR Computational Neurosceince Laboratories
Tel: +81 774 95 1230
Email: webmaster@atr.jp

BACKGROUND

The Advanced Technology Research Institute (ATR) was formed in 1985 and has nine divisions and two affiliated centers and ¥100 million in capital, all dedicated to understanding human-computer interactions, human-machine computing, and brain computing and robotics (Figure C.1).

Figure C.1. ATR units.

ATR activities have expanded to a high level of productivity and increasing patent activity in both basic and applied domains, in laboratories or centers for spoken language communication, adaptive communications, wave engineering, human information science, network informatics, computational neuroscience, intelligent robotics and communication, and brain activity imaging, among others.

RESEARCH AND DEVELOPMENT ACTIVITIES

The research scope of the ATR CNS (Computational Neuroscience) is broad but focuses on several thematic areas: (1) how computational properties emerge from biological mechanisms and circuitry; (2) how higher cognitive function emerges from these biologically-based computational capabilities; and (3) the hierarchical organization of motor centers and motor learning, and in the past several years, an increasing emphasis on noninvasive decoding of motor commands.

One major effort of the ATR CNS relevant to BCIs is the Noninvasive Neural Decoding Project. The thrust of this research effort is to use a combination of non-invasive recording and imaging methods, such as EEG, MEG, fMRI, and NIRS, to decode or classify brain representations of external events, emotional states, and movement plans. The essential assumptions are that these and all internal states are represented dynamically and in a distributed manner by large populations of neurons. Moreover, these representations change as a function of learning and adaptation by the organism. Thus, to decode the information represented internally by the brain requires technologies that can detect brain-neuron activity with high spatial and temporal resolution and that can account for the inherent nonstationarities, i.e., learning mechanisms, of the brain. For example, ATR CNS scientists have developed pattern recognition methods that can detect visual stimulus orientation based on fMRI of V1 (Figure C.2).

In addition, ATR CNS scientists have developed procedures to successfully classify intended hand movements (configurations of hand and fingers) that can, in turn, be mapped onto a robotic hand for a “rock, paper, scissors” task (Figure C.3).

Other pioneering work has focused on developing hierarchical Bayesian filters to estimate the current distribution from fMRI/MEG data, combined with NIRS input, to achieve high-resolution maps of spatiotemporal activity within the brain (Figure C.4).

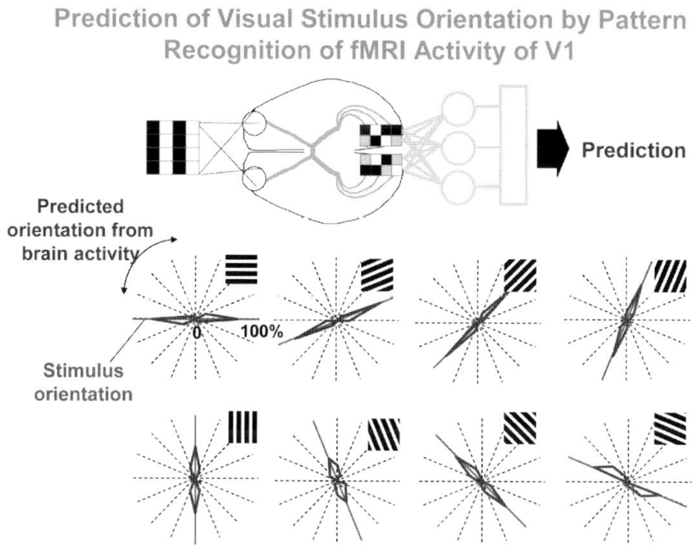

Figure C.2. Visual stimulus orientation.

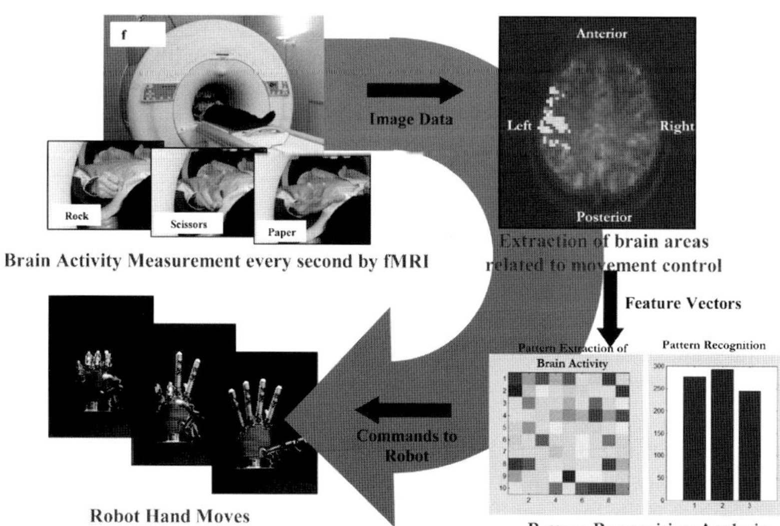

Figure C.3. Real-time fMRI decoding.

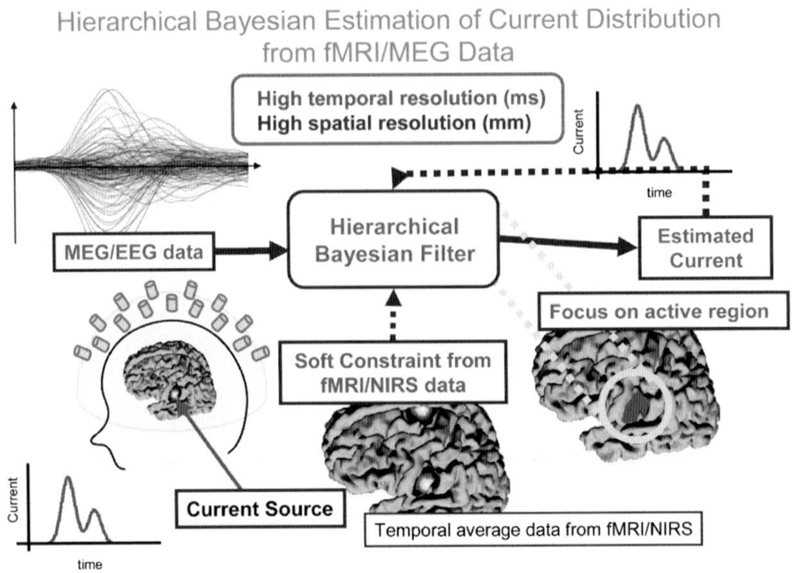

Figure C.4. Hierarchial Bayesian estimation.

This last paradigm of combining EEG/MEG/fMRI and NIRS data is important in an ATR distinction among

- BMIs or brain-machine interfaces: Invasive recording of neural activities by multiple electrodes with user/algorithm training
- BCIs or brain-computer interfaces: Noninvasive EEG activity recording with heavy user/algorithm training
- BNIs or brain-network interfaces: Noninvasive EEG/MEG recording constrained by NIRS/fMRI enabling higher spatial resolution with less intensive user training

The ATR CNS is clearly considering BNIs as the future class of technologies for noninvasive interactions with the brain, both for medical applications to repair brain injury and for nonmedical applications to assist and improve brain interactions with computers and other devices (Figure C.5).

In general, the panel was impressed with the vision of the ATR CNS with respect to possible applications of BCIs (or BNIs) and the strong commitment of the Director, Professor Mitsuo Kawato, and the ATR CNS Board of Directors to the development of BCIs for many future aspects of human existence.

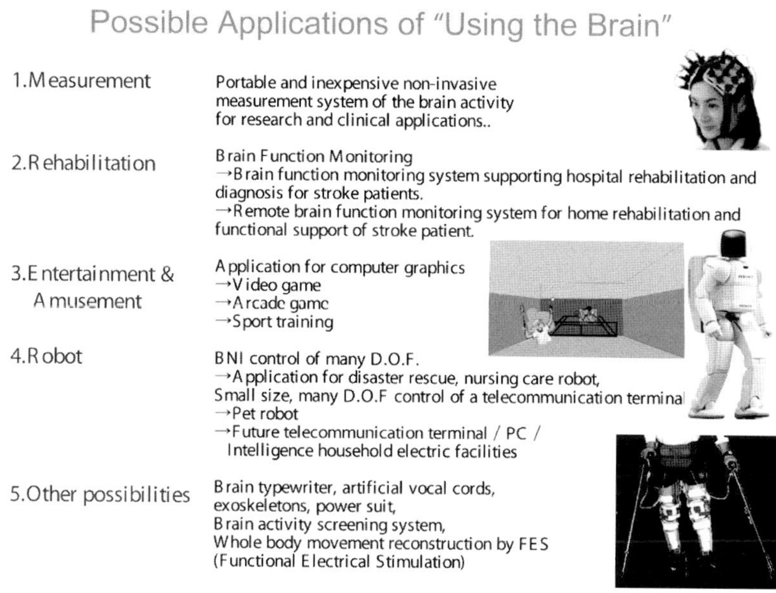

Figure C.5. Possible applications of BCIs.

FUNDING SOURCES AND COMMERCIALIZATION

Both private and public sources fund the ATR CNS. Major support comes from the National Institute of Information and Communications Technology (NICT) and the Honda Research Institute Japan Company.

Industrial Collaboration. ATR CNS has a collaboration with Shimadzu Corp.

Academic Collaboration. ATR CNS has extensive academic collaborations throughout the world, including in the United States with MIT, Haskins Labs, AT&T Labs, Bell Labs, Central Michigan University, Illinois University, Stanford University, the Santa Fe Institute, and the University of Southern California, as well as with Universidade Federal de Minas Gerais (Brazil), Buenos Aires University (Argentina), University of Sydney and Griffith University in Australia, Otago University (New Zealand), the Chinese Academy of Sciences (P.R. China), ETRI and KAIST in the Republic of Korea, the Indian Institute of Technology, Israel Institute of Technology, the Weizmann Institute of Science (Israel), the Sweden Royal Institute of Technology and Stockholm University, University of Karlsruhe, University of Madrid, University of Cambridge, and University of Glasgow.

More specifically, collaborations of ATR CNS and other major researchers related to BCI, BMI, and BNI in connection with "liberating the brain society" include the following:

- *Brain Research*: Masao Ito, Yukiyasu Kamitani, Yoshio Sakurai (rat), Shigeru Kitazawa (eye movement), Toshio Ijima (monkey), Ichiro Fujita
- *Engineering*: Takafumi Suzuki (electrodes), Yasuharu Koike, Osamu Shimizu (Shimazu Manufacturing Ltd, NIRS development), Shiro Ikeda, Hiroshi Yokoi, Kazuhiko Sagara (Hitachi)
- *Clinical*: Youichi Katayama (DBS), Amami Kato (ECoG), Fujikado Takashi (artificial neural networks), Ichiro Miyai, Kazunori Ikoma, Akimoto Sato
- *Theory and Modeling*: Osamu Sakura, Atsushi Iriki, Hideaki Koizumi, Takashi Tachibana.

SUMMARY AND CONCLUSIONS

The ATR CNS clearly has developed one of the most sophisticated BCI programs in the world, particularly when BCIs are considered in the broadest possible perspective, both medical and nonmedical. The ATR CNS administrators and researchers have decided to pursue noninvasive BCI technologies and have managed to employ the most advanced imaging methods available for visualizing brain-spatiotemporal dynamics. This choice has implications for future embodiment of BCI systems, given the obvious difficulty of miniaturizing such systems. Nonetheless, the ATR CNS is poised to develop the fundamental understanding of the relation between brain-state dynamics, modeling tools, and robotics to a degree not seen by the panel at any other site. The panel also was strongly impressed by the extent to which the ATR CNS has considered the societal and ethical consequences of broadly-distributed development and use of BCIs. This is a significant program.

REFERENCES

Atkeson, C.G., J. Hale, F. Pollick, M. Riley, S. Kotosaka, S. Schaal, T. Shibata, G. Tevatia, S. Vijayakumar, A. Ude, and M. Kawato. 2000. Using humanoid robots to study human behavior. *IEEE Intelligent Systems: Special Issue on Humanoid Robotics* 15:46–56.

Burdet, E., R. Osu, D. Franklin, T. Milner, and M. Kawato. 2001. The central nervous system stabilizes unstable dynamics by learning optimal impedance. *Nature* 414:446–449.

Exploratory Research for Advanced Technology. Kawato Dynamic Brain project website: http://www.jst.go.jp/erato/project/kgd_P/kgd_P.html (accessed March 21, 2007).

Gomi, H., and M. Kawato. 1996. Equilibrium-point control hypothesis examined by measured arm stiffness during multijoint movement. *Science* 272:117–120.

Haruno, M., T. Kuroda, K. Doya, K. Toyama, M. Kimura, K. Samejima, H. Imamizu, and M. Kawato. 2004. A neural correlate of reward-based behavioral learning in caudate nucleus: A functional magnetic resonance imaging study of a stochastic decision task. *Journal of Neuroscience* 24:1660–1665.

Imamizu, H., T. Kuroda, S. Miyauchi, T. Yoshioka, and M. Kawato. 2003. Modular organization of internal models of tools in the human cerebellum. *Proceedings of the National Academy of Sciences USA* 100:5461–5466.

Imamizu, H., S. Miyauchi, T. Tamada, Y. Sasaki, R. Takino, B. Puetz, T. Yoshioka, and M. Kawato. 2000. Human cerebellar activity reflecting an acquired internal model of a novel tool. *Nature* 403:192–195.

Imamizu, H., T. Kuroda, T. Yoshioka, and M. Kawato M. 2004. Functional magnetic resonance imaging examination of two modular architectures for switching multiple internal models. *Journal of Neuroscience* 24:1173–1181.

Kawato, M. 1999. Internal models for motor control and trajectory planning. *Current Opinion in Neurobiology* 9:718–727.

Kamitani, Y., and F. Tong. 2005. Decoding the perceptual and subjective contents of the human brain. *Nature Neuroscience* 8:679–685.

Osu, R., S. Hirai, T. Yoshioka, and M. Kawato. 2004. Random presentation enables subjects to adapt to two opposing forces on the hand. *Nature Neuroscience* 7:111–112.

Sato, M., T. Yoshioka, S. Kajiwara, K. Toyama, N. Goda, K. Doya, and M. Kawato. 2004. Hierarchical Bayesian estimation for MEG inverse problem. *NeuroImage* 23:806–826.

Schaal, S., D. Sternad, R. Osu, and M. Kawato. 2004. Rhythmic arm movement is not discrete. *Nature Neuroscience* 7:1137–1144.

Schweighofer, N., K. Doya, H. Fukai, J.V. Chiron, T. Furukawa, and M. Kawato. 2004. Chaos may enhance information transmission in the inferior olive. *Proceedings of the National Academy of Sciences USA* 101:4655–4660.

Shidara, M., K. Kawano, H. Gomi, and M. Kawato. 1993. Inverse-dynamics model eye movement control by Purkinje cells in the cerebellum. *Nature* 365:50–52.

Wolpert, D., and M. Kawato. 1998. Multiple paired forward and inverse models for motor control. *Neural Networks* 11:1317–1329.

Site: **Huazhong University of Science and Technology**
Department of Control Science and Engineering
Wuhan 430074, P. R. China
http://www.hust.edu.cn/english

Date Visited: October 24, 2006

WTEC Attendees: T. Berger (report author), G. Gerhardt (report co-author), G. Hane

Hosts: Prof. Yongji Wang
Department of Control Science Engineering
Tel: +86 27 87540014
Email: wangyjch@mail.hust.edu.cn

BACKGROUND

Huazhong University of Science and Technology (HUST) is a merger of the former HUST founded in 1953, Tongji Medical University founded in 1907, Wuhan Urban Construction Institute founded in 1954, and Wuhan Vocational College of Science and Technology founded in 1968. It was established on May 26, 2000. HUST is one of China's leading universities directly under the management of the Ministry of Education. HUST has 36 academic schools and departments and a variety of university programs, including 74 undergraduate programs, 200 master's-level programs, 139 doctoral programs, and 22 post-doctoral research centers. The university is home to over 50,000 full-time students, including 12,000 master's-level candidates and 4,400 doctoral candidates. The university has over 10,000 faculty members, of which 800 are full professors and approximately 1,200 are associate professors.

Dr. Wang's Department of Control Science and Engineering is located within HUST and is composed of Professor Wang and his research group.

FUNDING SOURCES AND COMMERCIALIZATION

Dr. Wang and his associates receive funding from the Chinese government. Funding originates from NNSF China and the Ministry of Education in China for the research on motor control of arm/hand movement. Funding for the spinal cord stimulation comes from NNSF China. Additional funding for the development of upper limb rehabilitative robotics systems (to be controlled by the motor cortical control systems being developed by Dr. Wang's group) is provided by NNSF China.

Industrial Collaborations

Spinal cord stimulation electrodes are made by AKM FPC Co., Ltd., Suzhou City, Jiangsu Province; then they are further processed by Wuhan Research Institute of Materials Protection, Hubei Province.

Academic Collaborations

Research funds from various government departments and enterprises amounted to 527 million Yuan (about U.S.$67,387,000) in 2004. There is a major collaboration between Dr. Wang's laboratory at HUST and Dr. Jiping He's laboratory at Arizona State University. All unit recordings from nonhuman primates are conducted by Dr. He's laboratory. Dr. Wang's laboratory is exploring a variety of algorithms for extracting information about the intended reach trajectory from the population motor cortical recordings. Thus, with respect to work on motor cortical control, part of the experimental work is being conducted at Arizona State University in the United States, and modeling/algorithm development is being conducted in China. Dr. He is also collaborating with Dr. Xu Qi on spinal cord stimulation research; in this case, the experimental work is being conducted onsite at HUST, and Dr. He acts as a consultant. There are also active collaborations between Dr. Wang's laboratory and other components of HUST, particularly with respect to upper extremity rehabilitative robotic systems and microfabrication of electrodes.

RESEARCH AND DEVELOPMENT ACTIVITIES

Professor Wang leads an active group concerned with extraction of information about upper-limb reach commands from 16-channel array recordings from motor cortex of nonhuman primates.

Dr. Wang's research represents part of an ongoing collaboration with Dr. Jiping He. All unit recordings from nonhuman primates are conducted by Dr. He's laboratory. The primary behavioral paradigm that Dr. He's group uses is a center-out task in which monkeys control a manipulandum in reaching from a center position to one of several peripherally positioned targets; the reaching movement is tracked continuously. Dr. Wang's laboratory is exploring a variety of algorithms for extracting information about the intended reach trajectory from the population motor cortical recordings. Some of the methods being applied include support vector machines (SVMs), support vector regression (SVR), Bayesian modeling, principal component analysis (PCA), artificial neural networks (ANNs), and non-linear systems identification.

For example, Dr. Wang's research has demonstrated that SVMs can be superior to learning vector quantization in terms of a higher accuracy of prediction, lower training time, and reduced data requirements for training. Dr. Wang's laboratory also has extensively studied the performance of SVR methods. His work has explored both linear and nonlinear regression approaches and has examined the effects of linear, radial basis function, and spline kernels on SVM regression. Importantly, Dr. Wang also has applied nonlinear systems identification methods for predicting the trajectory of hand position from multiple motor cortical unit recordings. Applications of this approach have not been reported previously. The majority of other researchers in this field, including those in the United States, have used only linear methods. In addition, the problem of mapping multiple-unit motor cortical activity ultimately to 3D hand position may be sufficiently complex that input-output methods such as nonlinear systems identification and ANNs, which require no prior knowledge of the system (motor cortex, spinal cord, and neuromusculature), may prove superior. Dr. Wang's results demonstrate that nonlinear models are superior to linear models in terms of prediction accuracy. Finally, Dr. Wang presented results of some preliminary research to develop spiking neural networks that include dynamics in the network connections. This is a novel approach that has been reported only once previously and may be a breakthrough approach for cortical control modeling and brain computer interfaces in general.

Dr. Xu Qi and Dr. Xu Jiang are actively involved with the development of epidural spinal cord stimulation approaches for restoration of individuals with spinal cord injuries. This is best applied to incomplete spinal cord injuries where a partial crush of the spinal cord has led to only partial loss of locomotion function. This work is done in conjunction with Dr. Jiping He. The scientific work embodied in this involves the development of microelectrode arrays that are flexible for spinal cord stimulation. The individuals stimulate approximately T10~L2 vertebral levels per animal using this approach and sustain studies for approximately two days. In terms of BCI, this is an interesting area that provides some technology development, as well as alternate approaches for reanimation after spinal cord injury.

Development of Indwelling Flexible Electrodes for Spinal Cord Stimulation

Dr. Xu Qi is part of the Department of Control Science and Engineering at HUST. Researchers in this department are working on the development of epidural spinal cord stimulating electrodes, specifically in T10~L2 implants. These are polyimide-based electrodes with silver stimulating sites. They currently have funding from NNSF China. They are capable of carrying out studies for several days in laboratory cats in order to investigate the mechanism of ESCS modulation on the energy metabolism during locomotion. The ultimate goal is to combine effective electrical stimulation of the spinal cord with the identification of motor cortical command signals to develop BCIs for patients with spinal cord transection to restore brain control of the neuromuscular system. This technology should be applicable to a number of scientific thrusts for overall development of better bioimplantable electrodes. It is a unique technology and may have applicability not only to laboratories within China, but also to those in the United States and Europe that are working on invasive BCI technologies.

SUMMARY AND CONCLUSIONS

Competitive Advantages Compared to the World

The advanced algorithms that are being developed by Dr. Wang for processing of motor cortex electrophysiological information could aid in the development of better BCI technology in other laboratories. These low-cost approaches for the development of data processing of BCI technology are sorely needed, and it is clear that many of the approaches being developed at HUST are cutting-edge and lead the field. Further collaborations with this group and others throughout the world would be beneficial for development of modern BCI technology.

The flexible electrode technologies being developed by Dr. Xu Qi may have applicability to a number of BCI applications and the development of better electrodes for many types of invasive recording technologies.

Site: **NTT Communication Science Laboratories**
3-1 Wakamiya-Morinosato, Atsugi-shi,
Kanagawa, Japan
http://www.kecl.ntt.co.jp/

Date Visited: October 26, 2006

WTEC Attendees: T. Berger (report author), G. Gerhardt (report co-author), G. Hane

Hosts: Dr. Yoshinobu Tonomura
Laboratory Director
Email: tonomura.yoshinobu@lab.ntt.co.jp
Dr. Takehiro Moriya
Executive Director
Human and Information Science Laboratory
Email: moriya@ieee.org
Dr. Hiroaki Gomi
Parasitic Humanoid Project
Tel: +81 0462 40 5221

BACKGROUND

Nippon Telegraph and Telephone Corporation (NTT) Communication Science Laboratories are dedicated to the pursuit of the communications of the future through pioneering research in communication science. Under the directorship of Dr. Yoshinobu Tonomura, this group focuses on developing the science and technology that will enable human-computer interfaces of the future. There are three main subdivisions of the NTT Communication Science Laboratories: (1) the Innovative Communications Laboratory, which is involved with innovative communication environments where human and information systems can coordinate, interact, and collaborate in the cyber and real worlds; (2) the Media Information Laboratory, which seeks to enrich peoples' lives by developing computer systems that can "see" objects, "hear" sounds, "feel" surfaces and shapes, and "talk" with humans; and (3) the Human and Information Science Laboratory, which is a laboratory focusing on BCI-related technology. The Human and Information Science Laboratory aims for a comprehensive understanding of human information processing and the establishment of relevant computational theory expected to enhance a wide variety of technologies.

FUNDING SOURCES AND COMMERCIALIZATION

NTT appears to be operating independently on the basis of company resources, yet it may be receiving funding from the Japanese government.

Industrial Collaborations

NTT collaborates with other industrial partners for the design and manufacture of some of the sensors and actuators included in NTT systems. Other significant collaborations with respect to the theoretical or scientific basis of the Parasitic Humanoid system were not clear at the time of the visit.

RESEARCH AND DEVELOPMENT ACTIVITIES

Directed by Dr. Maeda and Dr. Kashino, a major focus involving NTT BCIs is the Parasitic Humanoid (PH) project. The foundation of the PH project is a wearable robot for modeling nonverbal human behavior. (See Figures C.6–C.8; all these figures are also available online at http://www.brl.ntt.co.jp/people/parasite/) This anthropomorphic robot composed of a suite of sensors and actuators senses the behavior of the wearer and has an internal computer-processing capability to continuously learn the process of the human sensorimotor integration of the user. When the reliability of predicting future motor movements of the user is sufficient (based on the current stream of multimodal sensory inputs), the PH outputs the errors from the actual behavior so as to drive future actual motion of the wearer. Through this symbiotic interaction, the internal model and the process of human sensorimotor integration approximate each other asymptotically. The computer system begins to predict the next behavior of the wearer using the trained, internal learning modules.

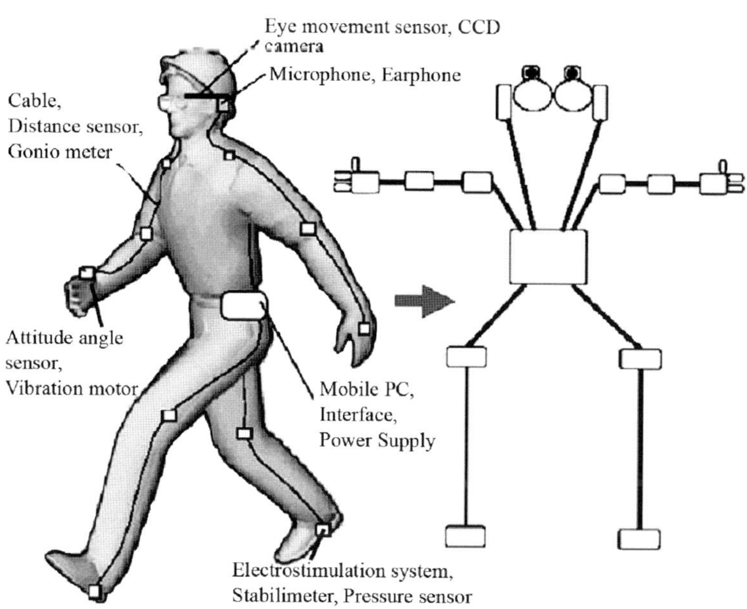

Figure C.6. Wearable sensory devices construct a wearable humanoid without muscle or skeleton.

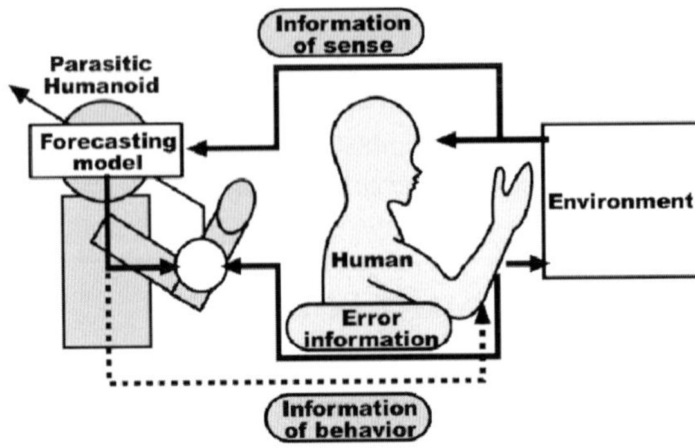

Figure C.7. The symbiotic relationship between the wearer and the Parasitic Humanoid (PH).

One application and scenario for interaction with the Parasitic Humanoid is capturing and retrieving movement patterns. For example, in playing golf, a user may want to capture and retrieve all of the complex motor dynamics that constitute the "best" golf swing. Imagine that the user may be able not only to measure those motor dynamics, but store them and later download that information. Thus, for sporting activities, one may be able to replay and learn the motor patterns for achieving the optimal golf swing.

A second class of applications involves system "dual consciousness." For example, one "dual consciousness" application involves taking advantage of avoidance of oncoming cars or motorcycles by predicting an approaching motorcar movement pattern. Through vestibular and other inputs, the individual may be directed away from the oncoming vehicle. This involves the humanoid being linked to traffic information. In addition, an approach may be to link the PH to traffic information to avoid and suppress traffic jams.

The sensors of a prototype of the Parasitic Humanoid are listed in Table C.1. As itemized in the table, this system consists of vibration motors and electrodes for stimuli that provide outputs to the user, and sensors that provide inputs to the PH. The total weight of motors, electrodes, sensors, and wiring is less than 500 g. The performance of typical mobile PCs is sufficient for logging sensory data, controlling vibrators, and training the internal neural network models. The wearable prototype itself is shown in Figure C.8.

Table C.1 Sensors in a Parasitic Humanoid Prototype

Type of Sensor	Configuration of Sensors	# of Signals
Three-Axis Postural Sensor	Head 1, trunk 3, each limb 3	16*3=48
Fingernail Sensor	Bending and touching of 3 fingers in each hand	3*3*2=18
Eye Movement Sensor	Each eye, two-axis motion and size of pupil	3*2=6
Shoe-Shaped Sensor	Pressure: 5 points, and 1 impact sensor in each foot	6*2=12
Audio and Visual Sensor	2 CMOS cameras (120 Hz) and 2 microphones	2 video, 2 audio

Figure C.8. A prototype of the Parasitic Humanoid.

DEVICES FOR THE PH SYSTEM

Wearable Limb-Motion System

For measurements of limb motion, there is a three-axis sensor manufactured by NEC Tokin Corp. This sensor includes three-axis gyroscopes, two-axis acceleration sensors detecting the direction of gravity, and two-axis magnetic compasses detecting the terrestrial magnetism. This sensor has an advantage that the wearer has no devices attached to the joints, because the sensor does not measure relative directions of the limbs, but absolute directions in space. The influence on behavior by the presence of motion sensors is minimized. The actuating electrodes stimulate muscles, and each actuator set is sent through an electrode that is attached on the skin with a supporting band.

Wearable Eye-Movement Measurement System

The goal is to measure the ocular position in three-dimensional space from the center of both eyes up to the hand's reaching limits, according to the angle of convergence and within an error of 20 mm. This device is positioned in a frame-type system analogous to glasses worn by the individual.

Fingernail Sensor to Measure Direct Touch with Fingertip

NTT has developed a novel fingernail sensor for detecting the touching and bending of the finger. The sensor is placed on the fingernail to avoid hindering the direct touch of the fingertip. The sensor consists of miniature light-emitting diodes and photodetectors that extract a force vector due to the various color patterns responding to the contact force direction (Figure C.9). This instrument is wearable and practical for daily use. NTT has already applied this technology to SmartFinger, which is a new type of display that provides supplementary tactile sensation for augmented reality (see Figure C.10).

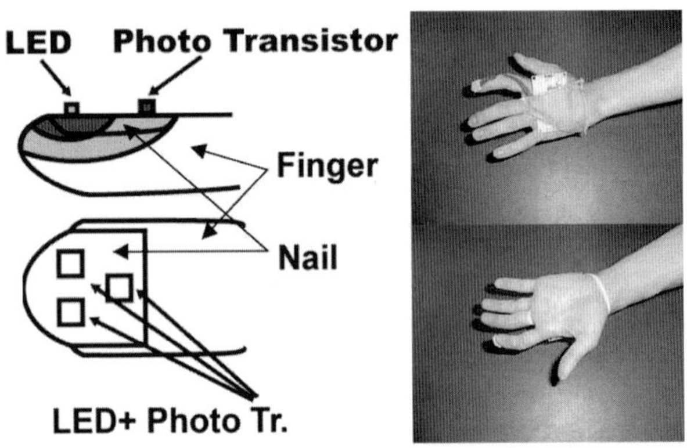

Figure C.9. Fingernail sensors.

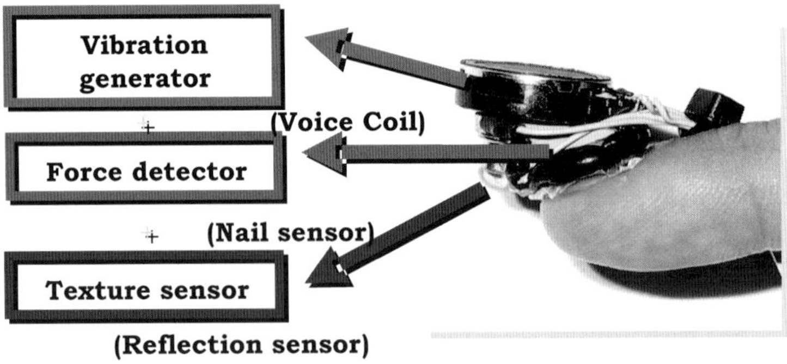

Figure C.10. Components of SmartFinger.

Shoe-Shaped Sensor to Measure Walking and Standing

The NTT labs have developed a shoe-shaped sensor designed to produce a specific walking cycle. An individual walking cycle is measured as pressure on the sole of the shoe. A vibration motor attached on the instep stimulates the foot with cyclic vibration. NTT investigators have found that stimulation during walking with cyclic vibrations does not obstruct normal walking movement, and if the cyclic vibration is similar to that of the walking cycle, the vibration influences the walking rhythm in a normal manner (see Figure C.11).

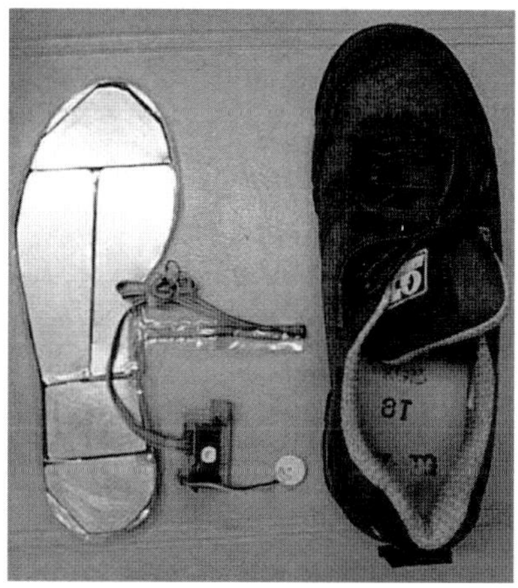

Figure C.11. Shoe-shaped sensors.

Galvanic Vestibular Stimulation System

When Galvanic Vestibular Stimulation (GVS) is delivered to the mastoid through electrodes during human walking, human subjects respond by deviating towards the hemispheric side of stimulation. The WTEC panel participated in an onsite demonstration of this device: electrical stimulation induced an unmistakable sense of walking on a moving surface, e.g., a ship at sea. Upon application of the current, an individual walking straight will deviate markedly to one side, experiencing the change in forces associated with disruption of the vestibular system (see Figure C.12).

Such a device could be used in conjunction with games to simulate environments such as rollercoaster rides and aircraft simulation and could be applied to real-world applications such as a device that would sense an oncoming vehicle to allow an individual to move out of its path (see Figure C.13). This is but one application of these interesting technologies that may be coupled in the future to BCI-type technology.

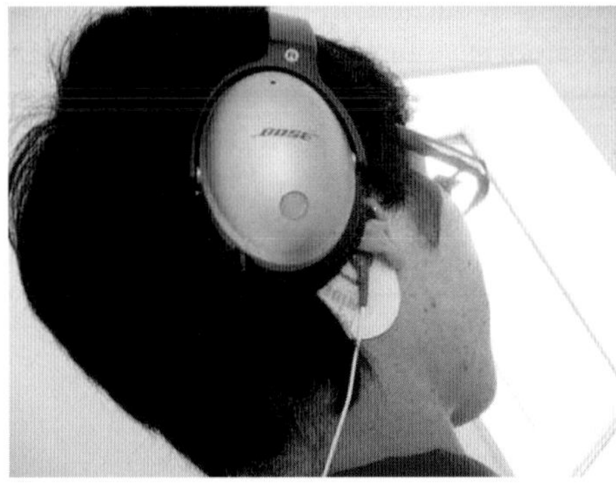

Figure C.12. An electrode on mastoid for galvanic vestibular stimulation.

Figure C.13. Example: an investigator unconsciously avoids a motorbike coming from behind.

SUMMARY AND CONCLUSIONS

The parasitic humanoid project is involved with the development of a variety of sensors and actuators that may be exceptionally useful for providing feedback for BCI applications. These approaches may be useful in feedback to individuals in conjunction with movement control devices. In particular, the eye-tracking vestibular systems and the photodiode tactile electrode system would be very useful components of new BCI technologies. In addition, the technologies may be useful for training neurologists and other healthcare professionals with respect to neurological disorders. Some of the sensory experiences produced by these devices are novel and could contribute to a better experiential understanding of the human sensory system.

REFERENCES

Ando H., T. Miki, M. Inami, and T. Maeda. 2002. The nail-mounted tactile display for behavior modeling. *ACM SIGGRAPH Conference Abstracts and Applications*, San Antonio, TX, 264.

Bent, L.R., B.J. McFadyen, V.F. Merkley, P.M. Kennedy, and J.T. Inglis. 2000. Magnitude effects of galvanic vestibular stimulation on the trajectory of human gait. Neurosci. Lett. 279(3):157–160.

Grillner, S. 1985. Neurobiological bases of rhythmic motor acts in vertebrates. *Science* 228:143–149.

Jacobsen, S. 2001. Wearable energetically autonomous robots. DARPA Exoskeletons for Human Performance kick-off meeting.

Japan Science and Technology Corporation and NTT Communication Science Laboratories. Parasitic Humanoid website, http://www.brl.ntt.co.JP/people/parasite/ (accessed on December 20, 2006).

Kansaku, K., S. Kitazawa, and K.Kawano. 1998. Sequential hemodynamic activation of motor areas and draining veins during finger movements revealed by crosscorrelation signals from fMRI. *Neuroreport* 9:1969–1974.

Kohonen, T., J. Hymminen, J. Kangras, and J. Laaksonan. 1996. SOM PAK: the self-organizing map program package. Technical Report A31, Helsinki University of Technology.

Mascaro, S., and H. Asada. 2000. Photoplethysmograph fingernail sensors for measuring finger forces without haptic obstruction. IEEE Transactions on Robotics and Automation 17(5):698–708.

Matsuoka, K. 1987. Mechanisms of frequency and pattern control in the neural rhythm generators. *Biological Cybernetics* 56:345–353.

Nomura, Y., and T. Maeda. 2001. The study of fingernail sensors for measuring finger forces and bending, *Journal of the Virtual Reality Society of Japan* 6(3) (in Japanese).

Tachi, S., H. Arai, and T. Maeda. 1988. Tele-existence simulator with artificial reality (1)—design and evaluation of a binocular visual display using solid models, *IEEE International Workshop on Intelligent Robot and Systems (IROS'88)*, October, Tokyo, Japan.

Taga, G. 1995. A model of the neuron-musculo-skeletal system for human locomotion. *Biological Cybernetics* 73:97–111.

Site: **RIKEN Brain Science Institute**
2-1 Hirosawa, Wako-shi
Satiama 351-1098 Japan
http://www.bsp.brain.riken.jp/

Date Visited: October 26, 2006

WTEC Attendees: G. Gerhardt (report author), T. Berger, G. Hane

Hosts: Shun-ichi Amari
Director, Brain Science Institute
Tel: +81 48 462 1111 x 7541
Email: Amari@brain.riken.jp
Andrzej Cichocki
Laboratory Head, Lab for Advanced
Brain Signal Processing
Tel: + 81 48 467 9668

BACKGROUND

The RIKEN Brain Science Institute (BSI) was founded in 1997. Its mission is to be "a global interdisciplinary and international center of excellence in the field of brain science." The RIKEN BSI budget for 2005 was ¥9.8 billion per year (U.S.$83 million). There were 504 staff members as of April 2005. The BSI has an organization structure consisting of the President, Director, and six major components: (1) understanding the Brain, directed by Masao Ito, Keiji Tanaka and Susumu Tonegawa; (2) protecting the Brain, directed by Nobuyuki Nukina, Tadafumi Kato, and Hitoshi Okamoto; (3) creating the Brain, directed by Tomoki Fukai and Shun-ichi Amari; (4) nurturing the Brain, directed by Katsuhiko Mikoshiba, Takao K. Hensch, and Keiji Tanaka; (5) advanced Technology Development Group, directed by Atsushi Miyawaki; and (6) Research Resources Center, directed by Chitoshi Itakura. The Director has a special advisor, Masao Ito. In addition, embodied in the BSI organization of RIKEN are an advisory council, a research review committee, and a search committee. Also embodied in this organization are an information center, a neuroinformatics Japan center, and a brain science promotion division. The WTEC panel initially met with Dr. Shun-ichi Amari, Director of the BSI. Dr. Amari clarified for the panel the important role of BCIs in providing a synthetic platform. The division we visited was headed by Dr. Andrzej Cichocki, head of the Laboratory for Advanced Brain Signal Processing, a subdivision of the BSI encompassing approximately ten other scientists. This group is primarily involved with noninvasive forms of brain-computer interfaces.

FUNDING SOURCES AND COMMERCIALIZATION

Currently, the funds supporting the BCI venture are encompassed in the BSI under the direction of Dr. Shun-ichi Amari. The general budget for RIKEN is provided by the Japanese Government.

Industrial Collaborations. Some laboratories are collaborating with Sony Corporation, Honda Research Institute, and a number of pharmaceutical companies. Collaboration with Toyota Motor Company is being started.

Academic Collaborations. The following is a partial list of RIKEN's academic collaborations with a Memorandum of Understanding (MOU) in place:

- Massachusetts Institute of Technology Picower Institute of Learning and Memory
- University of California, SanFrancisco, Neuroscience Program
- University College London, Department of Neuroscience
- University of New Castle (UK), Neuroscience
- École Normale Supérieure, Paris
- Queensland University, Australia, Queensland Brain Science Institute
- Bernstein Center for Computational Neuroscience, Germany.

RESEARCH AND DEVELOPMENT ACTIVITIES

Short- and Long-Term Scientific Goals

Dr. Cichocki has an extensive background in signal processing, mathematics and designs, machine learning algorithms, and tools for analyzing EEG signals for non-invasive BCI devices. Together with Professor Amari (2002), he published a monograph about blind source separation and independent component analysis. This group has perhaps the greatest collection of EEG hardware the team observed. Its researchers are capable of recording EEG signals from as many as 256 passive or active gel-type electrodes for EEG processing. One of their major thrusts is to extract the hidden information from the brain signals; one of the areas of focus by Dr. Cichocki and his group is the extraction of EEG information for the potential early diagnosis of Alzheimer's disease. This preliminary work was published in *Clinical Neurophysiology* (Cichocki et al., 2005). The work involved age-matched controls and 22 patients with mild cognitive impairment who proceeded to develop Alzheimer's disease. The team was able to develop an analysis approach involving filtering based on blind source separation (BSS) to diagnose and predict potential Alzheimer's disease patients versus controls. They were able to improve the percentage of correctly classified patients from 59% to 73% for Alzheimer's disease patients and from 76% to 84% for controls. The authors noted that their method is general and flexible, allowing for a variety of improvements and potential applications to other BCI applications.

This extensive research group has implemented a large number of commercially available devices for EEG monitoring of the CNS. Its members were poised to write a critical review of the available instrumentation and underscore the reality that there is a weakness in the development of electrodes for EEG signaling recording. In particular, the potential development of dry active electrodes for improved signal-to-noise and real-world applications is sorely needed. It is recommended that this group write a review article in the context of the available instrumentation in order to summarize the capabilities of many com-mercially available EEG recording systems and the needs for the field. In particular, this analysis is necessary for the entire field of noninvasive BCI. The group did suggest that in the context of real-world applications, newer electrode designs with active wet-type electrodes may be more applicable to real work situations involving recordings in the field, rather than in a Faraday cage, for optimum signal-to-noise behavior. Another application this group is exploring in its data analysis is transformation of EEG signals to sound. This interesting application has implications for biofeed-back, representation of brain responses, and a newer way of looking at EEG information.

SUMMARY AND CONCLUSIONS

Dr. Cichocki has an outstanding group of investigators focusing on noninvasive BCI applications at the RIKEN Brain Science Institute (Figures C.14 and C.15). They are poised to make significant impact concerning data analysis, manipulation of BCI signals, and developing multicommand, fully-online BCI systems with neurofeedback. As of October 2006, they did not appear to have extensive colla-borations with other investigators, which may be beneficial in the context of moving their data analysis approaches into other applications of BCI. Their under-standing of available instrumentation is outstanding and should be shared with other investigators in this field as well as with industry. They are poised to aid in the development of dry EEG electrodes with active recording technology to further improve the abilities of noninvasive EEG methods to be used in BCI applications. In addition to Dr. Cichocki's laboratory, Dr. Atsushi Iriki and Dr. Naotaka of the Fuji Laboratory for Symbolic Cognitive Development are working on invasive BMI using monkeys.

Competitive Advantages Compared to the World

This group has an exceptional breadth of understanding of current EEG methods. Its mathematical skills and ability to process EEG information are extensive and should be shared with other laboratories focusing on noninvasive BCI in Europe and the Wadsworth Institute in the United States.

Figure C.14. (L) WTEC panel with members of the RIKEN Brain Science Institute; (R) RIKEN BCI equipment.

Figure C.15. WTEC panel and Dr. Shun-ichi Amari, Director, RIKEN Brain Science Institute.

REFERENCES

Cichocki, A. Recent publications of Dr. Andrzej Cichocki (webpage), http://www.bsp.brain.riken.jp/%7Ecia/recent.html#bci (accessed on January 25, 2007).

Cichocki, A., and S. Amari. 2002. *Adaptive blind signal and image processing: learning algorithms and applications.* West Sussex, England: Wiley.

Cichocki, A., and P.G. Georgiev. 2003. Blind source separation algorithms with matrix constraints. *IEICE Transactions on Fundamentals of Electronics, Communications, and Computer Sciences* E86-A(1):522–531.

Cichocki, A., S.L. Shishkin, T. Musha, Z. Leonowicz, T. Asada, and T. Kurachi. 2005. EEG filtering based on blind source separation (BSS) for early detection of Alzheimer's disease. *Clinical Neurophysiology* 116(3):729–737.

Hoya, T., T. Tanaka, A. Cichocki, T. Murakami., G. Hori, and J.A. Chambers. 2005. Stereophonic noise reduction using a combined sliding subspace projection and adaptive signal enhancement. *IEEE Transactions on Speech and Audio Processing* 13(3):309–320.

Li, Y., A. Cichocki, and S. Amari. 2004. Analysis of sparse representation and blind source separation. *Neural Computation* 16(6):1193–1234.

RIKEN Brain Science Institute. Laboratory for Advanced Brain Signal Processing website: http://www.bsp.brain.riken.jp/index.php (accessed on January 25, 2007).

Site: **Shanghai Institute of Brain Functional Genomics**
The Key Laboratories of MOE and SSTC
East China Normal University
Shanghai 200062, P. R. China
http://sbg.ecnu.edu.cn/english/intro.asp

Date Visited: October 25, 2006

WTEC Attendees: G. Gerhardt (report author), T. Berger, G. Hane

Hosts: Dr. Longnian Lin
Email: lnlin@brain.ecnu.edu.cn

BACKGROUND

Dr. Lin is a productive neuroscientist working on freely-behaving mice. He has trained and collaborated with Professor Joe Tsien at the Center for Systems Neurobiology, Department of Pharmacology and Biomedical Engineering, Boston University. Dr. Lin's facility in the Shanghai Institute of Brain Functional Genomics is impressive, encompassing numerous laboratories, ample office space, and populated with a variety of students, technicians, and associated personnel. The facility is an ultramodern neuroscience research environment capable of *in vivo* studies in awake-behaving rats and mice. The facility contains a large variety of instrumentation, including advanced Plexon recording systems capable of single-unit discrimination from awake-behaving animals.

FUNDING SOURCES AND COMMERCIALIZATION

Funding Sources

This group receives funding from the major basic research sources of China, the Chinese Ministry of Education (MOE) and the Shanghai Science and Technology Commission (SSTC).

Commercialization

Technologies employed in Dr. Lin's laboratory can possibly be considered for commercialization, especially the high-density ensemble recording microdrive system used for recording from 96–128 channel electrodes. There is no indication at this time that the individuals are interested in commercialization of this device, but the technology has been reported in *Journal of Neuroscience Methods*.

Academic Collaborations

There is a major, apparently *ongoing*, collaboration with Professor Joe Tsien at Boston University's Departments of Pharmacology and Biomedical Engineering.

RESEARCH AND DEVELOPMENT ACTIVITIES

The major goal of this laboratory is to record large-scale neural ensemble recordings in the brains of freely-moving rats and mice (Figure C.16). The lab's mass-recording technologies, involving up to 128 channels of electrodes that involve the use of 13 μm-diameter electrodes for tetrodes and 25 μ-stereotrode, is an exceptional accomplishment, developed in part through further collaboration with Dr. Buzsaki in the United States. Dr. Lin has published a series of high-profile papers regarding the organization of real-time memory encoding in ensembles of neurons. He has identified hippocampal neurons that exude properties that are associated with the animal's identification of a nest and that share many of the properties of place cells of the hippocampus that were identified over 30 years ago. This technology is state-of-the-art and represents an enabling technology that would be useful to a number of laboratories in China, as well in the United States, Europe, and possibly Japan.

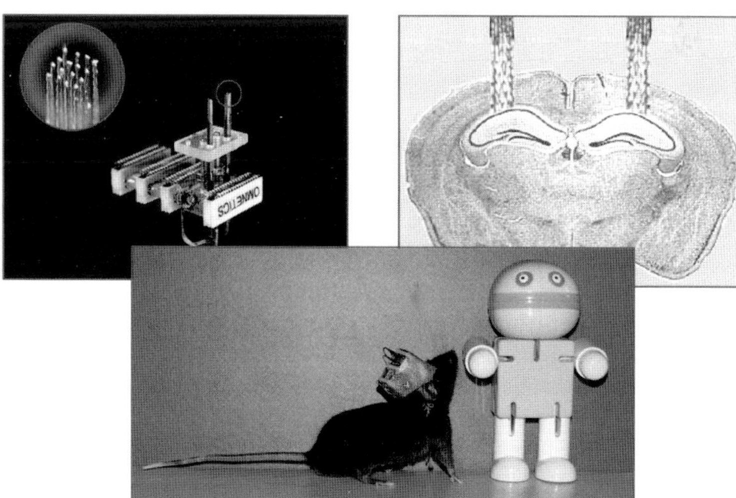

Lin, L et al., J Neuroscience Methods 2006

Figure C.16. 96-Channel ensemble recording in free-behaving mice (© Elsevier B.V.).

The Shanghai Institute investigators have developed a practical recording system for high throughput of multichannel neuronal information from laboratory mice that is also applicable to freely-moving rats. The recording system is truly advanced, encompassing minimal restraint of animals and maximum capability of recording large numbers of neurons in the awake-behaving animals (Figure C.17).

Figure C.17. Recording neurons in awake-behaving mouse.

To date, Dr. Lin and his collaborators have not investigated the possibility of employing transgenic or knockout animals to further understand network-level coding in these important animal models. This is an area that could be pursued in future experimentation. It is clear that Dr. Lin and his team are capable of carrying out these recordings on a daily basis. As such, a variety of higher-level studies concerning neural network activity are likely forthcoming from this group.

SUMMARY AND CONCLUSIONS

Dr. Lin has an outstanding group well able to address issues surrounding BCI and ensemble neuronal activity in a variety of animal models. His techniques are capable of recording routinely from a large number of neurons in freely moving rats and mice. Preliminary descriptions in high-impact papers support that the group is addressing state-of-the-art issues surrounding neural network properties. Currently, his group is not involved directly with BCI-related issues. However, the techniques are directly applicable to issues surrounding invasive approaches to using BCI for control of external devices. There is a definite opportunity for Dr. Lin to consider collaboration with other key investigators in the Shanghai area, such as the investigators at the Institute for Laser Medicine and Biophotonics at Shanghai Jiao-tong

University, who are working on retinal prosthesis projects, and investigators such as Professor Boming Sun, a neurosurgeon in the department of neurosurgery at Rui-Jin Hospital. Professor Boming Sun is one of the leaders in China on deep-brain stimulation and its applications.

Competitive Advantages Compared to World

This laboratory possesses the ability to record from large ensembles of neurons routinely from awake-behaving mice using a newly described, multiwire recording microdrive. This unique capability allows for investigations of neural network activity and issues surrounding BCI. This group is poised to collaborate with a number of leading groups throughout the world working on issues of invasive BCI recordings. This may be a potential untapped resource for technology and potential advancement of invasive recording techniques to be applied to BCI.

REFERENCES

Lin L., G. Chen, K. Xie, K.A. Zaia, S. Zhang, and J.Z. Tsien. 2006 Large-scale neural ensemble recording in the brains of freely behaving mice. *J. Neurosci. Meth.* 155:28–38.

Lin L., R. Osan, S. Shoham, W. Jin, W. Zuo, and J.Z. Tsien. 2005. Identification of network-level coding units for real-time representation of episodic experiences in the hippocampus. *PNAS* 102(17):6125–6130.

Lin, L., R. Osan, and J.Z. Tsien. 2006. Organizing principles of real-time memory encoding: neural clique assemblies and universal neural codes. *Trends Neurosci.* 29(1):48–57.

Site: **Shanghai Jiao-Tong University**
Institute of Laser Medicine and Biophotonics
Shanghai 200031, P. R. China
http://www.sjtu.edu.cn/english/index/index.htm

Date Visited: October 25, 2006

WTEC Attendees: T. Berger (report author), G. Gerhardt (report co-author), G. Hane

Hosts:
Prof. Qiushi Ren
Dept. of Biomedical Engineering
Tel: +86 213420 4080
Email: renqsh@sjtu.edu.cn
Prof. Xinyu Chai and Prof. Shanbao Tong
Dept. of Biomedical Engineering
Prof. Baoliang Lu and Dr. Banghua Yang
Dept. of Electrical Engineering
Prof. Boming Sun
Dept. of Neurosurgery
Rui-Jin Hospital

BACKGROUND

The Institute of Laser Medicine and Biophotonics resides within Shanghai Jiao-Tong University (SJTU) and is subordinate to the Ministry of Education. As a key university in China, it is jointly run by the Ministry and by Shanghai Municipality. This university is formerly Nang-Yang public school and was founded in 1896; it is one of the oldest universities in China. It has 20 academic schools and 60 undergraduate programs, 152 master's degree programs, 93 Ph.D. programs, 16 postdoctoral programs, 16 state key doctoral programs, and 14 state key laboratories and national engineering centers. Its enrollment is over 38,000 full-time students, including over 18,000 undergraduates and 18,100 candidates for masters and doctoral degrees.

We visited the Institute of Laser Medicine and Biophotonics, which was a new facility with exceptional laboratory space and capabilities. We saw surgery rooms for basic animal science and development of the technology, engineering facilities, data processing facilities, and analysis groups all housed on a single floor of a new building. In addition, we were part of a state-of-the-art symposium on BCI, which was organized especially for our visit. It was held in a new auditorium with sophisticated audio-visual capabilities. Clearly, this group is advanced in its focus on the development of BCI-related technology.

FUNDING SOURCES AND COMMERCIALIZATION

The institute's program on visual prosthesis, directed by Dr. Qiushi Ren, is a "China 973" [1] Project. Approximately $3.2 million has been provided for this project by China's Ministry of Science and Technology. Additional funding comes from the Science and Technology Commission of the Shanghai Municipality and the Natural Science Foundation of China.

Industrial Collaborations

There are a number of key technologies involving the development of the neural prosthesis and other BCI-related technology.

Academic Collaborations

The C-Sight project consists of contributions from multiple laboratories and is organized on a scale typical of a multiyear NIH project or an NSF ERC. The contributors to the project include four faculty members from Shanghai Jiao-Tong University's Biomedical Engineering Department (Professors Ren, Liang, Zhu, and Wang), Professor Zhao from the Shanghai Institute for Microsystems Fabrication, Professor Zhuang from the Shanghai Institute for Science and Technology, and Professor Li from Beijing University Medical School. There are substantial scientific collaborations within Shanghai Jiao-Tong University (SJTU), and between SJTU and other universities in China. These collaborations are of a high caliber, similar to ones in the United States.

RESEARCH AND DEVELOPMENT ACTIVITIES

One of the major programs directed by Professor Ren is the development of an advanced retinal prosthesis device based on direct stimulation of the optic nerve. Development of such a prosthesis is particularly significant, given that China has a blind population of over 5 million, with 25% due to retinal degenerative diseases. The "C-Sight" project is an eight-subdivision program that encompasses implantable camera technology, retinal coding, MEMS electrode design and fabrication, biocompatibility, optical nerve stimulation, signal processing, and clinical evaluation. The core concept of the program is the development of a microimplantable camera to bypass the retina and directly feed electrical stimulation of visual information to the optic nerve by digital signal processing of the video information. The signals will be transformed into electrical impulses and directly stimulate layers of the optic

---

[1] China's 973 Program is China's national keystone basic research program, approved by the Chinese government in June 1997 and organized and implemented by the Ministry of Science and Technology.

nerve by an advanced indwelling electrode, which will be a circular variant of a Utah array-type device with varying levels of penetration of the electrode into the optic nerve. This is an exciting and demanding area of study that is being addressed by a number of key collaborative investigators under the direction of Professor Ren. Although visual prostheses are not a major focus of this BCI overview, the technologies developed within this program are likely of value to a variety of investigators working in other BCI applications. Examples are the work involving the development of the penetrating electrodes highlighted below, the novel methods in neural signal processing, the hardware implementation of signal processing algorithms, and the outstanding work of a neurosurgeon, Professor Sun, working in conjunction with this group. Professor Sun is world-renowned for his work on deep brain stimulation.

Development of Optic Nerve Electrodes for Retinal Prosthesis

One of the major foci of the team headed by Professor Ren is the development, involving eight teams, of a retinal prosthesis (Figure C.18). The device will have a novel type of optic nerve recording electrode positioned in a circular fashion and at a variety of depths within the optic nerve in order to stimulate the nerve bundles and to produce appropriate signals in the visual cortex.

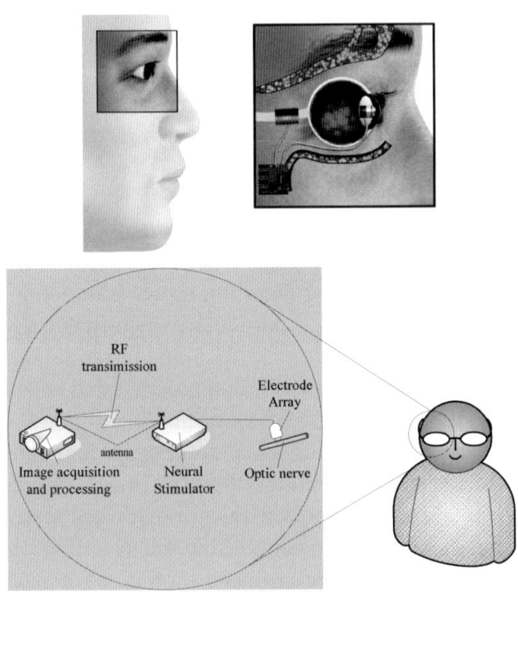

Figure C.18. Miniaturized camera, signal processing, neural stimulator, and optical nerve cuff for SJTU retinal process.

This is a new venture, still in its beginning stages; however, one of the particularly exciting aspects of the program is its well-conceived focus on optical nerve stimulation. The core faculty has engaged in a highly sophisticated analysis of the cytoarchitectural organization of the optic nerve. Through 3D reconstructtion of the organization of axonal bundling within the optic nerve, the investigators have been able to achieve an understanding of the spatial arrangement of the optical nerve fasciculi that has guided the design of the geometrical properties of their stimulating nerve cuff electrode. The novel circular and flexible electrodes will incorporate many aspects of electrodes that have been developed in part for spinal stimulation and cortical stimulation. The variable depth design proposed is a variant of that developed by the University of Utah. Unfortunately, the time constraints of the symposium allowed for little information exchange about the actual experimental details and the technology that has been developed. This is a challenging and novel area of study that should yield important results for other investigators in the field.

Visual Image Signal Processing

A major effort in visual image capturing and visual signal processing is being led by Professor Jia Wang and his colleagues. The essentials of their strategy are: (1) image capture through a miniature camera implanted in the eye, (2) a signal processing stage at the level of the camera, (3) RF transmission of the processed signal to an electrical stimulation system impanted under the skin near the eye, and (4) hard-wired connections to the optic nerve cuff electrode. The group has successfully implemented a working prototype of the miniature camera and signal processing hardware. Excellent examples of edge detection using their hardware-based algorithms were presented. Additional feature extraction and image enhancement algorithms have been developed and implemented using DSPs. Power and data telemetry hardware are being developed, and an overall system design for image-driven optical nerve stimulation was presented. In total, the WTEC team was impressed by the effort for visual neural prosthesis that is in progress at SJTU.

Brain Computer Interfaces for Monitoring Vigilance

Professor Bao-Liang Lu described the beginnings of an exciting research program to develop BCIs for a variety of conditions involving monitoring the degree of cognitive vigilance. One example includes monitoring vigilance during the operation of automobiles. Detecting attention level during driving is a long-standing problem of major safety significance; any significant progress would have a substantial impact both on reducing physical injury and rehabilitation costs, and on auto and health insurance costs. Professor Lu's group is developing novel approaches to real-time assessment of wakefulness, using measurement of both facial expression and/or multisite EEG. The project proposes a very interesting video monitoring of facial expression through cameras mounted in the steering wheel of the operator's car. By associating different facial profiles with different levels of wakefulness, an onboard system could monitor and detect critical levels of vigilance in real time during automobile operation. Also being investigated is the possibility of using EEG electrodes embedded in the headrest.

Over-Complete Feature Extraction for BCIs

Professor Liqing Zhang reviewed a series of studies in mathematical modeling and algorithm development for EEG-based BCIs. The problem is framed in terms of multiple sources of neural activity in the brain that are "convolved" in some un-known manner to produce the observed EEG. Dr. Zhang is investigating the application of independent component analysis (ICA) to EEG, and in particular, combining ICA with multichannel nonlinear adaptive filtering to decompose the EEG into separate sources. An interesting aspect of Dr. Zhang's approach is to focus on the temporal structure of the EEG signals in training ICA models. His laboratory has developed new methods for removing noise and artifacts from EEG signals. Solving the "inverse problem" with respect to EEG generators has remained an unsolved and difficult problem for decades, but Dr. Zhang's approaches and results show considerable promise in the area of enhancing imaginary hand movement detection and in geometrical pattern classification based on visual evoked potentials (VEPs) (Figure C.19).

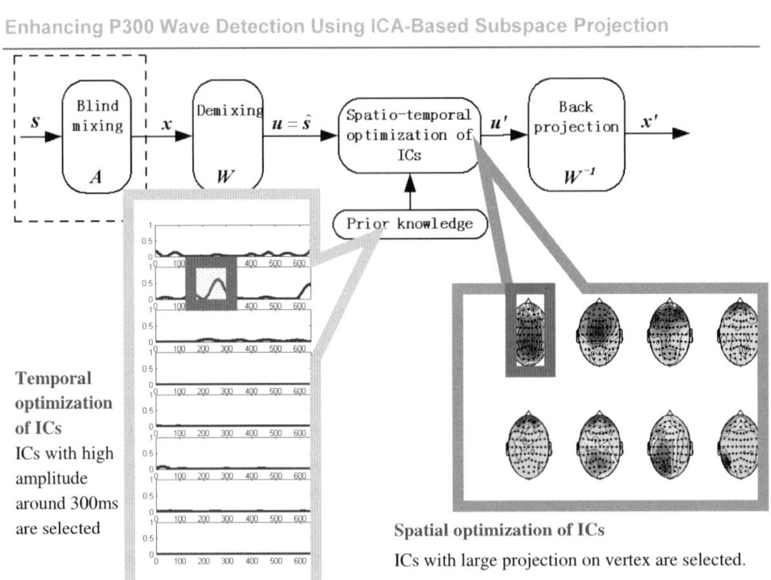

Figure C.19. Enhancing imaginary hand-movement detection using the overcomplete feature selection method.

Deep Brain Stimulation Surgery

Professor Bomin Sun, in the Department of Neurosurgery at Rui-Jin Hospital, leads a surgical team that has implanted approximately 400 subjects with deep-brain stimulators (DBS). The approach has been applied to patients with Parkinsonism, Tardive dyskinesia (TD), dyskinesia, dystonias, and other movement disorders. It is estimated that Dr. Sun and his team have completed approximately 40% of all DBS implants that have been carried out in China. This is an exceptional track record and attests to the skills and capabilities of this team. This technical group should be able to provide a wealth of information to investigators in the area of invasive BCI. In addition, this is an invaluable source of information regarding chronic electrode implants and potential side effects of implantation of chronic stimulating electrodes.

SUMMARY AND CONCLUSIONS

This is a highly accomplished group of scientists in an apparently well-funded environment that encompasses a training group and a developmental group for technologies for other BCI applications. The technologies involving electrode development, electronics and information processing, indwelling microelectrodes, and clinical deep brain stimulation could be of tremendous benefit to a variety of groups working on BCIs throughout the world.

Site: **Tsinghua University Department of Biomedical Engineering**
School of Medicine
Beijing 100084, P. R. China
http://neuro.med.tsinghua.edu.cn/

Date Visited: October 23, 2006

WTEC Attendees: T. Berger (report author), G. Gerhardt (report co-author), G. Hane

Hosts: Prof. Shangkai Gao
Department of Biomedical Engineering, and
Director, Institute of Neural Engineering of the
Institutes of Biomedicine
Tsinghua University
Tel: (+86) 10 6278 5472
Email: gsk-dea@tsinghua.edu.cn

BACKGROUND

Professor Shangkai Gao's laboratory is in the Department of Biomedical Engineering (BME), School of Medicine, Tsinghua University. It is also the primary component of the Institute of Neural Engineering, which is a part of the Institute of Biomedicine in Tsinghua University. The original BME program in Tsinghua University was created in 1979 in the Department of Electrical Engineering. Started with master-degree curriculum, the program established the overall tracks of undergraduate-to-graduate (master's and doctoral) degrees in 1988. In 2000, when the School of Medicine was founded in Tsinghua University, the BME program was moved to the school and became the BME department. Now approximately 30 undergraduates and 30 graduates are admitted to the BME department each year. There are eleven professors, seven associate professors, and eight lecturers in the department. The main research interests in the department include neural engineering, medical imaging, biomedical signal processing, biochips, medical instrumentation, and bioinformatics.

The WTEC team met Professor Gao and her team members in a beautiful, newly constructed building dedicated to the Institute of Biomedicine. The space and facilities were the equal of any biomedical engineering department or program in the world and included excellent animal facilities for invasive brain research. Clearly, a major thrust of the new institute is neural engineering and BCI research.

FUNDING SOURCES AND COMMERCIALIZATION

The funding for Professor Gao's extensive studies of EEG-based BCIs and newly developing electrophysiological studies in animals is currently provided by NNSF China, the China High-Tech Research and Development Program 863, and five-year National Science and Technology Support Projects. Funding through these sources may not be used to pay for faculty salaries. Support for graduate students is allowed, but that support is partial, not full; the panel understood that tuition costs are paid by the government support. The laboratories reimburse the university for space through an indirect-like charge. Support relationships with industry are possible through SBIR-like mechanisms.

Industrial Collaborations

Professor Gao's laboratory was beginning the process of developing industrial partners for what should be several readily commercializable BCI systems that the WTEC team observed in operation (see below). Professor Gao had already applied for four patents for BCIs and had been issued one U.S. patent (Patent No. U.S. 7,123,955 B1). She indicated that commercialization of her laboratory's BCI systems is planned for the near future. More specifically, her steady-state, visual-evoked potential (SSVEP) system has been tested for use in a web-page browser application.

Academic Collaborations

Professor Gao and her laboratory team collaborate with other laboratories at Tsinghau, including the Department of Mechanical Engineering, for the development of artificial limbs that would use Professor Gao's algorithms as control systems. Professor Gao's team also interacts with other BCI research laboratories throughout the world. In particular, her laboratory has been highly active in international competitions for algorithm development and highly visible in its success. Professor Gao also is very involved in leadership roles in IEEE societies and journal editorial boards. She has recently been elevated an IEEE Fellow for contributions to the study of brain-computer interfaces.

RESEARCH AND DEVELOPMENT ACTIVITIES

Professor Gao's laboratory is the premier BCI laboratory in China, and it holds a superior position with respect to BCI research throughout the world. Professor Gao has pioneered the development and applications of algorithms for noninvasive, EEG-based computer control and has led the effort to demonstrate the practical application of BCI systems. The lab's interests are broad and include combining the high temporal resolution of EEG methods with the high spatial resolution methods of fMRI and other imaging methods to achieve bioengineered systems for BCI, diagnosis of neurological diseases, and neurobiological understanding of cognition and perception (Figure C.20).

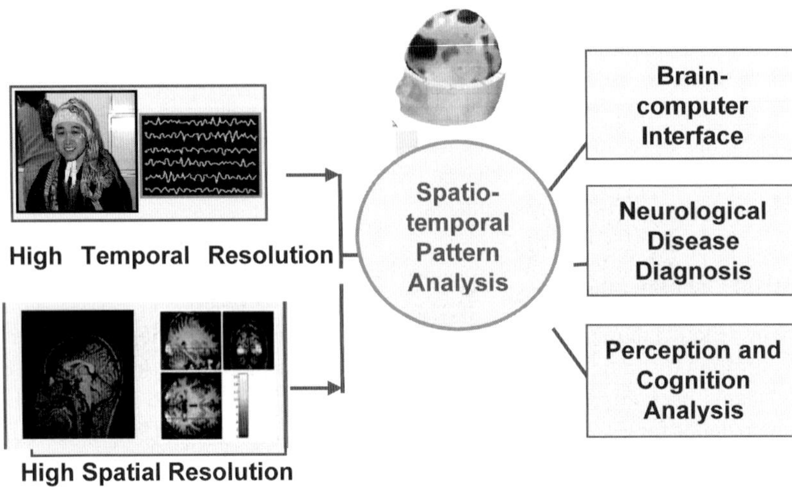

Figure C.20. Research interests of Professor Gao's laboratory.

With respect to BCI applications, the driving motivation for Professor Gao's research derives from the realization that China's large population requires BCI systems that are relatively low in cost, can be used in the home with relatively little effort on the part of the patient, and must be noninvasive for widespread use. These constraints have led Professor Gao to focus on what she sees as the primary obstacle for BCIs, advanced signal processing algorithms to extract maximum information from the low-amplitude, noisy signals of EEG.

Professor Gao's laboratory has pioneered new signal-processing methods for feature extraction and pattern classification. Her laboratory was first in the international BCI data competition in 2002/2003 for enhancing P300 signals using ICA techniques and also was first for classifying single-trial EEG data. In the 2004/2005 competition, Professor Gao's laboratory placed first in three of the seven data set categories—a significant achievement, particularly given that in these international settings, Professor Gao's laboratory was competing against many premier BCI laboratories the WTEC panel visited in Europe. The WTEC panel witnessed an alphabet selection and telephone-dialing BCI technology based on a straightforward, SSVEP approach that functioned with high accuracy and with very little setup specialization or fine tuning of the system. The SSVEP system utilizes a 3×4 array of flashing characters and requires very little training. Information transfer rates for this system approach 50 bits/minute. Professor Gao's laboratory has been able to develop systems that can use differences of 0.2 Hz in SSVEP-based frequencies to identify subject-directed choices. Repeated applications of this approach to both spelling and telephone calling were provided onsite with no difficulties; the panel agreed that it was an impressive demonstration (Figure C.21).

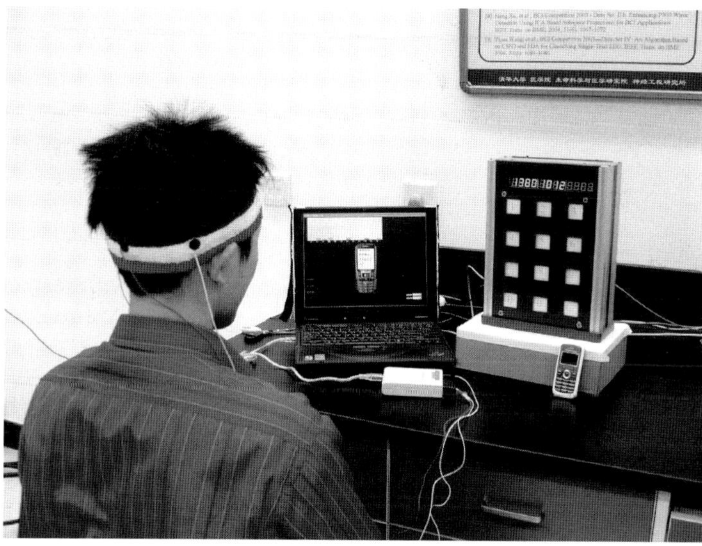

Figure C.21. SSVEP-based BCI for making a phone call.

Not only was the approach easily implemented (tasks were readily completed by multiple subjects with very few errors), but because of the high information transfer rates, tasks also were completed rapidly—much more rapidly, for example, than the P300-based spelling tasks the WTEC panel witnessed in Europe. In addition, the equipment required was relatively "low tech" and inexpensive. A very simple head-strap recording array was used to detect and classify the SSVEP. The panel interpreted this high level of performance in a context of ease of use and low-cost system requirements as evidence of the power of the underlying algorithms. Professor Gao's laboratory has expanded the application of this approach to include multiple functions that would be extremely useful to quadriplegic patients: typing (spelling), cursor control, appliance control, and telephone calling (Figure C.22).

Professor Gao's laboratory also had made significant progress in the development of motor-imagery-based BCIs. This approach is designed to empower the paralyzed patient with control of external devices through algorithms that can identify different spatiotemporal patterns of neocortical activity associated with the patient imagining movements of different limbs.

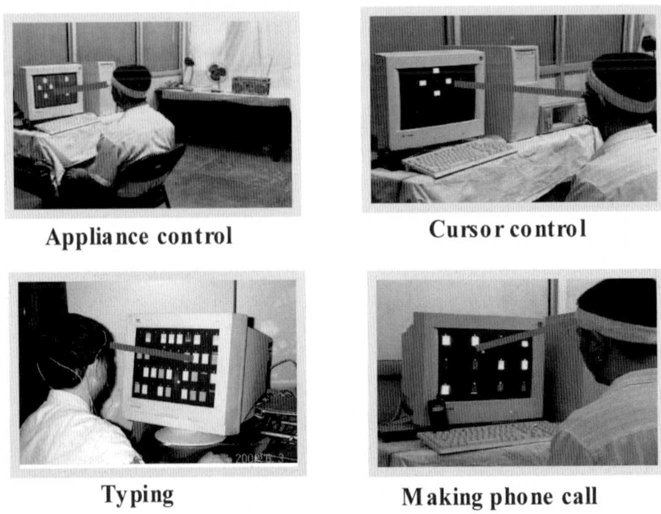

Figure C.22. The applications of SSVEP-based BCI.

Professor Gao showed the BCI panel videos of her group using this methodology to control teams of robotic dogs that competed against each other in a soccer-like game (Figure C.23). The users controlled the dogs through imagining left-hand, right-hand, and foot movements; 10 channels of a 128-channel EEG headset recording array were used by the subjects. Although trained users were able to successfully move the robotic dogs on the playing field and were able to maneuver them around barriers, this demonstration could not match the high performance levels of the SSVEP system.

Figure C.23. Motor imagery-based BCI for robotic dog control.

The BCI panel was intrigued with an early-stage project utilizing phase synchrony measurements to classify single-trial EEG patterns during mental imagery of motor movements. This approach has shown great promise for "active" rehabilitation training of stroke patients to regain motor control of paralyzed limbs (vs. passive rehabilitation movement) (Figure C.24).

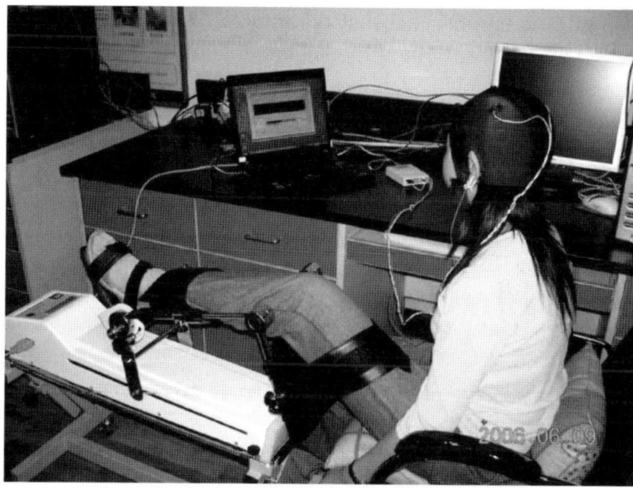

Figure C.24. "Active" rehabilitation training.

Finally, the BCI panel was shown new electrophysiological equipment that will provide Professor Gao's laboratory with the capability of recording from the rat brain. In particular, it will provide the capability of comparing EEG-level, intracortical field potential, and intracortical single-unit measurements of neural function. Although the Gao laboratory was just beginning the electrophysiological effort, it was clear that their facilities were first-rate. Future development of subhuman primate recording facilities was already in planning stages. In addition, the Gao lab has already established working collaborations with the epilepsy clinic associated with Tsinghau University. Given the established U.S. expertise in invasive electrophysiological studies of brain function, both at the subhuman and human levels, this new direction into invasive neural investigations represents a significant opportunity for U.S.-China collaborations that should be encouraged.

SUMMARY AND CONCLUSIONS

Professor Gao's laboratory is a world-class effort in the development in BCIs. The WTEC panel was extremely impressed with the algorithm development of the Gao lab and the apparent practical application possibilities of the systems under development. The superiority of the Gao algorithms is evident in their success in worldwide competitions and in the on-site demonstrations to the WTEC panel. The next-phase expansion of the laboratory into invasive electrophysiological studies of brain function is quite exciting. The obviously first-rate support for the Gao lab and for other associated laboratories at the Tsinghau University involved in Biomedical Engineering is evidence of a high-quality environment for this next developmental stage.

Again, it should be emphasized that the established U.S. expertise in invasive electrophysiological studies, both at the subhuman and human levels, represents a significant opportunity for U.S.-China collaboration. The WTEC panel also felt that commercialization possibilities for the BCI systems emerging from the Gao laboratory are likely to be many and fruitful.

Site: **Tsinghua University Institute of Microelectronics**
Beijing 100084, P. R. China
http://www.ime.tsinghua.edu.cn/english/index.htm

Date Visited: October 23, 2006

WTEC Attendees: G. Gerhardt (report author), T. Berger, G. Hane

Hosts: Zhihua Wang
Professor of Electronic Engineering and
Deputy Director of the Institute of Microelectronics
Tsinghua University
Tel: +86 10 62789251
Fax: +86 10 62771130
Email: zhihua@mail.tsinghua.edu.cn

BACKGROUND

The Institute of Microelectronics of Tsinghua University (IMETU) was founded in September, 1980, with the approval of the Ministry of Education of China. IMETU is located on the campus of Tsinghua University and encompasses approximately 10,000 square meters of space. It has 110 faculty (21 full professors) and staff. IMETU has played an important role as the Northern China base of microelectronics research and development and is strongly supported by the government. The researchers at IMETU are mainly focused on micro- and nanoelectronics, IC design and development, device physics, and processing techniques. IMETU has an advanced IC design capability and a semiconductor processing laboratory. IMETU is part of the China National Information Science Laboratory. Over the last 26 years, IMETU has trained over 1,000 bachelor's graduates, 400 master's-level graduates, and 110 Ph.D. graduates. In 2005, 26 patent applications were submitted, and 14 patents were granted. In 2005 it published 18 international journal articles, 88 domestic journal articles, 81 international conference articles, and 18 domestic conference publications.

FUNDING SOURCES AND COMMERCIALIZATION

IMETU receives a variety of national government funding, including from the National Basic Research Key Program termed "973"; the institute has four such grants. In addition, there are national high-tech research and development programs termed "863"; at the institute, these include 10 research programs, 1 key technology R&D program, 16 National Natural Science Foundation Projects, 14 programs from the Ministry of Education and other Ministries of State, 5 programs supported by

Tsinghua Basic Research Fund, 16 programs with international collaborations, and 19 industrial contract projects. The total research funding of these projects in 2005 was about 25 million Yuan (~$3,094,000). In the context of technology development, of the 14 patents that were granted in 2005, 10 of these patents were for IMETU inventions .

RESEARCH AND DEVELOPMENT ACTIVITIES

BCI-related Projects at the Institute of Microelectronics at Tsinghua University

Professor Wang directs a large group of engineers and investigators at IMETU whose work is focused on BCI development.

Background of Bidirectional Wireless Monitoring System for Orthopedic Implants

This work encompasses a low-power integrated circuit design of devices to assess implant wear and/or incipient failure of total knee replacement and total hip replacement orthopedic devices. The approach of embedding sensors within the devices is to (1) provide new *in vivo* diagnostic capabilities; (2) reduce clinical complications; and (3) improve implant materials and designs. This work has been in collaboration with the University of Nebraska Medical Center. The implantable devices that were shown involved types of surgical implants that would be driven by electricity generated from piezoelectric materials. This is an ingenious approach to generate power from the potential pressure in the joint. The investigators showed an impressive architecture design and circuit block diagram composed of analog parts and digital components. This is a unique technology that WTEC panelists did not see elsewhere. Also, this shows distinct systems-level integration among the engineers and clinical investigators.

IC Design of Digital Wireless Endoscopic Capsule System

The IC design of a digital wireless endoscopic capsule system involves development of an endoscope that would be wireless and capable of self-contained imaging. The miniaturized device would have a built-in camera, battery system, and antennae to transmit information. This novel design involves telemetry applications with both internal and external hardware. We were shown an external recording system and a mock-up of the internal system. Basically, this approach for potential wireless endoscopic evaluation in humans would be less invasive than many currently used standard endoscopic approaches . At present, the design incorporates bidirectional communication, imaging compression, low power, and successful verification of the component layout of the device. We were not shown a mock-up of the actual microendoscopic device, but the program appears to be well-financed and under development at a high level.

CMOS Chip Design Study of Neurorecording and Signal Processing in Implantable BCI

The CMOS chip design study involves development of a microelectrode array technology somewhat analogous to the invasive electrode approaches that are being pursued primarily in the United States. The electrode chip has onboard electronics that have been developed through a CMOS process and currently contains a 16-channel, 2:1 analog select array, a low-power, low-noise preamplifier filter buffer array, and a 9:1 analog multiplexer and digital control unit. Our hosts showed a 4×4 microelectrode array that has four electrode recording sites on a silicon substrate with a tip-extension of 2 mm and a shank diameter of 100 μm. They exhibited a complete layout of the integrated system architecture and a preamplifier design. A prototype device for recording electrical systems in the sciatic nerve of the bullfrog was shown, along with the microelectrode array mounted on a PC board for testing (Figure C.25). These investigators are working at an advanced level and are capable of making strides in the invasive electrode development technology that may be useful to investigators in the United States and abroad who are working on implantable electrode technology.

Figure C.25. Microelectrode array system for 16-channel recordings.

Multimode, Multichannel Cochlear Implant

We were shown a picture of a cochlear implant chip device embodying a miniature speech processor. It was a multimode, multichannel cochlear implant that our hosts claimed encompassed five patents. Although no performance specifications or further information about this implantable cochlear implant device were provided, the implant was very small and clearly at a high stage of development, representative of the capabilities of this outstanding institute of micro-electronics.

IMETU DIVISIONS

IMETU is organized into four basic units: (1) Integrated Circuit Processing and Technology Division; (2) IC and System Design Research Division; (3) Micro/Nanodevices System Division; and (4) CAD Technology Division.

Integrated Circuits Processing Technology Division

The Integrated Circuits Processing Technology Division is an important part of North Microelectronics R&D Base, Beijing, China. Its researchers are involved in novel SiGe microwave power devices and special types of IC technologies. A VLSI Pilot Line is under way to establish research and development to manufacture types of ICs. Main R&D fields involve: (1) research fields on deep seven micrometer VLSI manufacturing and process, (2) research on SiGe technologies, (3) nonvolatile semiconductor memory technologies, (4) research on high-voltage and power IC technologies under way involving CMOS in high-voltage power devices, and (5) research on virtual and automatic IC fabrication.

IC and System Designs Research Division

The IC and System Designs Research Division is engaged with education, research, a system-on-a-chip (SoC) design and testing, including theory, development, methodology, and systems architecture. Major research areas involve: (1) general processor development, (2) analog and mix signal circuitry, and (3) application of SoC technology.

Micro/Nanodevices System Division

The Micro/Nanodevices System Division involves new types of micro/nanodevices and systems consisting of these devices. The research scope of this division includes: (1) MEMS and smart sensor development, (2) nanoscience and technology, (3) nonvolatile memories such as MRAM and RRAM, (4) microdevices for biological and chemical applications such as microdevices in silicon-based and nonsilicon-based biochips, (5) fuzzy controller technology, (6) new materials for micro/nanoelectronic applications, and (7) packaging and assembly of ICs and MEMS.

CAD Technology Division

The CAD Technology Division involves Computer-Aided Design (CAD) research involving semiconductor device physics and VLSI CAD. The main research topics are: (1) carrier transport models and scaled-down MOS devices, (2) CAD software development for micro- and nanodevices, (3) new structures for micro- and nanodevices, (4) methodology for developing IP library and development IP cores for memory use and SoC designs, (5) layout-based extraction and verification of RF circuits, and (6) computer-aided manufacturing systems for IC fabrication and development of integrated biologic sensors.

This impressive facility has a number of collaborative efforts, including those with Professor Shangkai Gao from the Department of Medical Engineering, School of Medicine, and Tsinghua University. We were shown an EEG-based amplifier device that was small, very portable in design, and possibly could be manufactured for a low price for investigators in China and perhaps abroad. This potential, coupled with collaborations such as that with Professor Gao, will lead to marketable devices to further strengthen the economic growth of this institute. It is apparent that there is a need for translational collaboration from other investigators to take advantage of the advanced capabilities of this facility.

SUMMARY AND CONCLUSIONS

Competitive Advantages Compared to the World

Professor Wang manages an impressive group of investigators with ample funding and infrastructure to develop a variety of integrated circuits, micro/nanodevices, and CAD technologies. This institute could play a major role in the development of noninvasive BCI-type devices for other laboratories in the world. We were shown a device developed in conjunction with Professor Gao that was remarkable in its size and capability for control of BCI linked devices. The fusion of laboratories such as Professor Gao's with the Institute headed by Professor Wang, could develop BCI devices that are state of the art, highly useful, and potentially quite affordable. They could be mass-produced by the Institute to further increase the growth of BCI-based technologies. This institute is making an impact in China and is poised to make a major world impact, possibly in the area of BCI-based technologies.

REFERENCES

Tsinghua University, Institute of Microelectronics. 2006. Annual Report.

Tsinghua University, Institute of Microelectronics website, http://www.ime.tsinghua.edu.cn/english/index.htm (accessed on November 14, 2006).

Site: **Waseda University**
Department of Electrical Engineering and Bioscience
Takashi Matsumoto Laboratory
3-4-1 Okubo Shinjuku-ku
Tokyo 169-8555, Japan
http://www.matsumoto.elec.waseda.ac.jp/english/

Date Visited: October 27, 2006

WTEC Attendees: T. Berger (report author), G. Gerhardt (report co-author), G. Hane

Hosts: Takashi Matsumoto
Tel: +81 3 52863377
Email: takashi@mse.waseda.ac.jp

BACKGROUND

Professor Matsumoto's laboratory is part of Waseda University's Department of Electrical Engineering and Bioscience, School of Science and Engineering. This school is one of the largest research schools in Japan, with an enrollment of 2,589 graduate students, including 398 doctoral candidates. The school employs 231 professors and assistant professors, 67 professors and assistant professors (including guest professors) in other institutions, and 410 adjunct faculty members and other affiliated lecturers—a total of 708 faculty members. The Department of Electrical Engineering and Bioscience supports a forward-looking interdisciplinary combination of biological sciences, information sciences, electrical engineering, and electronics engineering (optoelectronics, materials science, etc.), as shown in Figure C.26. It therefore offers training and research opportunities ideal for the development of BCIs. Undergraduate students enroll in an interesting mixture of core engineering courses (e.g., systems analysis, computer architecture, control theory) and courses in biological and medical sciences (e.g., neurobiology, brain sciences, systems biology, medical devices). Graduate students continue with advanced engineering (e.g., solid-state electronics, photonics) and "bioengineering" courses (e.g., informational-based learning, intelligent control research, signal, and information processing). Highlighted research areas include superparallel image processing chips based on the vertebrate retina, studies of cytoskeletal actin and tublin of hippocampal neurons, and hovercraft-like mobile robots. The department's strength in core engineering and bioengineering areas is considerable.

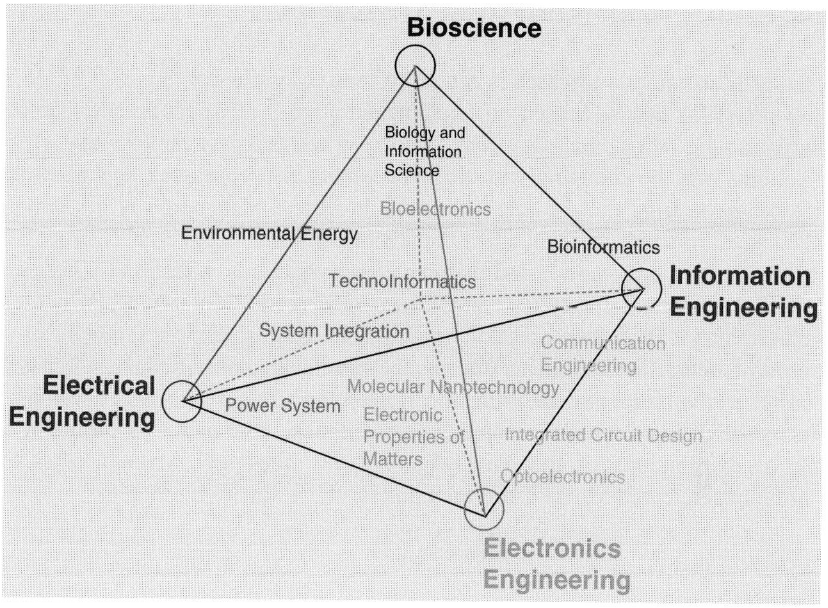

Figure C.26. Department of Electrical Engineering and Bioscience interdisciplinary structure.

FUNDING SOURCES AND COMMERCIALIZATION

Industrial Collaborations

There are numerous collaborations with industry. One such endeavor is the Waseda-Olympus Bioscience Research Institute, established as a combined system to integrate the research endeavors of industry and academia. Based in Singapore, a strategic location for information exchange with Asian and Western countries, the institute aims to create a global network of researchers in bioscience and to transmit intellectual information worldwide. Many of the participants are from the Department of Electrical Engineering and Bioscience.

As of October 2006, the Takashi Matsumoto Laboratory had collaborations with NHK (the Japan Broadcasting Corporation), Toshiba, Cool Design, and Apple Doctor (the latter two are venture companies).

Academic Collaborations

Multiple collaborations exist among the large faculty, though there are relatively few "wet-lab" neuroscience laboratories that could offer fruitful partnerships for BCI research and development.

RESEARCH AND DEVELOPMENT ACTIVITIES

Professor Matsumoto's research is central to the successful development of BCIs. The work of his lab is focused on fundamental studies of detecting and learning patterns in time series data (data from sensors), when that data can be noisy, when the patterns may be changing with time (nonstationary processes), and when the processes generating the data are unknown. Professor Matsumoto is developing mathematical methods that will allow models of nonlinear, nonstationary processes that provide predictive power and that can classify patterns in the data. These are many of the essential properties that mathematical models underlying BCIs must have in order for BCIs to successfully manipulate external devices on the basis of patterns of neural activity generated by the nonlinear, nonstationary, stochastic processes in the brain. One of Professor Matsumoto's major efforts is developing methods for online Bayesian prediction of "next-step" data based on estimates of posterior distributions of the data (Figure C.27). Professor Matsumoto is an advocate of sequential Monte Carlo methods for this estimation problem.

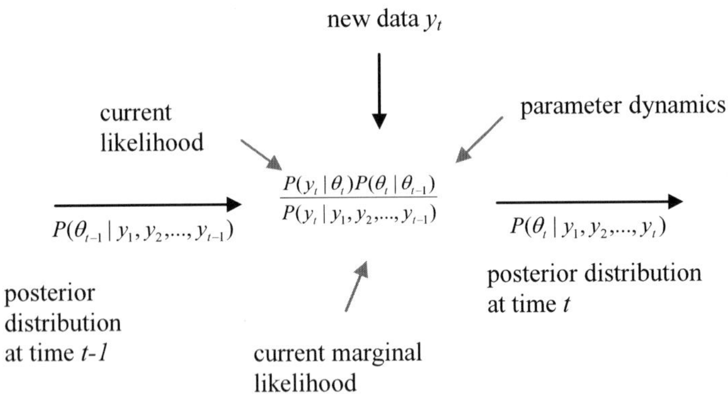

Figure C.27. Bayesian prediction of "next-step" data.

During the review of Professor Matsumoto's work, we were shown multiple examples of the successful applications of his methods using simulated data. He demonstrated successful (low-error) online regression and predictions for a smoothly varying process, as well as a process that exhibits an abrupt change in its dynamics, i.e., equivalent to a state change. Dr. Matsumoto also demonstrated the application of a Monte Carlo method that provided sequential marginal likelihood estimation for detecting online change detection. Other applications were demonstrated for online time series prediction, online pattern prediction, and online clustering. It was agreed that these apparently powerful methods need to be applied to real biological data, and Dr. Matsumoto indicated an eager willingness to initiate such studies. His laboratory and the relevance of his work to BCIs provide an excellent example of a rich possibility for U.S.-Japanese collaboration.

SUMMARY AND CONCLUSIONS

Dr. Matsumoto and his laboratory team conduct cutting-edge research into mathematical modeling methods fundamental to the development of BCIs. His laboratory could provide the basis for powerful collaborations with U.S. scientists. By leveraging the basic studies being conducted at Waseda University, progress on algorithms required for real-time analysis of neural data could be greatly accelerated. In discussions with Dr. Matsumoto, it was evident that he would welcome such collaboration, and that Waseda University offers additional opportunities with respect to BCI research that can be identified and encouraged.

REFERENCES

Ito, K., Y. Nakada, and T. Matsumoto. 2006. Online Bayesian learning for multivariate Gaussian mixture with Rao-Blackwellized sequential Monte Carlo. *Annual Conference of the Japanese Neural Network Society (JNNS2006)* 56–57.

Matsui, A., S. Clippingdale, and T. Matsumoto. 2006. A sequential Monte Carlo method for Bayesian face recognition. *Joint IAPR International Workshops On Structural and Syntactic Pattern Recognition (SSPR 2006)* and *Statistical Techniques in Pattern Recognition (SPR 2006)* 578–586.

Matsumoto, T. 2007. Online Bayesian modeling and prediction of dynamical systems: Sequential Monte Carlo approach. *Methods of Information in Medicine* 46:96–101.

Matsumoto, T., and K. Yosui. 2007. Adaptation and change detection with a sequential Monte Carlo scheme. *IEEE Transactions on Systems, Man and Cybernetics* 37(3):592–606.

Muramatsu, D., M. Kondo, M. Sasaki, S. Tachibana, and T. Matsumoto. 2006. A Markov chain Monte Carlo algorithm for Bayesian dynamic signature verification. *IEEE Transactions on Information Forensics and Security* 1(1):22–34.

Nakada, Y., and T. Matsumoto. 2006. A Bayesian change detection for unknown nonlinear systems: Online sequential Monte Carlo approach. *Joint 3rd International Conference on Soft Computing and Intelligent Systems* and *7th International Symposium on Advanced Intelligent Systems (SCIS & ISIS 2006)* 20–24.

Nakada, Y., Y. Mouri, Y. Hongo, and T. Matsumoto. 2006. GibbsBoost: A boosting algorithm using a sequential Monte Carlo approach. *IEEE International Workshop on Machine Learning for Signal Processing* 259–264.

Ohya, J., Y. Nakada, K. Yamada, and T. Matsumoto. 2006. Monte Carlo localization with change detection model. *2006 IEICE Annual Convention (IEICE2006)* 24–27.

Takashi Matsumoto Laboratory website, http://www.matsumoto.elec.waseda.ac.jp/english/ (accessed on December 20, 2006).

Site: **Wuhan University**
College of Chemistry and Molecular Science
Luojia Hill, Wuhan 430072, P. R. China
http://www.whu.edu.cn/en/

Date Visited: October 24, 2006

WTEC Attendees: G. Gerhardt (report author), T. Berger, G. Hane

Hosts: Prof. Yvonne Yanwen Duan
College of Chemistry and Molecular Science
Tel: +86 27 68752858
Email: yduan@whu.edu.cn
Prof. Juntao Lu
Department of Chemistry

BACKGROUND

Professor Yvonne Duan directed an Australian-based research group focusing on microelectrode technology and electrochemical measures of impedance *in vivo*. Her prior work and published studies have been focused primarily on flexible intracochlear electrodes and the tissue interface. She had recently moved to Wuhan University as a professor and is interested in development of invasive BCI electrode technologies and performance evaluation of these devices. Professor Duan has a growing research group. At the time of the WTEC visit, it was composed of Professor Duan, Professor Juntao Lu, and several students: Ms Xianhong Li, Mr. Zhengxu Cai, Ms. Wenjie Xiao, Mr. Binyin Liu, and Mr. Yi Lu.

Funding Sources and Commercialization

- National Natural Science Foundation, China
- National High-tech R&D Program, China
- Industry partners and potential commercial partners that are developing medical diagnosis instruments

Industrial Collaborations

Professor Duan has worked on intracochlear electrodes and is interested in the development of intracranial electrodes and their characterization. She collaborates closely with a high-tech company, Wuhan Greentek Scientific Pty., Ltd. The initial market of the company is neural scientific research and medical research instruments. Products include recording and stimulation electrodes and stimulators. The company will also develop medical instruments such as medical electrodes and nerve stimulators.

Academic Collaborations

Professor Duan continues collaborations with the investigators at the Cooperative Research Center for Cochlear Implant and Hearing Aid Innovation of the Bionic Ear Institute in Melbourne, Australia. In addition, she is developing ties with other investigators at Wuhan University and Tsinghua University.

Research and Development Activities

Professor Duan is pursuing the development of silicon-based microelectrode arrays for stimulating targeted neuronal populations in the central nervous system and recordings of single-unit activity using iridium oxide surfaces and platinum. She is currently developing simple and reliable surface modification methods for iridium oxide electrodes and microporous Pt electrodes. The latest newly developed porous Pt electrodes (0.0075 mm^2, 100 μm in diameter) have achieved a charge injection capacity of 3.1 mC/cm^2 (mean, n=6, using 200 μs biphasic charge balanced current pulse) that increases charge capacity 22 times the average compared to normal Pt electrodes. With the modified porous Pt electrode, the limited current amplitude increased from average 45 to 990 μA. The impedance of the electrodes was reduced by about 80% from 33 to 5.7 kΩ at 1 kHz in phosphate-buffered saline Her main expertise is in the area of electrochemistry and use of electrochemical methods to investigate impedance at the electrode/tissue interface. She completed "A study of intracochlear electrodes and tissue interface by electrochemical impedance methods *in vivo*," published in *Biomaterials* (Duan, 2004). Her prior work was carried out in cats and involved studies extending to six months *in vivo*, which is a very good testbed system for understanding potential changes in electrodes as a function of time *in vivo*.

Professor Duan has been less concerned with actual electrophysiological measures, which are critical for understanding the performance properties of the device *in vivo*. The neural electrode research group at Wuhan University is focusing on electrode-neural interface biocompatibility.

Electrode-Neural Interface Biocompatibility

Tissue responses that cause deterioration of function of implanted electrode arrays are a major problem, including an increase in stimulation threshold and electrode impedance as well as noise. Professor Duan's team proposes a model of the electrode-neural tissue interface and a modification strategy, shown in Figure C.28, which guides its research toward a neural tissue-friendly electrode array.

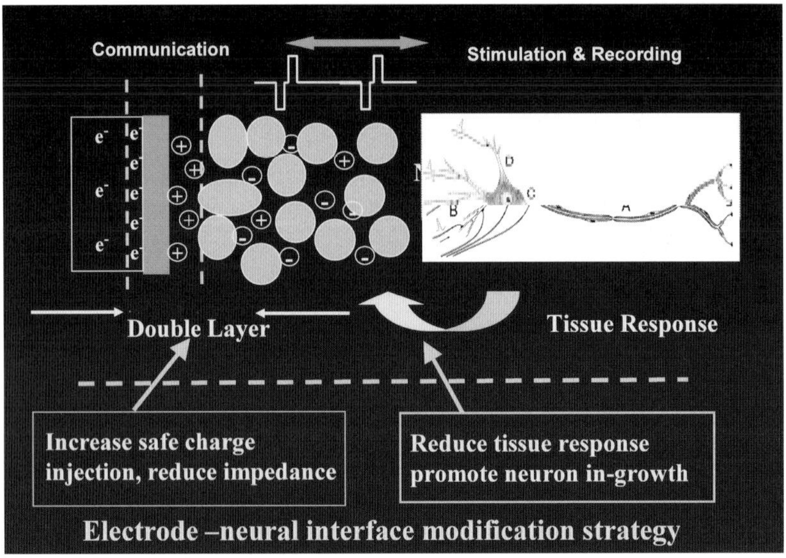

Figure C.28. An electrode-neural tissue interface model and modification strategy.

Currently, porous Pt and Iridium oxide electrodes coated with biocompatible hydro-gel film are being studied in the laboratory and will be applied to MEMS-based microelectrode arrays.

Safe Electrical Stimulation Study

Safe electrical neural stimulation has two criteria: (1) it does not cause extra cellular fluid electrolysis, and (2) it minimizes electrode corrosion. Electrode design, stimulation waveform, and stimulator design all contribute to this issue. Understanding the effects of the factors and increasing the safety margin are the goals of current study. The Wuhan researchers are developing a technique of real-time monitoring of electrode potential during electrical stimulation using a pseudoreference electrode in chronic animals.

Surface Biopotential Sensor

A sintered Ag/AgCl electrode with a special post treatment is being developed for applications in EEG, as well as applications in new EEG feedback systems. New conductive gels used with the sensor are also under development. The sensor system enables provision of high signal-to-noise ratio and long-term stable measurements. Figure C.29 shows the sensors (in EEG cap) being used in EEG-feedback rehabilitation equipment in collaboration with Tsinghua University.

Figure C.29. The sensors in the EEG cap are used in EEG-feedback rehabilitation equipment.

SUMMARY AND CONCLUSIONS

Professor Duan leads a relatively new research group on flexible microelectrodes that it has applied primarily to intracochlear applications. Her interests are in the development of intracranial electrodes and the development of better microelectrodes for long-term implantation. Her electrochemical technologies are good for determining potential changes *in vivo*, with further application in electrophysiological recordings. This could be a powerful approach for determining ways to improve the longevity and performance properties of indwelling electrophysiological electrodes and stimulating electrodes for BCI applications. Professor Duan's work is complementary to additional work that is going on at Wuhan University, directed by Professor Wang and the HUST Group. In particular, the flexible electrode technology that Professor Duan has been working on in conjunction with collaborators may be applicable to the spinal cord stimulation work that is being carried out in conjunction with Dr. Wang and team.

Competitive Advantages Compared to World

Professor Duan has excellent electrochemical skills for studies of potential changes of electrodes *in vivo*. These technologies could be applied with others capable of *in vivo* recordings to develop better materials and electrode designs with greater longevity for *in vivo* recordings involving BCI technology. Professor Duan may be a potential collaborator for investigators in the United States working on development of microelectrode arrays and BCI technology.

REFERENCES

Duan, Y.Y., G.M. Clark, and R.S.C. Cowan. 2004. A study of intracochlear electrodes and tissue interface by electrochemical impedance methods *in vivo*. *Biomaterials* 25:3813–3828.

APPENDIX D. GLOSSARY

A/D	Analog to digital
ADHD	Attention deficit hyperactivity disorder
Ag	Silver
ALS	Amyotrophic lateral sclerosis
AM	Amplitude modulated
AR	Autoregressive
ARM	Assistive robot manipulator
ARTS	Advanced Robotics and Technology Systems Laboratory (SSSA, Italy)
Au	Gold
BCI	Brain-computer interface
BMI	Brain-machine interface
BOLD	Blood-oxygen-level-dependent (signal)
CE	Mandatory quality mark for certain product types marketed in the European Union (it may have originally have meant Communauté Européenne or Conformité Européenne)
CMOS	Complementary metal oxide semiconductor
CNS	Central nervous system
DBS	Deep-brain stimulation
DOF	Degree(s) of freedom
ECoG	Electrocorticography, electrocorticogram
ED	U.S. Department of Education
EDP	Ethylene diamine pyrocatechol water
EEG	Electroencephalography/electroencephalogram
EPFL	Ecoles Polytechniques Fédérale de Lausanne (Switzerland)
EPSP	Excitatory postsynaptic potential
ERD	Event-related desynchronization
ERS	Event-related synchronization
ESEM	Environmental scanning electron microscope
EU	European Union
FES	Functional electrical stimulation
FFT	Fast Fourier transform
FIRST	Fraunhofer-Institut für Rechnerarchitektur und Softwaretechnik (Germany)
fMRI	Functional magnetic resonance imaging
FPGA	Field programmable gate array
IC	Integrated circuit
ICA	Independent component analysis
IDA	Intelligent Data Analysis research group of Charité University of Medicine, Berlin, and the Technical University of Berlin (Germany)

IDIAP	A Swiss research institute, at one time referred to as "Institut Dalle Molle d'Intelligence Artificielle Perceptive"
Ir	Iridium
IRCCS	Scientific Institute for Hospitalization and Treatment of National Importance and High-level Specialisation in Neuromotor Rehabilitation (Italy)
LFP	Local field potentials
LMS	Least mean squares
MEA	Microelectrode array
MEG	Magnetoencephalography/magnetoencephalogram
MEMS	Microelectromechanical system(s)
mEP	Movement-evoked local field potential
MRA	Multivariate regression analysis
MRAM	Magnetic random access memory
MRCP	Movement-related cortical potential
NIBIB	National Institute of Biomedical Imaging and Bioengineering (U.S.)
NIDRR	National Institute on Disability and Rehabilitation Research (U.S.)
NIH	National Institutes of Health (U.S.)
NINDS	National Institute of Neurological Disorders and Stroke (U.S.)
NIRS	Near-infrared spectrum
NNSF China	National Natural Science Foundation of China
NSF	National Science Foundation (U.S.)
PDF	Probability density function
PECVD	Plasma-enhanced chemical vapor deposition
PETH	Peri-event time histograms
PNS	Peripheral nervous system
Pt	Platinum
Pt-Ir	Platinum-iridium
PVD	Physical vapor deposition
REM	Reflection electron microscopy
RF	Radio frequency
RRAM	Reconfigurable or redundant read only memory
SCI	Spinal cord injury
SCP	Slow cortical potential
SEM	Scanning electron microscope
SMR	Sensorimotor rhythm
SoC	System on a chip
SS	Stainless steel
SSSA	Scuola Superiore Sant'Anna (Italy)
SSSEP	Steady-state somatosensory-evoked potential
SSVEP	Steady-state visual-evoked potential

TATRC	Telemedicine and Advanced Technologies Research Center (U.S. Army)
TTD	Thought translation device
VLSI	Very-large-scale integration/integrated (electronic circuit)